Genda's Blade

Japan's Squadron of Aces

343 Kokutai

三四三海軍航空隊

TO SUE MASSEY
WITH BEST WISHES!
THANKS FOR BEING MY
FRIEND.
HENRY SAKAIDA
11-25-03

Genda's Blade

Japan's Squadron of Aces

343 Kokutai

三四三海軍航空隊

Henry Sakaida & Koji Takaki

CLASSIC

An imprint of
Ian Allan Publishing

Henry Sakaida, a third-generation Japanese-American, was born in Los Angeles, California, in 1951 and has researched the exploits of Japanese fighter pilots since he was 13 years old. He graduated with a degree in advertising from San Jose State University and wrote his first book, *Winged Samurai,* a photo-biography of the famous Zero ace, Saburo Sakai, in 1985. His five subsequent books – *Pacific Air Combats WWII - Voices From The Past* (1993), *Siege of Rabaul* (1996), *Japanese Army Air Force Aces 1937-45* (1997), *Imperial Japanese Navy Aces 1937-45* (1998), and *B-29 Hunters of the JAAF* (co-authored with Koji Takaki) (2001), all deal with Japanese combat aviation. He specializes in uncovering rare first-hand accounts from the men who flew on both sides of the Pacific air war.

He is a wholesale nursery plant grower by profession and he writes and researches in his spare time.

Koji Takaki was born in Saeki City, Oita Prefecture, Kyushu, Japan in 1933. His hometown was located at the mouth of the Bungo Straits, and he remembers one occasion in March 1945 when, as a boy, he watched American carrier-borne Corsairs attack Saeki Airfield before he jumped into the safety of his family's 'foxhole'!

A graduate from Kyoto University School of Economics, he worked in the textile business for forty years before retiring as President and Chairman of a leading textile trading firm in 1997.

In 1992, he wrote his first book *Ashizuri no Umi to Sora* (*The Sky and Sea of Ashizuri*), dealing with the combat history of the Imperial Army's 56th *Sentai,* a Tony fighter group, based at Itami near Osaka and Kobe, which intercepted B-29s over North Kyushu, Nagoya, Osaka, Kobe, and the Bungo Straits.
He has subsequently collaborated with Henry Sakaida in writing articles for *The Hook* magazine as well as acting as co-author and translator on various other projects connected with the air war over Japan.

First published 2003

ISBN 1 903223 25 3

Produced by Chevron Publishing Ltd
Project Editor: Robert Forsyth
Jacket and book design by Colin Woodman Design

Published by Classic Publications

an imprint of Ian Allan Publishing Ltd, Hersham, Surrey KT12 4RG.
Printed by Ian Allan Printing Ltd, Hersham, Surrey KT12 4RG.

Visit the Classic Publications website at www.classic-books.co.uk

Contents

三四三海軍航空隊

Authors' Introduction and Acknowledgements

三四三海軍航空隊

The heroic battles of the *Shiden-Kai* fighters of the Imperial Japanese Navy's 343 *Kokutai* (air group) first became widely known to the Japanese people through the well-known memoir written by Gen Minoru Genda. Genda, then a member of the House of Councilors, wrote his work under the title of *Kaigun Kokutai Shimatsu Ki* (The Rise And Fall Of The Navy Air Force) in 1962, 17 years after the end of the Second World War. Their outstanding activities impressed and attracted many people with the notion that there was an élite fighter group equipped with Kawanishi Aircraft's masterpiece fighter *Shiden-Kai* ('Violet Lightning Remodeled') flown by a group of veteran pilots as its mainstay.

The history of the 343 *Kokutai* resembles that of the famous *Luftwaffe* squadron *Jagdverband* 44 (JV 44), commanded by the charismatic Gen.Lt. Adolf Galland, a respected fighter leader. Known as the 'Squadron of Experts' after the veteran aces who wore the coveted Knight's Cross and who filled the unit's ranks, JV 44 was equipped with the latest German fighter – the jet-powered Me 262. As the Allied forces advanced across the Reich, Galland endeavored to stem the onslaught with an élite unit, and win back air superiority. Galland and JV 44 went down fighting and gained the respect and admiration of their people and former foes.

Capt Minoru Genda was the strategist who masterminded the attack on Pearl Harbor. Like his *Luftwaffe* counterpart, he was displeased with the existing air war situation. As the Allied forces closed in on Japan, Genda, like Galland, proposed the establishment of an élite unit, equipped with the most modern fighter and manned by veteran pilots. The 343 *Kokutai* also went down fighting, but very little is known about it outside of Japan. The American pilots who had fought the 343rd had no idea who their opponents were, except that they were skilled and very aggressive. After so many 'turkey shoots' in and around Japan, it came as a great shock to the Hellcat and Corsair pilots who first engaged the unit. Some of them did not survive.

Trained on a basis of formation tactics, the *Shiden-Kai* group tackled powerful American fighters and bombers to the last with a high level of commitment, losing a number of pilots in action, including all four squadron commanders in succession. The Air Group fought almost every type of American aircraft during the last stage of the Pacific War – carrier-based Grumman F6F Hellcats; carrier or land-based F4U or other models of Corsairs; land-based P-47 Thunderbolts and P-51 Mustangs. The bombers included carrier-based SB2C Helldivers and TBM Avengers, as well as large B-24 Liberators and B-29 Superfortresses. Patrol aircraft such as PBM Mariners and PB4Y-2 Privateers also crossed paths with the *Shiden-Kais*. In fact, the only formidable American fighter aircraft that the 343rd did not fight was the P-38 Lightning.

The 343 *Kokutai* commenced air operations with a big mêlée against the carrier squadrons of Task Force 58 over its base at Matsuyama in Shikoku on 19 March 1945. It ended the war with an organized interception against heavy bombers with their fighter escorts over Northern Kyushu on 8 August 1945. In its brief eight-and-a-half-month existence, the unit participated in more than 15 large air battles. Vestiges of those aerial encounters can be traced back with the help of Genda's memoir, books, and articles by surviving pilots and aviation historians, and what remains of the unit's records. The combat history of the 343 *Kokutai* as seen by the Japanese side was fragmented and incomplete. This was because 88 pilots – all first-hand witnesses – did not survive. None of the four squadron commanders lived to tell the tale. In addition, most of the Air Group's combat reports and other official documents were destroyed at the time of Japan's defeat. Such being the case, it did not seem that the whole story could be told. What was left was the legend of the 343 *Kokutai*, encrusted with myth and half-truths. There were just too many blank spots in its history.

On the other hand, in the United States, most of the official American aircraft action/mission reports have been preserved and are available. Starting his research in 1979, co-author Henry Sakaida realized that the 343rd's activities as seen by its US counterparts were, to a considerable degree, preserved. He started grappling with the task of identifying specific American fighter units which had fought the 343 *Kokutai*. Although he had some success in corresponding with the surviving veterans of the 343 *Kokutai*, he was greatly hampered by the language barrier and the distance.

Co-author Koji Takaki joined this project in 1997 after the research remained stalled for more than ten years. He was able to piece together the actions of the unit from the written US records by supplying information from Japanese sources. He conducted interviews with various veterans by phone and letters. Fortunately, some important individuals were still alive during the critical research period.

Each co-author conducted original research in their home countries, and then analyzed and shared their findings. Much like police detectives going over cold case files, many events of long ago were finally pieced together. Thanks to fax and email, there was daily communication – and monumental expense!

It was not always easy to pick the 343 *Kokutai's* combat out of the US reports. This was because the *Shiden-Kai* was often misidentified by US pilots. The new fighter made its debut during the last period of the war and the Americans were not familiar with it. *Shiden-Kais* were misidentified in most cases as Jacks, Franks, Tojos, or Zekes. The co-authors were able to determine that they were 343 *Kokutai's* aircraft based on the following: their American-style formation fighting by divisions or sections, the American pilots' observation about their extreme aggressiveness, and recollections from the Japanese veterans concerning the time, location, and circumstances of their engagements. Surviving American veterans were contacted about their aircraft and combat experience so that there could be some basis for making certain assumptions and presenting certain scenarios. In this way, the co-authors believe that the fighting history of the 343 *Kokutai* has been revived to a considerable degree from the records and witnesses of the opposite side. It was pure detective work, utilizing critical analysis and process of elimination.

The co-authors also examined the figures of the 343 *Kokutai* pilots' victory claims. As accurate records on personnel or aircraft losses have been kept on both sides, true figures of the losses are available in most cases only if the opponent squadrons are identified. In the post-war years, aviation historians have revealed that pilots' victory claims on each side were generally overestimated. The 343 *Kokutai* has proved not to be an exception.

The 343 *Kokutai* claimed 52 aerial victories in the first big melee over Matsuyama on 19 March 1945. In-depth research on all the action reports of 27 squadrons from US Task Force 58 shows that 321 aircraft flew on the Kure raid. The co-authors reached a conclusion that a total of eight US fighters were shot down by the 343rd's *Shiden-Kais*, four were shot to pieces and consigned to the sea upon their return to the carrier, and two were forced to ditch. On the other hand, about 50 fighters were claimed shot down by six US fighter squadrons (80 aircraft) that apparently fought *Shiden-Kais*, while the 343 *Kokutai* (which launched more than 60 fighters) lost 15 fighters and 13 pilots to the US fighters in the air. The claims and losses for both sides were remarkably similar.

The exaggerated claims can be attributed to mass confusion, poor observation, multiple counting, and excitement in the heat of battle. The greatest factor was 'tunnel vision'. So intent on scoring a victory, a pilot is totally focused on his prey, oblivious to his comrades also shooting at the same enemy aircraft. Gun camera film and a wingman would further "confirm" such a victory for each pilot!

It was common to both sides that confirming crashes became difficult as dogfights intensified. Individual victories were encouraged by the Americans by awarding acedom. Shooting down five enemy aircraft brought instant glory. The fiercely competitive nature of fighter pilots contributed to overclaiming. In those days, the Japanese adopted a policy to regard aerial victories as a result of collaboration in formation fighting and had no special merit system for individual victories. Even under such a policy, however, it appears that overclaims could not be avoided. It is evident that the American pilots reported accurate figures in some encounters against *Shiden-Kais*, while the *Shiden-Kai* pilots claimed more than actual.

The 343 *Kokutai's* powerful performance under difficult circumstances as well as their many proclaimed victories have inspired Japanese readers. This book may cause umbrage. It is the co-authors' view that the time is now ripe to know the facts and objectively observe the strength of the *Shiden-Kai* group. There was never any attempt to devalue the prestigious actions of the 343 *Kokutai*. The valor and excellent flying skills of its pilots are expressly stated in the commendatory remarks in the US action reports as quoted in this story. But most importantly, the co-authors believe that the surviving *Shiden-Kai* pilots want to know what really happened. They do not wish to be given false praise.

The co-authors have introduced to the readers new facts about missing pilots, including the 343rd's four missing squadron commanders, based on information obtained from the US official reports, and have tried to make assumptions on their deaths with the support of their research. One of the most dramatic actions was a series of encounters and engagements on the intercept mission against US carrier aircraft over the Bungo Straits on 24 July 1945. On this day, Lt Takashi Oshibuchi, the senior air commander, and Ens Kaneyoshi Muto, the Japanese Navy's top marksman, did not return. A submerged *Shiden-Kai* was subsequently discovered at the north part of Sukumo Bay in Southwestern Shikoku, 34 years after the war's end. At that time, the identity of the pilot of the ditched *Shiden-Kai* was a mystery and the incident was left unsolved. The co-authors dared to tackle this difficult problem and have presented one hypothesis in this book.

History is often an assembly of hypotheses. It is written on a basis of existing documents, historical materials including real evidence, and relevant information. When new facts are discovered or new information is obtained, a new hypothesis is presented. As the new hypothesis is generally accepted, history is rewritten or reconstructed to make a step forward to the truth. It does not mean that the preceding researches are devalued because new hypotheses cannot be conceived without them.

As 'forensic historians', the co-authors have considerably clarified many aspects or circumstances of the *Shiden-Kai* actions that were hitherto unknown. To read that LCdr Robert Weatherup shot down CPO Shoichi Sugita over Kanoya Airfield on 15 April 1945 may invite skepticism. How is it that historians are able to reveal the names of the specific participants? The answer is that it can be done through painstaking research. Most aviation historians will not go to these lengths simply because they do not have the resources and/or the time to do so.

The attempt to reproduce each main combat of the 343 *Kokutai* based on the American records is, of course, supported by the memoirs of Gen Minoru Genda and LCdr Yoshio Shiga, other pilots' recollections, and published researches by aviation historians. Those works are mentioned with great respect and appreciation in the bibliography. In particular, books by distinguished aviation historian Yoshiro Ikari were indispensable. Likewise, Frank Olynyk's computerized listings of all Second World War victory credits were invaluable. Each credit was listed by the pilot's full name and rank, his unit, date, time, location, and enemy aircraft type.

The co-authors owe much to the 343rd's surviving pilots who kindly provided valuable information and advice. In particular, LCdr Yoshio Shiga was always available for consultation. Gen Ryoichi Yamada kindly supplied information in response to frequent inquiries and critiqued the work in progress. Mr. Jiro Yoshida, Public Relations Director and goodwill ambassador of the Zero Fighter Pilots Association, worked tirelessly to support and encourage the co-authors. His personal communications with Gen Genda and others on behalf of the co-authors were of invaluable assistance.

Our thanks to...

In Japan:

343 Kokutai

HQ: Gen Minoru Genda (Ret.), Tadashi Nakashima, Yoshio Shiga

S701: Gen Ryoichi Yamada (Ret.), Toru Kurita, Saburo Sakai, Teruichi Sato, Yoshio Takahashi, Takumi Takaishi, Masayuki Tashiro

S407: Hideo Harada, Minoru Honda, Goro Ichimura, Hisaichi Kubo, Shokichi Suzuki, Yoshinori Matsumoto, Takao Mitsumoto, Shigehisa Osawa

S301: Chitoshi Isozaki, Tomokazu Kasai, Gen Masaji Matsumura (Ret.), Yoshikazu Miyamoto, Isamu Miyazaki, Tsuneharu Tamura, Seiichiro Yamamoto

S401: Bokugun Tsuchiyama

Maintenance: Wakashige Kondo

Radio: Tadashi Kakegawa

Jiro Yoshida of the Zero Fighter Pilots Association, Hisao Morikawa of the Navy Academy 70th Class Association, Toshio Imamura of the Army Air Force Akeno Fighter School, Ryosuke Yamagishi, Army 56th *Sentai*, Gen Goro Takeda (Ret.), Army 244th *Sentai*, Maj Saiji Matsuda (Ret.), *Kempei Tai*.
Aviation historians: Yoshiro Ikari, Dr. Yasuho Izawa, Yasuo Kumoi, Shigeaki Mori, Kazuhiko Osuo, Yoji Watanabe.
Maru Magazine (*Kojinsha*): Masao Kikuchi, editor
Organizations: 343 *Kokutai* Association, NHK-TV Matsuyama
Jiro Date, Toru Fukubayashi, Haruyoshi Hayashi, Kikuyo Hayashi, Yayoi Ishikawa, Susumu Koyama, Haruko Mihara, Kiyoko Muto, Shigeru Nohara, Hiroko Sakuragi, Mayumi Shirakata, Shori Tanaka, Hisashi Tashiro.

In the United States:
Marine Air Group 22: LtCol Russel D. Baade (Ret.), Col Elkin S. Dew (Ret.), Willard Pedrick, Hardy C. Wilcoxon
VBF-1: Robert M. 'Apple' Applegate
VBF-10: Roy D. 'Eric' Erickson
VF and VBF-17: Capt Marshall U. Beebe (Ret.), Jack 'Chili' Crawford, Harry 'Chief' Hanna, Cdr Tommy S. Harris (Ret.), Forrest McCormick, James L. Pearce, C.E. 'Bill' Watts, Cdr Charles Weiss (Ret.)
VF-46: Cdr Robert A. Weatherup (Ret.)
VF-49: Jack A. 'Hoot' Gibson, George M. 'Willy' Williams
VF-82: Phil Perabo Jr.
VF and VBF-85: George M. Blair, William R. Clarke
VF-88: Adm Malcolm Cagle (Ret.)
VMF-113: Alson C. Frazer, Russell Hunchar
VMF-123: Col Archie Clapp (Ret.)
VMF-224: Joseph P. Lynch
VMF-311: Col Mike Yunck (Ret.)
VPB-21: Ray Armknecht, John Martinn
VPB-109: George D. Fairbanks, Val Higgins
VT-17: Robert H. Williams
USAAF 19th Fighter Squadron: Vernon Wischer
USAAF 342nd Fighter Squadron: Edward S. Popek
Aviation historians: Dr. Eric Bergerud, Tom Britton, Josh Curtis, Hill Goodspeed, Bill Hess, Larry Hickey, Jack Lambert, James Lansdale, Robert Mikesh, Gary Nila, John Stanaway, Don Sweet, Osamu Sam Tagaya, Barrett Tillman, Col Raymond F. Toliver (Ret.), Sallyann Wagoner, Ron Werneth.
Organizations: American Fighter Aces Association, Marine Corps Aviation Association, National Archives, National Museum of Naval Aviation at NAS Pensacola, Naval Historical Center, Tailhook Association, USAF Historical Research Center, Veterans of Foreign Wars.
Alexander R. 'Sandy' Amell, Maurice Ashland, Bonnie Dunham, Rick Dunn, Cynthia A. Smith Garlock, Junichi Gondai, Raymond 'Hap' Halloran, LCdr Harvey Herzog (Ret.), Olga Hunchar, Don Roberts, Tadashi Sakaida, Ronald Sathre, James C. Sawruk, Kurt Stauffer. And to all others who contributed to this monumental task during the last 21 years, deep appreciation and gratitude is expressed.

From Henry Sakaida:
Lastly, but most importantly, I wish to thank my ever-suffering wife, Donna Butsumyo Sakaida, who toiled endlessly on the computer, scanning photographs and maps, for her computer-illiterate husband. I promise you that I will never ask you to do anything like this again...until the next time.

And to my son, Eren Tadashi Ganta, who consented to model the JNAF flight gear, and daughter Allison Tokie – much appreciation for taking pity and rescuing me as I fumbled along pathetically on the home PC.

Koji Takaki, Takarazuka, Japan, and Henry Sakaida, Los Angeles, California

Foreword

Yoshio Shiga

Former Lieutenant Commander and Executive Officer, 343 *Kaigun Kokutai*

三四三海軍航空隊

This is a detailed combat story based on the facts, of the fighter squadrons equipped with the Kawanishi N1K2-J *Shiden-Kai* fighter, recognized as the Japanese Navy's last superior fighter aircraft and given the Allied codename "George" during the last phase of the Second World War.

Not a few articles and books have been written about the operations of the 343 *Kaigun Kokutai* (343rd Naval Air Group) over the past half a century since the war. So much has been written in Japan that it seemed nothing new could be told. But the accounts were all one-sided and could not convey the true story. Additionally, to our regret, some of the actual circumstances of our combats could not be fully told. This is because a greater part of the air group's official action reports on the main air battles were lost in the confusion of defeat and many pilots, including four commanders of three squadrons under the air group, were killed in action.

Mr. Henry Sakaida is a well-known aviation historian and writer in America, who has taken note of the combat activities of the *Shiden-Kai* fighters from the 343 *Kokutai* since 1979. He has spent many years gathering relevant materials, including mission reports from the US Navy, Marines, and Army Air Force squadrons and also interviews with former American pilots who fought our *Shiden-Kais*. He firmly believed that the combat activities of the 343 *Kokutai* could be documented from the American side through their own aircraft mission reports and he has been proven right.

During the 1980s, I offered to support Mr Sakaida's efforts as far as possible and, in return, he promised to write a definitive history of our air group. He made steady progress in the work, thanks to the assistance and arrangements of Mr Jiro Yoshida of the Zero Fighter Pilots Association in Tokyo. The timely collaboration of Mr Koji Takaki, an aviation author and retired businessman, was the key to completing this monumental project. As co-author, Mr Takaki, well-versed in the Japanese and English languages and Second World War combat aviation history, brought his great analytical ability to this task. Both of the authors have unearthed a number of historically significant and interesting facts and episodes, which would have been lost to history. It is not an overstatement to say that they have achieved much more than supplementing the insufficient official data on the Japanese side. It took the joint efforts of a Japanese and American historian, combing their resources and talent, to finally tell our story.

Now, we have *Genda's Blade* published by Classic Publications in the UK, over 55 years after the end of the war. It is my pleasure and honor to have an English book in which the operations of the *Shiden-Kais*' old battles are going to be made known to people, not only in England and America, but also the world over. I believe that my feelings would be shared by the 161 victims who died in the air and on the ground during the battles, those who passed away after the war, including Captain Minoru Genda, the commander of the air group, as well as surviving members of the unit and their families. Our brave pilots helped write an interesting chapter in the history of the Imperial Naval Air Force. I am honored to have served with them.

Genda's Blade is recommended as a definitive air group history of the 343 *Kaigun Kokutai*.

(Yoshio Shiga)

Chapter 1

"Outstanding from the start"
Birth of the 343 *Kokutai*

精鋭戦闘機隊

The Japanese Navy began the Pacific War on 8 December 1941 with the strongest air force in Asia. Its pilots had gained extensive combat experience in the China War, beginning in 1937. At the time of Pearl Harbor, the JNAF seemed invincible. However, by June 1942, the tides of war began to change. During the Battle of Midway, the Imperial Navy lost four fleet carriers, over 280 aircraft, and up to 200 veteran pilots and air crewmen. From then on, the Japanese began fighting on the defensive.

Rabaul. Before the war, hardly anyone had heard of this sleepy port town on New Britain Island in the Southwest Pacific, known only for its copra export, but the Japanese quickly seized it in January 1942.

The Navy pilots staged their raids in the Solomons from this base and the various air groups that operated there were known as the Rabaul Air Corps. Their hard fighting brought them fame in the struggle for Guadalcanal. It represented the last proud moments of the Japanese Naval Air Force.

By 1943, Rabaul was known as the "Graveyard of Fighter Pilots." What the Russian Front was to the Germans, Rabaul was to the Japanese. Allied aircraft and submarines cut off supply efforts. Rabaul was abandoned by the JNAF in February 1944 and bypassed by the Allied forces. The Japanese were pushed further back toward the homeland, but attempted to make their stand in the Philippines. In a desperate attempt to destroy the American fleets, they resorted to *Kamikaze* suicide attacks. The Japanese threw men and equipment into the giant meatgrinder with very little success.

Capt Minoru Genda of the Naval General Staff was painfully aware of the deteriorating war situation. Two and a half years earlier, his plans for the Pearl Harbor raid culminated in a resounding victory.

A 1923 Naval Academy graduate, Genda had completed fighter training in 1929. From 1936 to 1937, he attended Naval War College and developed a keen taste for strategy. It was during this period that he submitted a radical thesis to his instructor: "*Our Naval armament should be reorganized based upon the idea that the core should be a land-based air force with carrier air groups supported by submarines… Destroyers and cruisers will be kept as far as they are needed for the purpose of supporting the air forces. The existing main ships such as battleships and fast battleships should be scrapped or used as a substitute for piers.*"

Capt Minoru Genda

Genda was way ahead of his time, but the instructor asked his student sarcastically: "And when do you think your idea will materialize? In 50 or 100 years?"

At the time the China War broke out, Genda was on the staff of Air Flotilla 2 and flew missions from land bases. During the Battle of Britain, he was an assistant Naval air attaché in London. The lessons of the air war between the *Luftwaffe* and the British Royal Air Force would prove to be an invaluable education for the young commander.

"*In late 1944, I thought thoroughly about the unfavorable war situation,*" Genda wrote in his memoir. "*Our battles were being lost because of defeats at sea. Defeat in battle was caused by the fact that we were overwhelmed in the air war. We were losing the air war because we had failed to secure air superiority with our fighters. In short, we were losing the war because our fighters were being defeated.*"

After the Battle of Leyte in the Philippines, Capt Genda proposed a new strategy to Naval GHQ during a high-level staff meeting. He lectured that Japan was losing the war because the enemy had control of the air. A radical idea was brought forth which envisioned an élite unit,

MINORU GENDA

Minoru Genda will forever be remembered by the Americans as the mastermind behind the Pearl Harbor attack. Because he was a brilliant strategist, the young commander was selected to draft the plan for the surprise attack against US forces in the Hawaiian Islands.

Genda was born in Hiroshima on 16 August 1904, the second son of a farmer. Two brothers were graduates of Tokyo University, another graduated from Chiba Medical College, and his youngest brother entered the Army Academy. Graduating from the 1st Hiroshima Middle School, Minoru entered the Navy Academy with the goal of becoming a fighter pilot.

Following distinguished service in the Second World War, Genda's military career came to a temporary halt. The Imperial Navy was formally dissolved in November 1945. Many professional military officers struggled to make ends meet. However, Capt Genda did not meet with such destitution because he was financially supported by a wealthy businessman. He was active in the movement to protect the lineage of the Imperial Family as it was feared that the Emperor would be designated as a war criminal. Gen Douglas MacArthur and the US Government decided to keep the Emperor in place.

At the start of the Cold War, the United States needed Japan's support to counter the advance of Communism on the continent of Greater Asia. When the Korean War started in 1950, Japan's rearmament became necessary. It is believed that Capt Genda had a role in policies involving the rebuilding of the Japanese military. He entered the Defense Agency in 1954.

Maj Gen Genda was appointed as Chief of Staff of the Japanese Self Defense Air Force in 1959. When the decision came to arm the JSDAF with advanced jet fighters, Genda went to the United States and test-flew various Lockheed fighters even though he was almost 60 years old! The Japanese public was amazed at his stamina. They recalled the golden years of pre-war Japanese aviation when Genda had led a division of biplanes around the country, conducting demonstrations and aerobatics. Known as the "Genda Circus", his team was a part of the public relations campaign to promote Naval aviation.

The United States, once his foe, bestowed on him the US Legion of Merit degree of Commander in 1962 upon his retirement from the military. It was presented by Gen Lyman I. Lemnitzer, chairman of the US Joint Chiefs of Staff. Genda was recognized for his leadership in the Japan Self-Defense Air Force and his cooperation with US forces in Japan.

To the Japanese, Genda will always be remembered as the beloved commander of the 343 *Kokutai*, a general in the post-war Japan Self-Defense Air Force, and a member of the upper house of the Japanese *Diet* (parliament). Even in retirement, the diminutive gentleman exuded an aura of quiet strength and determination. He was quite active in politics and served many years in the National Defense Committee of the Liberal Democratic Party.

It is ironic that he died exactly 44 years after the Japanese surrender, on 15 August 1989 in Matsuyama. He was married and had three children.

1924: Graduated from the Naval Academy at Etajima and served on the heavy cruiser *Izumo*.
1928-29: Completed flight training at Kasumigaura with 160 flight hours.
1929: In Yokosuka *Kokutai* for two months.
1929: Served on the carrier *Akagi*.
1929-30: Three months training at Yokosuka under RAF Flight Leader Wingate and Squadron Leader Chappel.
1930: Served on *Akagi* from March to November.
1930-31: Flight instructor at Kasumigaura for one year.
1931-32: Combat aircraft instructor with Yokosuka *Kokutai* and served three months on the carrier *Ryujo*.
1932: Again served on the *Ryujo* until the end of the year.
1933-35: Instructor in Yokosuka *Kokutai*.
1936-37: Student at Naval War College.
1937-38: On staff of Air Flotilla No.2 in China area. Flew from land bases.
1938: Again instructed at Yokosuka *Kokutai* until the end of the year.
1939-40: Assistant Naval Attaché for Air in London; left in September 1940.
1940-41: Staff officer in Carrier Division No.1.
1941-42: Staff officer of the 1st Air Fleet (then including eight carriers).
1942: Senior air officer on carrier *Zuikaku* (July to September). Staff officer 11th Air Fleet at Rabaul (October to November).
1942-45: Naval General Staff, in overall Naval Air planning.
1945: CO of 343 *Kokutai*, a fighter group based first at Matsuyama (Shikoku), moved in April to Kanoya, then Kokubu, and finally to Omura. Over 3,000 flight hours at the end of the war, all in fighters.
1954: Joined the new Japan Self-Defense Air Force.
1959-62: Chief of Staff of the JSDAF.
1962-86: Member of the Japanese parliament.

RIGHT: Generalleutnant Adolf Galland was ousted from his position as General der Jagdflieger following disagreements with Reichsmarschall Göring over the leadership of the Luftwaffe fighter force. In 'exile' he commanded Jagdverband 44, a unique unit numbering several Experten and equipped with the revolutionary Me 262 jet fighter.

FAR RIGHT: Pilots of JV 44 with their Me 262s at Munich-Riem, April 1945.

commanded by Genda and led by veteran squadron leaders, and armed with the newest fighter – the *Shiden-Kai*.

Capt Genda's goal was to regain air superiority and try to turn the tide of war. He was finally given approval to put his plan into action. The birth of the special air group was remarkably similar to that of the *Luftwaffe's* JV 44, the élite jet fighter squadron organized by *Generalleutnant* Adolf Galland, its charismatic fighter leader. In the face of criticism from Göring and other quarters within the *Luftwaffe*, he had to fight an uphill battle to be given a chance to prove himself. There were so many aces brought in to fly the advanced Me 262s that the unit became famous as the 'Squadron of Experts.'

The newest JNAF unit was designated 343 *Kokutai* (2nd Formation) and established on 25 December 1944. It was attached directly to the 3rd Air Fleet on 10 February 1945. The first 343 *Kokutai* had been formed on 1 January 1944 at Kagoshima Air Base in Kyushu. It was bloodied in heavy fighting in the Marianas and was disbanded on 10 July 1944 due to the loss of so many pilots and aircraft.

Matsuyama Airfield on the island of Shikoku was selected as the new home base for the 343 *Kokutai*. Before the unit was formally organized, one of the group's new squadrons, the 301st, trained at Yokosuka. The busy airfield was not large enough to accommodate its needs. PO1/c Tomokazu Kasai once trained at Matsuyama before he went overseas and recommended this large airfield to Lt Naoshi Kanno, the squadron leader. Kanno immediately flew to the airfield alone to inspect it.

Matsuyama was indeed larger than Yokosuka and the locals were friendly. In addition, there were hot springs there for tired bodies. Kanno passed his hearty endorsement to Capt Genda, who was waiting to be appointed CO of the 343rd. Genda commented: "I thought that Matsuyama was appropriate for our base because, geographically, it was located a little inland and we would be able to gain good time against enemy attacks."

The base was formally selected on 25 December 1944 when the air group was established. Around this time, two other squadrons due to be assigned to the 343 *Kokutai*, Squadrons 701 and 407, were in Kyushu or on their way back to the homeland from the front. Squadron 407 began training with the *Shiden* at Izumi in mid-December. Most of Squadron 701's pilots were returning from the Philippines via Formosa. They began to regroup at Miyazaki.

As soon as orders to reorganize the 343rd were issued, pilots of Squadrons 301 and 701 began to assemble at Matsuyama. From the end of December to the beginning of January 1945, the air group started to take shape, even before its command structure was formally established.

Capt Genda was appointed as commanding officer of the 343 *Kokutai* on 15 January 1945 and arrived at Matsuyama on the 19th. The first order of business for the new CO was to assemble his staff. As a pilot with over 3,000 flight hours, all in fighters, he knew exactly the caliber of men he needed – veteran air officers who had fought in combat and distinguished themselves in a leadership position. LCdr Yoshio Shiga was the first he appointed and he became executive officer.

At the time of his nomination, LCdr Shiga was waiting to command the air group aboard Japan's last carrier, *Shinano*. On 29 November 1944, the carrier was sunk en route from Yokosuka to Kure to pick up

YOSHIO SHIGA

Yoshio Shiga was born in Tokyo in 1914 as the third son of R/Adm Kensuke Yotsumoto, who was the head instructor at the War College. In April 1931, he entered the Naval Academy at Etajima. After graduation, Ens Yotsumoto served on a cruiser and went overseas. The dapper and handsome young man was often teased by his peers for going to nightclubs in Shanghai dressed in brown suits rather than in uniform. Light on his feet, young Yoshio charmed women with his dancing talents while envious comrades looked on.

In December 1936, Yotsumoto commenced his aviation career when he was nominated as a flight trainee officer. He initially received his training at Kasumigaura *Kokutai* and his goal was to become a divebomber pilot. However, he soon switched to fighters because it afforded more excitement. In February 1938, the new fighter pilot was sent to China to join the 13 *Kokutai* stationed in Nanjing. Veteran NCO aces Toshio Kuroiwa and Sadaaki Akamatsu, amongst others, tutored the "green" officer and protected him on missions. This pair of eccentric flyers was famous for their daring exploits. Yoshio Shiga scored four victories against the Russians in China during 1938.

It was in September 1940 that Lt(jg) Yoshio Yotsumoto married Ms Yoshino Shiga and took on his wife's family name to become Yoshio Shiga. This was a common practice in Japan when the wife had no brothers to carry on the family name. A veteran of Pearl Harbor, Lt Shiga served as division leader on the carriers *Kaga*, *Junyo*, and *Hiyo*. After combat operations, he became a test pilot. He was one of the first test pilots to fly the new *Shiden-Kai* and knew the secrets of this fighter intimately.

equipment. The calamity was actually a blessing in disguise. While waiting around for a new assignment, Shiga was nominated to the 343 *Kokutai*. As far as he was concerned, he was the right person in the right place at the right time. There was no other officer pilot who was as familiar with the new *Shiden-Kai* as he. Shiga arrived at Matsuyama on 16 January.

The 343 *Kokutai* needed a vice-commander. The first person to fill this position was Cdr Tadashi Nakashima, who was appointed on 18 January. He arrived at Matsuyama Airfield on 9 February. As division leader aboard the carrier *Kaga* at the start of the China War in 1937, he participated in aerial combat for the first time on 4 September of that year. He and his wingman shot down three Curtiss Hawks. By the end of his tour of duty, he had achieved three or four victories but would score no more. Nakashima's greatest strength was his ability to organize and lead. He was a former squadron leader of the famed Tainan *Kokutai* at Rabaul in 1942, and later a CO of a *Kamikaze* unit in the Philippines.

LEFT: LCdr Tadashi Nakashima (front, center) at Rabaul in 1942 with the Tainan Kokutai. Sitting in the chair next to him is WO Toraichi Takatsuka (16 victories). Squatting to his right is the great ace PO1/c Hiroyoshi Nishizawa who claimed 86 victories before he was killed in action. (Y. Izawa)

ABOVE: N1K2-J Shiden-Kai No. 343-A-17 in the foreground of a line-up photographed at a snow-covered Matsuyama Airfield, probably January 1945. Just visible on 343-A-17's 'meatball' is the faint number '17'. The faintness of the number is proof that the number was applied in a water-soluble chalk.

ABOVE RIGHT: Lt Takashi Oshibuchi, Squadron 701.

Capt Genda also reviewed a list of squadron leaders (and their units) whom he wanted for his new air group. His nominees were Lts Takashi Oshibuchi, Yoshishige Hayashi, and Naoshi Kanno. They were all Naval Academy graduates with distinguished records of leadership and fighting ability.

Lt Takashi Oshibuchi was born in Nagasaki in 1919 to a medical doctor's family originally from Ojika Jima, the northernmost island of the Goto Retto chain, southwest of the island of Kyushu. He was always the top student through primary and middle schools, and he graduated No.1 in his class at the Naval Academy.

Oshibuchi arrived at Rabaul towards the middle of 1942 after serving as an instructor at Oita and became a member of the Tainan *Kokutai*. In March 1944, he was appointed S304 squadron leader under 203 *Kokutai* at Atsugi and posted to Northern Japan for air defense duty.

Squadron 304 moved to Okinawa in October 1944 to participate in the Battle of Taiwan. On 22 October, Lt Oshibuchi led 29 aircraft to Banban Airfield on Luzon. After losing pilots and aircraft in heavy fighting, S304 was attached to the 221 *Kokutai* based at Angeles. It was reported that Oshibuchi was wounded in the foot by anti-aircraft fire or a Hellcat and returned to the homeland in November. He arrived at Matsuyama Airfield on 8 January 1945 after being discharged from Beppu Naval Hospital.

"Oshibuchi's formation leading was splendid," remarked Genda. "I had almost nothing to criticize about him. His relationship with his men was wonderful. They seemed to be proud to die with their leader." He was known to have scored over six victories by the time he returned to Japan. Oshibuchi was appointed leader of the 701st.

RIGHT: Lt Yoshishige Hayashi (seated left), commander of Squadron 407, seen with his men in 1944. Hayashi brought those pilots who survived into 1945 with him to the 343 Kokutai. (Photo credit: Y. Izawa)

ABOVE: Lt Naoshi Kanno, commander of Squadron 301, photographed in front of a Zeke fighter while in training.

Squadron 701 was one of the three *Shiden* squadrons under the 341 *Kokutai* in the Philippines. Their pilots were accustomed to flying the *Shiden*, the predecessor of the *Shiden-Kai.*

Lt Yoshishige Hayashi was born in Kamakura in Kanagawa Prefecture on 17 July 1920. He was born to a Buddhist priest of Jomyoji Temple, one of the five famous Kamakura temples. He attended the Kanagawa Prefectural Shonan Middle School, an institution that sent many graduates to the Naval Academy before the war.

The mild-mannered Hayashi was known for his persistence. Once he set himself a goal, he set out to reach it, and there was no stopping him. His mother would visit the Shinto shrine to pray for her son. She started this habit when he entered the Naval Academy and continued to do so for eight years until he died. "There was a great mother behind a man of noble character!" remarked Capt Genda.

Hayashi graduated from the Naval Academy and was one term junior to Lt Oshibuchi. In April 1944, he was appointed to lead S407, equipped with *Zeke* fighters. He entered combat in the Philippines when his squadron was attached to the 221 *Kokutai* and achieved at least five victories.

Somewhat reticent and modest, his men respected and loved him dearly. Cdr Tadashi Nakashima later remarked: "There is a type of man who always smiles and seems to be unreliable, but he exhibits his real ability and does a big surprising thing at the critical moment. Yoshishige Hayashi was such a man."

Lt Kunio Iwashita, his classmate and one of the *Shiden* squadron leaders in the Philippines added: "Lt Hayashi was a quiet and reticent man in the Naval Academy days. It was impressive to me that he had become a dauntless-looking man with sharp eyes when I met him in the Philippines. It was also unexpected to me that he would lead a *Shiden-Kai* squadron as an able commander."

Hayashi continued his command of the 407th, which converted from *Zekes* to the new *Shiden-Kai.* After training at Kagoshima and Izumi following his return from the Philippines, he and his men arrived at Matsuyama in two Douglas transports on 26 January.

Lt Naoshi Kanno was born in Korea in 1921 and brought up in the Miyagi Prefecture. He was the shortest of the three squadron leaders in the 343rd, but he had a bulldog-like fortitude. When he was eight years old, he was involved in his first dogfight; he fought a big dog and stabbed it to death. Once, when he received a bullet in his thigh during combat in the Philippines, he refused a painkiller, and ordered the surgeon to dig out the bullet. Kanno endured the operation in silence.

Kanno started his military career when he entered the Naval Academy in 1938 as a student one term junior to Yoshishige Hayashi. After graduation in November 1941, the young ensign was posted as an anti-aircraft gunnery division officer aboard the battleship *Fuso.* The role of a fighter pilot seemed well suited to his aggressive and daring personality, so he decided to switch to aviation. He trained at Kasumigaura from July 1942 and proceeded to the fighter course at Oita from February to September 1943.

"Lt Kanno was one term junior to me and I was in charge to guide him as an instructor at Oita," explained Lt Kunio Iwashita. "He was reserved but I understood from my experience of frequent mock dogfights with him that he was full of fighting spirit. He was very competitive and challenging with reckless maneuvers. Once we nearly collided with each other. As he was a man of violent temperament, some of his instructors avoided mock fighting with him."

The young officer pilots heading into battle in 1944 were vastly different from those who had entered combat during the China War. Japan was now being pushed back to the homeland and fighting defensively. There were major supply problems and lack of replacement pilots and aircraft. Into this situation Lt Kanno became division officer in the 343 *Kokutai* (1st formation) in April 1944 and saw heavy action around Micronesia. The air group lost so many pilots and aircraft that it was disbanded in July 1944. He was then transferred to S306 with the 201 *Kokutai* and fought ferocious battles over Yap and in the Philippines.

Lt Kanno would have been the first leader of the *Kamikaze* team, but as fate would have it, the honors went to his classmate, Lt Yukio Seki. Cdr Asaichi Tamai, deputy commander of 201 *Kokutai,* had to select the team. Someone from the Naval Academy had to lead the first effort. Kanno held seniority, but was temporarily absent from Mabalacat. He was in Japan with some of his men to ferry replacement aircraft back to the Philippines. It is said that Kanno would have been appointed the *Kamikaze* leader if he had been

NOTE: 'Zeke'/'Zero': In Japanese, the Mitsubishi Navy Type Zero Carrier Fighter (Designation A6M2 to A6M8) was known as 'Rei Shiki Sentoki' (Type Zero [Rei] Fighter [Sentoki]), usually abbreviated to 'Rei-sen' or 'Reisen'. However, use of the term 'Zero' by the Americans became so popular in the post-war years, that even the Japanese adopted it and it was often used - confusingly - to describe virtually any Japanese fighter aircraft. However, the American codename for the Zero fighter was 'Zeke'. The authors and publishers have left unchanged any references to Zero or Zeke in any interviews or correspondence from veterans or official reports and have generally used the term 'Zeke' to describe the Zero Fighter in the text where appropriate.

RIGHT: Flight trainee Shoichi Sugita (standing, right) and his classmates pose with their instructor in front of the Type 93 Intermediate Trainer at Tsukuba Air Base in May 1941. Standing next to him is Takeo Tanimizu, who survived the war with 32 victories. (T. Tanimizu)

FAR RIGHT: NCO pilots of the 204 Kokutai at Rabaul, early 1943. Standing in the middle is CPO Shoichi Sugita. "I go after them, no matter what!" is the way he explained his aggressive approach to air combat. Ryoji Ohara, who claimed 48 victories and survived the war, is kneeling left. (R. Ohara)

at the base. Because of his absence, Lt Seki died in his place and Kanno resented it. He asked Cdr Tadashi Nakashima to send him out as a *Kamikaze*, but he was politely refused. "We can't make you a *Kamikaze*," explained Nakashima. "You should fly and fight!"

Capt Genda was so impressed with the tenacity of Lt Kanno that he was the first man he thought of when it was time to assemble his fighter team. When Lt Kanno arrived in Matsuyama, he had a personal score of over 30 victories. There were no other fighter pilots who started their combat career in 1944 who had attained such a score. It shows that the young leader was exposed to very intensive air battles. Kanno took control of S301.

LCdr Yoshio Shiga, the Executive Officer, was equally pleased with his commander's selection. "In brief, Lt Oshibuchi was like a resourceful admiral," added Shiga. "Lt Hayashi was like an admiral of virtue. And Lt Kanno was much like an aggressive admiral."

The 343 *Kokutai* was unique in that it had an aerial reconnaissance squadron. Heading this squadron was LCdr Toshio Hashimoto, who distinguished himself in the Battle of Midway. Capt Genda had always felt that the disaster at Midway was due to faulty reconnaissance.

Hashimoto was the leader of the 2nd *Chutai* torpedo bombers during the Battle of Midway under Lt Joichi Tomonaga. Escorted by only six *Zekes*, they attacked the *Yorktown*, which was initially hit by Val divebombers. Five Kates of the 1st *Chutai* led by Tomonaga went in from the starboard side and were all shot down by anti-aircraft fire and F4F Wildcats. But Hashimoto, whose 2nd *Chutai* attacked from the port side, delivered two fatal hits.

The veteran enlisted men who had fought at Rabaul and in the Philippines were valued for their wealth of combat experience. It was rumored that Capt Genda selected these pilots personally. However, this is incorrect. "We assembled outstanding fighter squadrons from the start," Genda pointed out. "There were veterans, of course, but there were also many unskilled pilots in those squadrons, too."

Some enlisted pilots came with their respective squadron leaders. Others were transferred to the 343rd upon orders from the Navy's personnel authorities in accordance with Capt Genda's needs. Survivors from the three *Shiden* squadrons under 341 *Kokutai*, which saw heavy action in the Philippines also, arrived at Matsuyama. All through the month of January, veterans and new pilots arrived at the airfield.

BELOW: PO1/c Tomokazu Kasai fell victim to Lt Kanno's wry sense of humor. He often flew as Shoichi Sugita's wingman. (T. Kasai)

The most famous NCO pilot to arrive was CPO Shoichi Sugita, who began his career with the 204 *Kokutai* in the South Pacific. He claimed his first victory when he rammed a B-17 over Buin on 1 December 1942. He was one of the original six *Zeke* escort pilots on the ill-fated Adm Yamamoto mission when they were ambushed by P-38s. On 16 August 1943, he was hit by F4Us near Ballale and parachuted with severe burns, returning to Japan for hospitalization. By March 1944, the popular ace with a rustic personality returned to the fighting in the Marianas and later in the Philippines where he became an indispensable comrade of Lt Naoshi Kanno in the 201st. By this time, he had claimed nearly 120 victories.

Also coming into the 343rd's 301 Squadron was 19-year-old PO1/c Tomokazu Kasai. He entered combat after March 1944 and became Shoichi Sugita's wingman. By Western standards, he was a double ace. After the Battle of Leyte, Kasai sometimes served as wingman to Lt Kanno. Once, he was the victim of his squadron leader's wry sense of humor. "I don't want any fat boys like you in my squadron! Quit fighters and go to the divebomber squadron!" snapped Kanno. That night, Kasai went to Kanno's room and pleaded "I beg you sir, please don't let me go!"

"Hey, can't you take a joke?" snapped Kanno.

WO Minoru Honda, division officer, was a veteran pilot who had served in the 407th since its start in April 1944. He gained minor fame as a pilot who "...came back from the dead." In late 1942 while attached to the Kanoya *Kokutai* in the Solomons, he had the distinction of being promoted two ranks posthumously when he failed to return from a mission. Ten days later, he returned to base with the help of friendly natives. His immediate superiors were not happy to see him because they had gone to the trouble of reporting his death and issuing a prestigious double promotion! To alleviate this unfortunate situation, Honda was sent out alone on virtual suicide missions, but he managed to return each time. When the senior commander heard about this, he put a stop to such nonsense. His double promotion was rescinded. Such an honor for a living, lowly enlisted man would have been unprecedented. His aerial victory claims were over 40 by this time and he no longer bothered to count them.

Ens Akio Matsuba was an old China War hand who scored his first victory in August 1937 over Shanghai. After the China War, he served aboard the carrier *Ryujo* and then with various air groups. In June 1944 over

Squadron 701 Ishin-Tai, at Matsuyama in early April 1945. Front row, left to right: WO Narunobu Ibusuki, Ens Saburo Sakai, Lt(jg) Hirohisa Yasui, Lt Takashi Oshibuchi, Cdr Takahide Aioi, Capt Minoru Genda, LCdr Yoshio Shiga, Lt Ryoichi Yamada, Lt(jg) Tetsuya Omura, Ens Misao Kishimoto, Ens Teruo Sasaki, and Ens Daijiro Nakajima (standing extreme right by banner). Second row: PO1/c Morita, CPO Yoshio Takahashi, PO1/c Takamoto Kashima, CPO Takeo Yamada, CPO Ryuji Yagi, CPO Hisao Nishimoto, CPO Yasuo Matsumoto, CPO Nakano, CPO Takumi Sugitaki, CPO Jiro Hatsushima, PO1/c Yanagida (first name unknown), and CPO Tomeshiro Hiro. Third row: PO2/c Masaru Morikawa, PO2/c Naruhiko Yamada, PO1/c Toru Kurita, CPO Jiro Funakoshi, CPO Kiyoshi Toyohara, CPO Akira Takahashi, CPO Toshiro Komabashiri, CPO Masamori Ono, CPO Sukeo Yoshioka, PO2/c Masao Hasegawa, PO2/c Kenzo Sakae, PO2/c Yukio Muramoto, PO2/c Masayuki Tashiro, and CPO Mitsuo Anraku. Back row: PO1/c Tsutomu Yamada, PO2/c Hiroyoshi Yoshida, PO2/c Hisao Sugita, CPO Junichi Miyake, PO2/c Ryoji Niwa, PO1/c Teruichi Sato, PO1/c Sanpei Shiono, PO1/c Tsutomu Kimura, PO1/c Tanezo Inoue, PO2/c Sueji Imamura, and PO2/c Kikuo Fujiki. (Y. Shiga)

Squadron 407 Tenchu-Gumi at Matsuyama, February 1945. Front row, left to right: PO1/c Tomio Yamamoto, LA Kinzo Kasuya, PO1/c Kazuo Yotsueda, unknown, LA Mitsuyoshi Hasui, LA Yoichi Saiki, LA Eikichi Nishitsuruzono, and unknown. Second row: LA Yoshimi Daito, Lt Kozo Shima, Cdr Tadashi Nakashima, Capt Minoru Genda, LCdr Yoshio Shiga, and Lt Yoshishige Hayashi. Third row: Unknown, unknown, LA Akira Hoshino, unknown, Ens Seikichi Aizu, unknown, Ens Haruo Takamure, Lt(jg) Sadao Oshio, Lt(jg) Itaru Kawabata, Ens Mitsuo Ishizuka, WO Minoru Honda and PO1/c Morio Kitsunezuka. Fourth row: LA Toshiharu Hara, CPO Sakae Masumoto, CPO Mitsuo Nakajima, CPO Hideo Nakao, CPO Shiro Endo, CPO Toshio Tanaka, CPO Seiji Noma, CPO Tadayoshi Aizawa, CPO Miyuki Shimotsuru, and PO2/c Kenshin Mizoguchi. Back row: LA Shokichi Kurumoto, PO1/c Hideo Matsui, LA Nobuyuki Nishimoto, LA Hitoshi Kotake, PO2/c Shigehisa Osawa, LA Ikuji Matsumura, PO2/c Kiyoshi Miyamoto, CPO Hide Kitagawa, PO2/c Zenji Asai, PO1/c Takashi Yanagisawa, LA Magoshichi Koma, LA Yoshio Takagi, and unknown. (Y. Shiga)

Iwo Jima, he claimed six Hellcats shot down during a two-day period. The following month, he was transferred to S701 and fought in the Battle of Taiwan and in the Philippines. Coming to the 343 *Kokutai* at age 31, he was one of the oldest operational fighter pilots in the unit. His personal victory tally was well over 12.

When Ens Matsuba arrived at Matsuyama as one of the last pilots of S701 to escape the Philippines, Genda assembled all the pilots the following day for a special lecture. Matsuba talked for over two hours, telling his comrades about his experiences with the *Shiden* and the invaluable lessons he had learned. He had led four *Shidens* from S701 and destroyed four Hellcats in October 1944 off Taiwan.

These talented pilots of various ranks and backgrounds were thrilled to become a part of Genda's élite unit, but there was some criticism and resentment from others who felt their squadrons were being robbed of irreplaceable veterans. "We needed these experienced pilots to help instruct and lead our beginners," complained one *Zeke* pilot.

The veterans helped train the new pilots. Great emphasis was placed on quick take-offs and formation flying. Each squadron conducted mock dogfighting, using eight against eight. All the squadrons' pilots training with the new *Shiden-Kai* started at Matsuyama.

To support his new unit and also to promote morale, Capt Genda named his air group *Tsurugi Butai* (Blade Force). According to Genda's recollection, this title was probably suggested by Lt Naoshi Kanno. The

Squadron 301 Shinsen-Gumi, January 1945, Matsuyama Airfield. Front row, left to right: Lt(jg) Tatsutoshi Hashimoto, Lt Naoshi Kanno, LCdr Aritsune Shimokawa, Capt Minoru Genda, LCdr Yoshio Shiga, Lt Masaji Matsumura, and Lt(jg) Isaburo Inoue. Second row: PO1/c Eiji Suzuki, CPO Mitsuo Hori, Ens Yoshio Murata, WO Isamu Miyazaki, WO Shoji Shibata, Lt(jg) Hiromu Tanaka, Lt(jg) Soji Sawada, CPO Shoichi Sugita, CPO Yasuharu Nikko, and PO2/c Takeshi Ishikawa. Third row: CPO Wataru Tomisugi, PO1/c Noriyoshi Kubo, CPO Yoshio Aoyama, CPO Itaru Shikano, PO1/c Teruo Kiriyama, PO2/c Toru Otsubo, PO2/c Eiji Harada, CPO Eiichiro Sakurai and PO1/c Toshinobu Shimizu. Fourth row: CPO Tomoichi Yokoyama, PO1/c Mahito Yoshihara, PO2/c Toyomi Miyazawa, LA Tsuneharu Tamura, PO2/c Mikio Mitsuishi, PO1/c Makoto Nishimura, PO1/c Eiji Mikami, LA Osamu Himei, LA Kiichi Miyama, and PO2/c Susumu Imai. Back row: PO1/c Rokuro Asama, PO1/c Osamu Omori, PO2/c Ken Okimoto, CPO Koichi Niizato, CPO Tomokazu Kasai, CPO Seiichiro Sato, LA Hideo Izawa, LA Mutsue Naka, and PO1/c Yoneyasu Tsuneishi. (Y. Shiga)

three new fighter squadrons each chose their own title: S701, *Ishin-Tai* (Imperial Restoration Unit); S407, *Tenchu-Gumi* (Heavenly Punishment Unit); and S301, *Shinsen-Gumi* (Elite Guard Unit). The reconnaissance squadron bore the designation T4 and was named *Kihei-Tai* (Commando Unit).

Each squadron made their own banner and a friendly rivalry was encouraged between the squadrons. The names of *Shinsen-Gumi* and *Tenchu-Gumi* held bloody connotations for the Japanese and they were somewhat inappropriate for the modern military. But Capt Genda encouraged his men to adopt the names to raise morale and heighten the solidarity of the squadrons. He made a conscious effort to promote an *esprit de corps* and LCdr Shiga's duty was to support him.

In February, Squadrons 401 (Lt Masaaki Asakawa, CO) and 402 (Lt Iyozoh Fujita) were added to the air group. On 1 March, Squadron 402 was transferred to the 601 *Kokutai* and Squadron 401 was moved out to Tokushima Airfield to serve as a pure training unit.

"The Engineering Department built the base according to Genda's specifications," explained Yoshio Shiga. "For example, we dispersed quarters for the commanding officers. Strenuous efforts were made by Service and Radio Departments, from morning to night, to complete dispersed servicing and supply depots. We tried to obtain better equipment and devices to strengthen the function of wireless communications, including collecting and analyzing information. In particular, serious consideration was given to more effective performances of fighter radios and 20 mm cannon by asking for the aid of Yokosuka *Kokutai* and the *Kugisho* (Naval Air Arsenal)."

Overall, the members of the 343 *Kokutai* numbered roughly 3,000. They included over 120 pilots, staff, groundcrew, and support personnel.

About ten pilots in each squadron held an "A" class rating ("A" class pilots were able to participate in day and night operations). These were the hardcore veterans. Most of the pilots were in the "C" class, which meant that they could participate in easy operations.

The following pilots made up the air group: Naval Academy 71st term graduates who completed flight training in 1944; *Kohi* 10th term graduates; *Toku Otsu* 1st and 2nd Classes; and Reserve Student 13th and Naval Academy 72nd term pilots. At the time the unit was organized, they had the highest degree of skill in the JNAF.

Co-author Koji Takaki has made an analysis of the experience levels in Squadron 301. Veterans comprised 15 per cent of the squadron, which included warrant officers and chief petty officers. They had received training during 1941 and 1942, and were division leaders.

About 30 per cent were middle-class pilots, which included *Kohi* 10 pilots who completed fighter training in 1943. They flew as 2nd and 3rd men in a division and usually had a minimum of 800 flight hours. Finally, 55 per cent were pilots in the *Kohi* 11 and *Toku Otsu* 1st and 2nd Classes who had completed fighter courses in 1944. Some had seen action in the Philippines.

LCdr Shiga observed the squadron leaders and their men, and found that they had something more than the older generation had to offer. Different from those who entered into combat after extensive training, these young pilots had been thrown into battle before their wings had been fully developed, then brought up with severe experiences. They survived with youthfulness and fighting spirit as well as being blessed with good luck.

Capt Genda was always thinking about the welfare of his men, and if they should fall, how they would be remembered. "No one knew when he would die, above all, the pilots," Genda wrote. "In most cases, combats would occur over the sea. There would be no remains. So I decided and ordered the men to cut their hair and make fingernail clippings and put them into a small box for their time of death."

RIGHT: Members of Squadron 301 pose at the flight line in March 1945. From left to right: Lt(jg) Tatsutoshi Hashimoto, Lt(jg) Hiromu Tanaka, CPO Seiichiro Sato, PO2/c Toyomi Miyazawa, PO1/c Rokuro Asama, CPO Nobuya Komeda. Squatting to front, from left to right: LA Mutsue Naka, LA Tsuneharu Tamura and PO1/c Makoto Nishimura.

Chapter 2

"Shikishima!"
Operations – March 1945

「シキシマ！」

Much had changed since LCol Jimmy Doolittle led his B-25s from the carrier *Hornet* to bomb Japan on 18 April 1942. Capt Marc A. Mitscher was the commander of the *Hornet* on this historic mission in the darkest days of the war in the Pacific. Nearly three years later, V/Adm Mitscher was leading the Fast Carrier Task Force in the final push toward the enemy's stronghold.

The Allied forces' objective was to take Okinawa, a prime piece of Japanese territory due to its strategic location. Firstly, it filled all of the necessary requirements and had enough space for airfields to handle B-29s and their escort fighters. Secondly, its harbor facilities were large enough to accommodate shipping and stockpiling of logistical supplies. Okinawa was the perfect staging area for the final invasion of Japan.

The second strike by American carriers on the sacred Japanese homeland was not a propaganda ploy. On 16-17 February 1945, Task Force 58 pilots swept airfields in the Tokyo and surrounding regions. They claimed 344 aerial victories, although the Japanese actually lost a total of 78 aircraft. The appearance of enemy aircraft over the Imperial capital sent shock waves throughout the island nation. With Pearl Harbor a distant memory, the war had now come dangerously close.

Capt Minoru Genda, Cdr Tadashi Nakashima, and LCdr Yoshio Shiga busied themselves in training their new unit for the air battles which lay ahead. Genda wrote: "*Once we had completed building up a powerful air group, and once the fighters were launched, we intended to mow down a throng of enemy fighters like a raging wave of the sea. So I planned not to use our fighters in combat before they completed their training, about the middle of May. That was our intention, but the enemy's plan was different.*"

Intercept strategies against anticipated American carrier strikes were developed in Genda's command building. Various scenarios were played out in war games using large maps. Movements of the enemy task force assembly at Ulithi Atoll in the West Caroline Islands had been detected by radio intelligence since the beginning of March.

The first scramble by 343 *Kokutai* fighters took place on 13 March when a warning was issued. Forty *Shiden-Kais* under the command of Lt Yoshishige Hayashi took to the sky. When the formation had reached 1,800 m above Hashira Jima, anti-aircraft shells exploded all around it. It seems that no one had alerted the battleship *Yamato* to the Japanese flight plan! The crew of the giant ship had not seen such a large formation of friendly aircraft for a long time. Nervous gunners opened up at the unidentified formation because the stout *Shiden-Kais* resembled Grumman Hellcats from a distance. Fortunately, no aircraft were hit, although many pilots had their nerves frayed. The enemy did not materialize however and the first combat mission was a classic "snafu." On the following day, 14 March, the massive American fleet weighed anchor from Ulithi Atoll for the homeland of Japan.

BELOW: Capt Minoru Genda (left) with LCdr Yoshio Shiga at Matsuyama in the spring of 1945. Even though both officers seldom flew, they wore their flight helmets to show solidarity with their men. (Maru)

Prelude to combat: 18 March

On 18 March 1945, the 16 carriers of TF58 steamed toward the southern coast of Kyushu Island. The huge armada did not go unnoticed. It was detected by radar and by patrol aircraft of the 5th Air Fleet at Kanoya. Despite their formidable names, neither the 5th Air Fleet nor the 3rd had any carriers but instead commanded land-based air groups.

A Nakajima C6N Myrt recon aircraft of the 343rd's 4th Reconnaissance Squadron spotted the Americans advancing toward the mainland. "Enemy task force sighted south of Tosa!" came the warning. A total of 72 *Shiden-Kais* and *Shidens* went airborne, commanded by Lt Takashi Oshibuchi. It was a magnificent formation, its pilots full of fighting spirit and eager to tangle with the enemy's Hellcats and Corsairs. The 343 *Kokutai* was the envy of all the other units of the Imperial Navy air groups.

Back on Matsuyama Airfield, off-duty pilots had assembled at the dispersal hut at one corner of the airfield. The men were cooking over a converted fuel drum stove. It seemed as though the war was a million miles away. Suddenly, four aircraft in the distance caught their attention. They were coming in too fast and low for a landing approach. They looked like *Shiden-Kais*.

The wings of the four unidentified aircraft twinkled like sparklers. Breakfast flew in every direction; the men bolted for cover. The Hellcats strafed the airfield and were gone in a blur. They were snoopers sent over to gather the latest information on potential targets. The men cursed the Americans for ruining their breakfast and vowed revenge.

Lt Oshibuchi and his flyers returned exasperated. It was another major disappointment. The large formation flew over the coastline between Capes Ashizuri and Muroto but saw nothing. While they were away, the airbase had been attacked. The physical damage was negligible, but the insult was humiliating. However, this may have been a blessing in disguise. When the snoopers strafed the airfield, there were only a few aircraft in sight. It was fortunate for the 343 *Kokutai* that their large force of fighters was airborne at the time and had not been photographed on the airfield.

TOP: CPO Shigeyori Ina of Squadron 407 in the cockpit of his Shiden-Kai at Matsuyama Airfield, March 1945.

ABOVE: CPO Ina poses by the tailplane of 343-B-03. Note the white band on the fuselage to right.

Based upon visual observation by its recon teams, TF58 issued the following report: "*Commander Task Force 58 decided to attack these units (a large number of major fleet units) the following day with the harbor installations at Kure and Kobe as alternate targets. Fighter sweeps against intervening airfields on Shikoku and Western Honshu were to precede and follow these strikes.*"

TF58 pilots had a field day on 18 March over the Japanese airfields on Kyushu. Their mission was to smash the staging areas for the dreaded *Kamikaze* suicide aircraft, which threatened their carriers in Okinawan waters. A total of 440 aircraft under the Japanese 5th Air Fleet were launched, including 108 Zero fighters of the 203 *Kokutai* at Kasanohara and Kagoshima.

The US Navy claimed 102 enemy aircraft destroyed in the air and approximately 275 on the ground. According to Japanese historian Masatake Okumiya, a total of 110 aircraft including 32 *Kamikazes* did not return. This sum was very close to the American aerial victory claims. The Japanese reported a total of 50 aircraft demolished on the ground during the four days from 18 to 21 March, which far exceeded the American claims for aircraft destroyed on the ground on the 18th alone.

With so many Japanese aircraft in the air, the Americans did not escape unscathed. A *Kamikaze* hit the *Enterprise* at the forward elevator, but its bomb failed to detonate. The *Intrepid* had a close call and was almost hit.

Back at Matsuyama Airfield, tension and anticipation filled the air. A showdown was imminent and all the pilots looked forward to the next day.

Baptism of fire: 19 March

During the early morning hours of 19 March, the 16 carriers of Task Force 58 were in position east of Kyushu and south of Shikoku. There was a flurry of activity aboard the ships while hundreds of aviators slept fitfully

BELOW: A crewman prepares to load a belt of .50-caliber ammunition into a Hellcat. (NMNA)

ABOVE: Topping off a 150 gallon drop tank under a Hellcat in readiness for a forthcoming mission. (NMNA)

LEFT: Feeding a ship's hungry crew was a 24 hour per day job. Here, US Navy cooks prepare automatic dough mixers in a carrier's bakery. (NMNA)

BELOW: Navy enlisted men formed up into a 'chow line' to receive their meals and ate in a mess hall. Unlike infantrymen, who ate their meals out of cans and boxes, Navy men serving aboard ships could look forward to hot meals straight from the galley. (NMNA)

LEFT: Though not describable as 'fine dining', Navy officers enjoyed slightly better accommodation. Served by black stewards in a wardroom, unlike enlisted men, they paid for their meals through an allowance in their monthly salary. Japanese Navy officers also paid for their meals. (NMNA)

ABOVE: Pre-strike planning involved the use of topographic maps. Here, Marine aviators of VMF-123 aboard the Bennington go over plans for the Tokyo raid scheduled for 16-17 February 1945. Their next major strike was against Kure on 19 March. In center: Capt William Cantrell (standing rear), Maj Tom Mobley Jr. (CO of VMF-123, stooping forward), 1/Lt Archie Clapp (kneeling), and Navy intelligence officer Joe Scott on map, pointing at target.

below decks in their narrow bunks. Aviation maintenance men fueled aircraft and attached belly tanks. The ordnance men pulled belts of .50 caliber ammunition out of storage cans and layered them into the six magazines in the wings of fighters. Each magazine in the F6F-6 held 400 rounds. Once loading was completed, tape was placed over the gun barrels to prevent damage from salt spray. Others on deck wheeled 500 lb bombs on low carts to the TBMs, SB2Cs, Hellcats, and Corsairs.

An army of men in the ships' galleys labored over hot stoves to prepare breakfast for more than 3,100 men serving on each of the large fleet carriers. Intelligence officers pored over charts and aerial photographs. At 00.30 hrs, five TBM-3Ds of VT(N)-90 began their take-off for a last-minute search of the Bungo and Kii Channels leading to the morning's targets.

Photo reconnaissance of Kure and Kobe Harbors the day before revealed concentrations of warships and logistical installations. Four carriers were assigned to attack Kobe, Osaka, and Itami, on the main island of Honshu. The remaining flattops were to launch against Kure Harbor and nearby airfields. The giant 64,000 ton battleship *Yamato*, the pride of the Imperial Navy, was serenely anchored off Hashira Jima near Hiroshima Bay, south-southwest of Kure. Other notable targets included the battleship *Haruna* (misidentified as the *Kongo*), three aircraft carriers, and the light cruiser *Oyodo*.

Ens Roy D. "Eric" Erickson, a Corsair pilot of VBF-10 from Lincoln, Nebraska, had "hit the sack" rather late. He would later recall: "I crawled into my upper bunk with freshly laundered sheets and I thought, '*What a way to fight a war. At least you live good and when you get killed you die quickly.*' I soon fell off into a deep sleep. The much-despised squawk box came to life at 0430. I thought *'Christ! It can't be that time already! I just got to sleep!'* "

In the cramped confines of the carriers, hundreds of groggy aviators made for the washroom to splash water on their faces, brush their teeth, and take care of other business. They went back to slip into their flight suits and then continued to the 'chow line' for the monotonous morning fare. The much-disdained scrambled eggs were slopped onto metal trays, along with bacon and sausages, Navy beans, and hash browns. Fastidious eaters picked out the small black specks from their toast. Most of the pilots had more important things to worry about than weevils in their bread. The 'chow' was washed down with plenty of coffee or hot cocoa.

There were always a few pilots who were perennially late. Lt (jg) Forrest McCormick of VBF-17 lagged behind. He had slept through the rumpus of comrades getting up and no one had bothered to wake him. When he arrived in the mess hall, there was nothing left to eat except cold scrambled eggs. With the morning off to a bad start, he headed for the ready room.

With their stomachs full, the aviators entered the ready room, which was located directly beneath the flight deck. Each man grabbed a pair of red-lensed goggles, which he wore during the briefings and just prior to going topside; these were to aid night vision since there would be no light on deck. Flight gear and shoulder holsters hung from hooks along the walls. For the most part, the aviators were quiet and serious. Some lit cigarettes as the briefing commenced. None dared to show their nervousness.

In the ready rooms of the aircraft carriers, the ACI officers revealed the locations of various targets and suspected anti-aircraft batteries. Aboard the *Hornet*, Lt Charles Weiss, who was considered the best aerial gunner in the squadron, was very excited. "I went to sea and I wanted to shoot down aircraft like everybody else," he recalled. "And I would lie awake at nights thinking that maybe today will be my chance. I was very eager. We were briefed by the intelligence people that we were going to run into the best that Japan had. The best aircraft, the best pilots. We knew that they were going to be good and I was just praying for this chance to shoot down my plane!"

The flight leaders gave out the final instructions and wished everyone good luck.

"Pilots, man your aircraft!" came the order over the speaker. Each man went to retrieve his gear from the wall. He strapped on his shoulder holster and made a final check of his .38 Smith and Wesson revolver. This sidearm provided a false sense of security and had very little practical value. If a pilot was about to be captured, it was the first piece of equipment disposed of or surrendered. Life jackets were slipped on, and finally parachute harnesses. As the men left the ready room, their night vision preserved, they threw their red-lensed goggles into a large box. Walking up to the deck, they parted from their comrades with a handshake, a thumbs up, or a nod.

They climbed into their cockpits, strapped themselves in, then started their pre-flight procedures. Ens Roy Erickson switched on the landing lights of his Corsair without thinking. An angry voice boomed from the megaphone: "*Turn off those Goddamn lights!*"

"Pilots, start your engines!" came the final command. The deafening roar, the salt wind, and the smell of exhaust signaled the start of a momentous day. Many pilots recited their prayers silently. Some would return as aces, many would have epic stories to tell, and some would never return; but defeat was never on anyone's mind.

Matsuyama Airfield

At Matsuyama Airfield, Capt Genda and his staff burned the midnight oil. Cdr Nakashima, LCdr Toshio Hashimoto (4th Recon Squadron Leader), and Radio Officer Lt Tsuyoshi Tsuzurahara had not slept. The flood of information pouring into the radio room demanded their fullest attention. They had to select and analyze messages necessary for their action plan. Staff officers labored over last-minute details and strategies.

The enlisted pilots slept side by side in large sleeping areas, with futons over straw mats on the wooden floor. Flight officers were afforded better accommodation and usually slept two or four to a room. In the rank-conscious military, each level of rank brought special privileges. At 04.00 hrs, all available pilots were roused from their sleep by the call of "*All pilots up!*" Quickly and silently, they prepared for the day's actions, folded up their futons and headed for the washroom.

The men slipped into their one-piece brown gabardine flight suits and put on their short black leather boots. The officers stepped out into the cold, dark morning and headed briskly to the mess hall while the NCOs and lower ranks took their breakfast in their quarters where they sat down at a long communal table. They were each given a prepared pack of food, which was taken by flight crews on their missions. This was a departure from their usual morning meal which consisted of hot rice topped with a raw egg, pickled vegetables, broiled fish, seaweed, and soybean paste soup. Inside the pack was *sekihan* – rice boiled with red beans – prepared for auspicious occasions. Green tea added the necessary caffeine to get their day started.

At 05.00 hrs, Capt Genda gave the command for all pilots to assemble. The flyers quickly finished their breakfast, tied on their life jackets and wound silk mufflers around their necks. Grabbing their leather flight helmets, they raced outside.

Lt Goro Ichimura of S407, formerly a home defense pilot flying Jacks with the 302 *Kokutai* at Atsugi, walked quickly to the apron of the airfield with his classmates, Lts Kozo Shima, Kunio Matsuzaki, Ryoichi Yamada, and Masaji Matsumura. Ichimura recalled: "We walked along the road in the darkness under a blackout but with a light step amongst many rank-and-file pilots."

Leading Airman Tsuneharu Tamura was appreciative of the men who serviced the fighters. "Leaving our hut, I heard the sound of engines under trial running in the dark. I ran to the airfield to thank the groundcrews for their labor."

The mechanics had been up since midnight, preparing the *Shiden-Kais* for flight. Due to the poor quality of aviation gasoline, ranging anywhere from 85 to 100 octane, the mechanics were constantly adjusting the Homare 21 air-cooled, 18-cylinder engine. Some of the fuel was derived from the roots of pine trees and blended with standard aviation gas to stretch the fuel supply.

It usually took the groundcrew half an hour to service each fighter. Fuel tanks were topped off, the two oxygen bottles behind the pilot's seat were charged, and the four 20 mm Type 99 Model 2 cannon loaded. Belts of 20 mm shells were layered in two trays in each wing – 200 rounds per gun. To complete the servicing, the windshields were cleaned.

Capt Genda felt proud at the sight of the leaders of the 301st, 701st, 407th, and the 4th Reconnaissance Squadrons at attention in front of their men. The rows of pilots looked confident and resplendent with mufflers around their necks.

The NCO airmen of the S301 showed their solidarity by sporting silk violet mufflers, each embroidered with the pilot's name. Since they flew the *Shiden* aircraft – *Shiden* translates as 'Violet Lightning' – it was decided that their mufflers should be violet. However, such silk was extremely difficult to obtain. Miss Kotoko Onishi, who "adopted" these men while they were stationed in Matsuyama, managed somehow to obtain enough silk for about 20 mufflers. Pupils at the local girls' high school were asked to embroider a pilot's favorite motto. CPO Shoichi Sugita's muffler, as well as those of his three division members, read: "*Smile sweetly and shoot him down without fail.*"

The Japanese had received reports that American carriers were lurking somewhere south of Kyushu. The previous day, the 18th, the enemy had raided airfields in Southern and Eastern Kyushu. Genda gave an assessment of the situation. He knew an aerial assault was imminent and surmized that warships in and around Kure Harbor would be the logical targets. "The enemy's invasion is inevitable this morning," he addressed. "We are going to intercept the enemy aircraft and deal a severe blow to them. Our target is the enemy's fighter force. Do not give your eyes to bombers, ignore them. Try to shoot down as many of the enemy fighters as possible!"

Cdr Tadashi Nakashima, the deputy commander, briefed his men about the probable movements of the enemy task force. LCdr Yoshio Shiga gave simple instructions on the intercept plan and ordered the pilots to remain on standby alert. When the pilots were dismissed, they boarded trucks for their respective flight lines.

At 05.45 hrs, upon Genda's order, three Nakajima C6N (Myrt) single-engined carrier reconnaissance aircraft took off from Matsuyama Airfield in search of the American task force. The search involved flying south in a fan-shaped pattern between Capes Muroto and Ashizuri. Each crew was to report the direction of enemy aircraft movement. Ens Mitsuru Takada's aircraft took off after a slight delay, probably due to engine problems. His assignment was to patrol the area to the east.

Capt Genda, widely known as the mastermind behind the Pearl Harbor attack, was counting on his reconnaissance team to spot the enemy. There would be nothing more disgraceful for an air force commander than to have his aircraft destroyed on the ground. He regarded it as a stain on his family honor, which would last forever. It was tantamount to a Samurai warrior being slain before unsheathing his sword.

Shortly after sending his Myrts aloft, one aircraft returned due to landing gear trouble. With only two scouts in the air, Genda's problem intensified. The prospects of an American-style Pearl Harbor payback weighed heavily on his mind.

LEFT: The 343 Kokutai's 4th Reconnaissance Squadron played an integral part in the unit's operations. Nakajima C6N Saiun ('Painted Cloud') aircraft are seen here on Matsuyama Airfield in early 1945. With a maximum speed of 329 knots at 6,100 m (379 mph at 20,015 ft), it was almost equal in speed to the Grumman Hellcat.

The carriers launch

The American plan was simple enough. The fighters, many loaded with rockets and bombs, were to be launched first, followed by SB2C Helldivers and TBM Avengers. The first fighters to go in were to engage enemy aircraft and clear them from the skies before the arrival of the bombers. If no enemy aircraft were met, they were to attack their assigned targets in and around Kure Harbor and Kobe.

Task Groups 1, 3 and 4 were assigned to attack Kure. TG1 consisted of the carriers *Hornet, Wasp, Bennington,* and *Belleau Wood.* TG3 was comprised of the *Essex, Bunker Hill,* and *Cabot.* The *Yorktown* and *Intrepid* of TG4, minus the *Enterprise,* completed the Kure Harbor attack force. The *Enterprise* was forced to retreat when its forward elevator was damaged by a bomb-hit on 18 March, the day before.

Cdr James J. Southerland, CO of VF-83, was given the honor of starting the strike. The *Essex* turned into the wind. At 05.40 hrs, he was catapulted off the deck. Of his 16 Hellcats, 12 carried High Velocity Aerial Rockets (HVARs), three under each wing. Two aircraft carried one 500 lb bomb each, and the remaining pair were loaded down with full ammunition. Their target was Matsuyama Airfield. The home base of the 343 *Kokutai* stood in the path of the carrier strike force heading toward Kure.

At 05.45 hrs, the *Yorktown's* VF-9 began their take-off for a fighter sweep of Kure, Matsuyama, and Saijo Airfields. The leader was 25-year-old Lt Leslie DeCew, on his second tour of duty with the squadron. The six-victory ace was to lead his 16 Hellcats to rendezvous with ten F4U Corsairs (VBF-10) from the *Intrepid.* Most of the aircraft from both squadrons carried HVARs, while the remainder carried bombs.

The Marines from the *Bennington* took to the air, starting at 05.45 hrs. Sixteen Corsairs from the 'Eightball Squadron' (VMF-123), led by Maj Thomas E. Mobley Jr., climbed into the dark for their assault against Hiroshima Airfield. All of the aircraft carried rockets. One Corsair turned back a few minutes after launch due to engine trouble. As they neared the end of a destroyer picket line, they were fired upon by trigger-happy gunners. Fortunately, their aim was dismal. The destroyer stopped shooting and the Leathernecks regained their composure and proceeded on to their target 148 miles away.

At 06.18 hrs, the first of 20 Grumman F6F-5s of VBF-17 began rolling down the flight deck of the *Hornet.* The strike leader was an ex-Flying Tiger pilot named John Francis Perry, who was flying under the *nom de guerre* of Edwin S. Conant. His task was to neutralize enemy aircraft and airfields at Iwakuni and Matsuyama prior to the arrival of the strike force against Kure. He had joined the American Volunteer Group under his friend's name before the war. The

BELOW: Task Force 58 launched strike aircraft from 05.45 hrs until 08.00 hrs on 19 March 1945. Here, a Corsair of VF-84 starts its journey toward Kure from the Bunker Hill. (NMNA)

false identity was to skirt international law, which forbade American military personnel from direct involvement against the Japanese. When the war started, he resigned from the AVG and rejoined the Navy, still posing as Edwin Conant. He would be court-martialed after the war for this duplicity.

At 06.30 hrs, the *Cabot* sortied seven TBMs and 12 Hellcats. In succession, hundreds of aircraft took to the air. As "Eric" Erickson recalled: "As the different groups rendezvoused over the Task Force, the sky was full of aircraft as far as the eye could see, all making their way toward the home islands of Japan. A warm feeling of pride filled my heart as I realized I was part of this gallant force."

ABOVE: Forming part of a huge American aerial armada, SB2C Helldivers and TBM Avengers escorted by Hellcats and Corsairs, head toward the Japanese homeland, 19 March 1945.

Standby on Matsuyama Airfield

The 343 *Kokutai* was about to go into battle suffering a serious disadvantage. When the air group was formed, each squadron was allotted 48 *Shiden-Kais*. However, due to the realities of war, this overly optimistic figure was not realized. As a result of around-the-clock operation in the Kawanishi Naruo Factory, production climbed from 35 aircraft in January, to 47 in February, and 56 in March. The completed aircraft were ferried from Naruo Airfield by the 343rd pilots in turns, using K11W1-2 *Shiragiku* crew trainers.

Each squadron was divided into two *buntais* (battalion). The 1st *Buntai* of Squadron 301 had already completed the transition to the *Shiden-Kai*, but the 2nd *Buntai* under Lt Masaji Matsumura had not. Thus, the 2nd *Buntai* was sidelined. Squadrons 407 and 701 were still converting.

Capt Genda decided to use eight N1K1-J *Shidens* (mid-wing early model George) as cover flights. He knew that the inferior *Shiden* could not work together with the *Shiden-Kai* in formation tactics. Squadrons 407 and 701 would each contribute four aircraft, led by Lts Goro Ichimura and Ryoichi Yamada. LCdr Yoshio Shiga was deeply impressed with Lt Yamada of the 701st. He had volunteered to fly the old *Shiden* fighter because no one else wanted to and because he was accustomed to flying this hard-to-control aircraft from his service with the *Shiden* unit, Squadron 402, under the 341 *Kokutai* in the Philippines.

At 06.30 hrs, Genda ordered the *Shidens* to take off and circle Matsuyama Airfield at 5,000 m. The base would be under their protection to insure that the rest of the interceptors could be launched in safety. The *Shidens* carried belly tanks to buy them additional air time. This would be the first – and the last – day that the 343 *Kokutai* would use the old mid-wing fighters in combat.

BELOW: Lt Goro Ichimura

Lt Ichimura understood the gravity of the situation. He recalled: "The result of our training for the past months would be proven in a few hours... The immense duty was on me and Lt Yamada with two four-aircraft divisions under the codename of *Hayate* (swift wind). We taxied to the take-off line on the coast side across the field in the darkness and finished assembling at the end of the asphalt runway." The wind was southwest. Aircraft were to take off individually from the seaward side toward the mountain.

As the dawn drew near, the vague outline of the mountains could be seen. The roar of engines broke the serenity of the cold morning. Ichimura watched as LCdr Shiga positioned himself in front of his aircraft and swung down the white flag. Ichimura opened full throttle and his aircraft screamed down the field, followed closely by his three comrades, CPO Tsuneo Ozeki (No.2 man), LA Magoshichi Koma (No.3), and LA Yoichi Saiki (No.4).

Lt Yamada was off to a bad start in more ways than one. The right landing gear of his aircraft broke while taxiing forward. This common occurrence which had plagued the *Shiden* did not deter the young officer from his task. He jumped out, ordered his No.3 man, CPO Sukeo Yoshioka, to vacate his mount, then jumped into that aircraft, and roared off. But now he was leading an irregular three-aircraft formation. His wingman now became LA Naruhiko Yamada, a novice pilot who was originally assigned to CPO Takumi Sugitaki in the

2nd section. Sugitaki, the original No.2 man, became the odd man and was ordered to cover the two Yamadas in the lead section.

CPO Sugitaki was a hardcore veteran who had fought with the 302 *Kokutai* at Atsugi just before his transfer to the 343rd. He was the pilot who caused a B-29 to explode on 3 December 1944 over Boso (Chiba) Peninsula when he shot it up with his 30 mm cannon-equipped Jack. Lt Yamada felt that in a combat situation, Sugitaki would be able to fend for himself without a wingman.

The 343 *Kokutai* recognized the strengths of the American fighter formations and formulated their own tactics; section leaders in a four-aircraft division piloted the first and second aircraft. The first flight would be covered by the second flight. The No.1 man was paired with No.3, while the No.2 pilot teamed up with No.4.

The JNAF originally flew a three-aircraft formation at the start of the war. At Rabaul, Japanese airmen noticed that the American four-aircraft division was more advantageous and dangerous. It was found that in a mêlée, the third man had no protection and was easily separated and exposed to danger. Lt Zenjiro Miyano, commander of the 204 *Kokutai* at Rabaul, recognized this fact and became the first to adopt the four-aircraft division in early 1943. The Navy formally changed to a four-aircraft division in 1944.

At the flight line of Squadron 301, LA Tamura climbed into the cockpit of aircraft 343-A-13, strapped on his parachute harness which was kept in the aircraft, and buckled himself in. The oxygen mask hung from his wireless receiver; the day's flight was not expected to reach more than 18,000 ft and he would not need it. He felt adequately prepared. In his leg pockets, he carried caramel candies and tube chocolate for a quick energy boost.

Looking forward to getting back into action was division leader CPO Shoichi Sugita, the highest-scoring ace in the air group with claims of over 120 victories. The division of Sugita, CPO Tomokazu Kasai, PO2/c Toyomi Miyazawa, and Tamura was a well-coordinated team and they thought of one another as brothers.

The Sugita Division had a newcomer that morning. The usual No. 2 man, 19-year-old CPO Tomokazu Kasai, had been replaced due to a bout of diarrhoea. Kasai had been Sugita's wingman since early 1944 when they had fought in the Marianas. Kasai's stand-in was CPO Satoshi Yokoshima, a recent transfer from Squadron 402. With the battle about to commence, Tamura did not like the idea of having a substitute on the team. He had never flown with Yokoshima before.

Tamura began to go through his pre-flight procedures and connected his earphones to the radio. Japanese aircraft radios were generally of very poor quality. Unshielded spark plugs caused too much static and made them virtually useless. However, the 343 *Kokutai's* R/T had been improved to the point where the enemy's transmission could be heard. However, its pilots still relied heavily on hand signals. Wingmen could use their radios only for receiving transmissions – except in emergencies, while only the formation or division leaders could transmit. The idea was to eliminate unnecessary chatter and confusion during combat. The Americans, on the other hand, relied heavily on their radios during combat to issue directions, warnings, and pleas for help in tight situations.

BELOW LEFT: LA Tsuneharu Tamura (standing to right) with comrades from Squadron 301. (T. Tamura)

BELOW: Taken a few minutes after the photograph to left, seen here, front row, left to right: CPO Itaru Shikano, CPO Yoshikazu Miyamoto, CPO Tesshin Kamei (on ground), CPO Shigeo Suzaki, and PO1/c Tsuneishi Yoneyasu. Standing left to right: PO1/c Mahito Yoshihara, PO2/c Susumu Imai, CPO Hiroshi Tanaka, CPO Eiichiro Sakurai (wearing sunglasses), CPO Tomokazu Kasai (leaning forward), LA Tsuneharu Tamura, and PO2/c Kiyoshi Arima.

Back at the airfield, LA Tamura switched on his radio. He was surprised to hear the Americans chattering on his frequency despite strict orders to maintain radio silence.

The radiomen at the command center heard the incoming American pilots conversing and were insulted by their arrogance. It was a breach of regulations and a sure sign of overconfidence. There must have been some exceptional reason why their conversations could be heard on Japanese radios.

Scramble!

The seven *Shiden* aircraft circled Matsuyama Airfield as the sun began to rise from the horizon in the east. It was a breathtaking sight. Lt Ichimura could see the entire fighter force of the 343rd below waiting to sortie.

In the command center below, Cdr Nakashima constantly checked radio communications with his fighters. "If by some chance the radio would be out, it would be all up for us!" he wrote. "The radioman was vital. For a trial run, I radioed the cover flight. A response came back from Lt Yamada in a lively voice. I thought to myself – *Good!* Then, the first report came in!"

At 06.50 hrs, a wireless message from Ens Takada's Myrt reconnaissance aircraft was received and it was electrifying: "Enemy task force sighted, 50 kilometers south of Cape Muroto!" Instantly, Capt Genda shouted to Cdr Nakashima: "All squadrons get ready to scramble!" Nakashima, in turn, barked the orders to the radioman, 17-year-old PO3/c Tadashi Kakegawa.

"Shikishima!" came the order over the aircraft radios – the code for "all aircraft start engines." Engines belched smoke, then whined as propellers started to spin. Groundcrews scurried under the aircraft to remove the wheel chocks.

A moment later, a second message came in from Takada: "Large enemy formation heading north over the south coast of Shikoku!" The enemy were now 128 km away. Cdr Nakashima was marking their advance on a map. Genda finally gave the scramble order.

"Sakura, Sakura, Niitakayama nobore!" ("Cherry Blossoms, climb Mt. Niitaka!") shouted Radioman Kakegawa into his microphone. [1]

Sixteen fighters of the 701st, with Lt Takashi Oshibuchi leading the formation, began to taxi forward. The senior squadron leader was also the air commander for all three squadrons. Capt Genda had full confidence in him and so did his men. The wind was blowing from the northwest. The asphalt runway would not be used for the formation take-off. The aircraft rolled in unison across the grass field toward the sea, picking up speed against the wind. Each pilot felt extremely confident as he opened full throttle, feeling the might of his aircraft's 1,990 hp engine.

From the wide, grassy fields, eight fighters roared off into the sky, line abreast. Rolling through clouds of dust, their groundcrews and base personnel cheered them on, waving their caps. Surviving pilots who participated in the sortie proudly remember the scene.

Lt Yoshishige Hayashi's 16 fighters from the 407th followed the 701st. He started with 17 aircraft, but one was forced to cancel take-off due to engine problems. The young leader had formed a strong bond of trust with his men. Hayashi commanded Squadron 407 from its inception, leading his subordinates into combat in the Philippines. Surviving that campaign had brought his men closer to him.

Lt Oshibuchi's formation ascended rapidly in a large left turn, steadily gaining altitude over Iyo Nada on the Inland Sea, west of Matsuyama Airfield. Hayashi's group raced after the first formation.

The last group to scramble was Lt Naoshi Kanno's 301st. The squadron's take-off was slightly delayed. Three fighters had mechanical problems and were forced to abort, leaving the formation with only 18 aircraft. "As soon as our squadron reached the take-off line," recalled Tamura, "the first division led by Lt Kanno, followed by Sugita's and Shibata's divisions, began formation take-off all at once in perfect order with throttle levers down, raising dense clouds of smoke. I was determined to die in the sky that day; while I lifted the landing gear, our squadron went up straight toward the sea and gained altitude rapidly. CPO Sugita looked back to us wingmen frequently. I felt relieved and encouraged by him. I concentrated my nerves desperately so that I would not disturb the order of the formation." It was understandable that Tamura would be nervous; he had less than 500 flight hours at this time.

It was now 07.00 hrs. With clouds of dust and smoke, the scramble of all three squadrons was over within ten minutes. Never had Genda's pilots moved so swiftly and as one. It was a breathtaking sight seldom seen since the early years of the Pacific War.

In the meantime, another urgent message came in: "A large group of fighters and bombers heading north over the Bungo Straits at altitude 3,000 meters." The Americans were approaching in roughly two directions from the south: along the eastern coast of Kyushu and up the straits along the western coast of Shikoku.

Messages continued to come in, one after another. The information not only came from the Myrt reconnaissance aircraft, but also from the radar site at Cape Ashizuri, watch posts along the coast, and even fishing boats at sea. Therefore, Capt Genda was aware of the enemy aircraft types, numbers, their direction, and altitude.

When the last aircraft left the field, Oshibuchi was climbing past 4,000 m. Lt Kanno was still climbing at 760 m when he apparently heard a radio warning that an enemy formation was heading north over the Bungo Straits. American carrier aircraft could be seen far in the distance as they headed toward Kure. At that moment, the 301st was in a vulnerable position at low altitude. Kanno ordered a right turn to avoid the enemy formations and headed eastward.

VBF-17 versus the Georges

Lt "Muscles" Conant was angry. His 20 VBF-17 Hellcats were supposed to rendezvous with the 16 Corsairs from the *Bennington* for a coordinated strike against airfields in the Kure area. However, the Marines of VMF-123 were nowhere to be seen. With a higher cruising speed, the Corsairs arrived ahead of their Navy comrades. Anxious to get into action, the Marines were not about to wait for their Navy counterparts and set course for the target area, droning onward, maintaining strict radio silence.

Capt Genda was relieved to have finished launching his aircraft by 07.00 hrs. By 07.05 hrs (Japan local time), Lt Oshibuchi's squadron was fully assembled and heading westward, hunting for the enemy which would be coming into Iyo Nada from the Bungo Straits. The advancing Hellcats of VBF-17 were not aware of their presence. Oshibuchi was planning an ambush and carefully calculated the interception point.

The pilots of the 343 *Kokutai* were fully prepared and ready. They had trained for months just for this moment. Lt Oshibuchi had full confidence in his division leaders. His 2nd division leader was Ens Akio Matsuba, a China War veteran with 3,800 flight hours and a double ace. The leader of the 2nd *Chutai* of another eight-aircraft group was Lt Kunio Matsuzaki, followed by his 2nd division leader, Lt(jg) Yukihiro Watabe, 21, the youngest officer pilot from the Naval Academy on this mission. His veterans had not seen combat in months and were keen to take on the Americans.

Lt Hayashi, commanding S407, who was chasing after Oshibuchi's group, was supported by Lt Kozo Shima, the 2nd *Chutai* leader who was one of the survivors from the Philippines. WO Minoru Honda, 2nd division leader, was a very stern disciplinarian whose heavy-handed ways helped his subordinates survive. By his own reckoning, he hit and downed over 40 enemy aircraft before he stopped counting.

The seven *Shidens* were circling the Matsuyama region and had just seen the last aircraft sortie from the airfield. Their assignment was now over, but they would enjoy only momentary relief.

Capt Genda, listening in on the radio, barked orders to his radioman. PO3/c Kakegawa transmitted a warning to Oshibuchi. "An enemy formation coming in at altitude 13,120 ft southwest of the airfield!"

"I've already caught them! We're going in!" responded the squadron leader. LCdr Shiga was pleased to hear Oshibuchi's strong and lively voice in his headphones. The XO had returned to the command center after sending off the 50 *Shiden-Kais*.

"*Hayate, Hayate,* join your squadrons!" crackled the voice of Cdr Nakashima over the headphones to the *Shiden* cover flight. Lt Ichimura was making a left turn over the southwest side of the airfield when he heard this transmission. The two *Shiden* divisions banked from each other to part and started to chase after their squadrons.

Lt Yamada spotted Hellcats far on the horizon. He had been watching the Oshibuchi and Hayashi squadrons advancing westward 1,000 m below his *Shiden* group, when a formation of enemy aircraft came into his view. His eagle-eyesight may have gained advantage for the Japanese side. "The enemy was ten nautical miles southwest of Matsuyama Airfield. They did not seem to have noticed us," he recalled. Yamada quickly got on his radio and shouted a warning.

"At last they came," recalled Nakashima. "I was frustrated as I could not see the sky because I was busy listening to the incoming messages and sending orders. I continued shouting with tension. Then one man rushed into the radio room, saying *'Sir, I can't identify if they are enemy or friendly!'* I jumped out of the center and saw that they were Corsairs and Grumman F6Fs. *'You fool, they're enemy aircraft!'* I shouted to him. Again, I jumped back to the wireless."

"In order to join our 407th Squadron which was in a climb about 1,000 meters below our altitude," Ichimura remembered, "we circled over the north of the airfield when I spotted a Grumman formation coming over the sea from the west."

Now he was close enough to see the white star markings on the Hellcats. The enemy was in combat formation at an altitude about 1,000 ft below his comrades. He had no time to join his squadron and prepared to jump into the fray with the 701st.

Conant's formation passed the western side of Shikoku Island and was about six miles from the coast, heading north over the Inland Sea. He was surprised to see about 40 fighters up ahead. It was way too early in the morning for that many 'friendlies' to be there. They could not be American aircraft because they were flying in the opposite direction, away from Kure. The unidentified aircraft were flying at about 13,000 ft in the familiar US Navy formation of four-aircraft divisions and two-aircraft sections. The *Hornet* aircraft were cruising at 12,000 ft. It was overcast 10/10 at 13,500 ft with visibility between ten and 15 miles. At that moment, 32 aircraft led by Lts Oshibuchi and Hayashi were flying near the ceiling (underside) of the clouds.

The VBF-17 pilots were unfamiliar with the new Kawanishi N1K2-J *Shiden-Kai*. They knew about the mid-wing George through flash card drills. In the furious combat to follow, the *Shiden-Kai* would be mistaken for the Japanese Army's Ki-84 *Hayate* (Frank), the Ki-44 *Shoki* (Tojo), and the standard Navy A6M Zeke fighter. To add to the confusion, some of the Grumman pilots would report fighting the Japanese Army's taper-nosed Ki-61 *Hien* (Tonys).

Lt Yoshishige Hayashi, leading the Squadron 407 behind the Squadron 701, had to make a quick decision. He could stay with Oshibuchi or take his men back to Matsuyama Airfield to defend it. He would opt for the latter. With no air defense over his home base, the enemy could cause untold damage.

The *Shiden-Kais* were flying on a south-easterly course and the Hellcats were approaching from the north-west. As the Japanese and American aircraft drew closer, the strict radio silence maintained by the Hellcat pilots fell apart.

Lt(jg) Robert A. Clark's heart jumped. He too had spotted a large group of aircraft coming in from the side and now realized they were not friendly. Breaking the radio silence, Clark's division leader, Lt Charles F. Weiss yelled out a warning.

The duel between the Hellcats of VBF-17 and the fighters of the 343 *Kokutai* was about to commence. As the two groups closed, Lt Oshibuchi took his men into the clouds. His intention was to get to the rear of the Hellcats. However, Lt Conant was not about to be that accommodating. He ordered his men into a 180-degree turn in order to meet the enemy head on from above. His Hellcats would go straight in and slaughter the opposition.

The Japanese came at the Americans with altitude advantage and the sun to their backs. The Grummans pulled up to face their challengers. The divisions led by Conant and Weiss took the brunt of the attack. "I met the first one that came down knifing through us," reflected Weiss who positioned his 2nd Division to Conant's side. "From then on, it wasn't like any dogfight I've ever been in. It was just a big swarm, like a giant hornet's nest."

BELOW: Conant Division. From left to right: Lt Edwin S. Conant, Lt(jg) Forrest McCormick, Lt(jg) Roger W. Karr, and Lt Fred Prinz.

ABOVE: Lt Fred Prinz crashes his Hellcat into a barrier upon his return to the Hornet.

The first Americans to be hit were Lt Fred Prinz and his wingman, Lt(jg) Roger W. Karr from Conant's division. Oshibuchi and his wingman, CPO Kikuichi Ishikawa, rocketed down with their cannon hammering. Karr went down and was never seen again. Prinz was hit in the port wing, which rendered his guns inoperative and destroyed his hydraulic system. He hauled out quickly and returned to the carrier alone, where he made a no-flap landing; he crashed into the aircraft barrier, but escaped injury.

Conant and McCormick fired at the incoming enemy aircraft. Conant hit Oshibuchi's wingman in the engine. CPO Ishikawa parachuted out of his stricken aircraft but did not survive. He is said to have shot down one Hellcat.

Lt Conant rushed in to help his wingman. He followed a George down to 10,000 ft and made a highside run, stitching the enemy machine from the cockpit to the engine. It immediately burst into flames and went down. Conant recovered only to find three fighters bracketing him, one on either side and one above. He scissored down to 3,000 ft, snapped out a burst, damaged one and escaped up to 12,000 ft; he never rejoined his wingman.

"On their first run knifing through our formation, they shot down or disabled about half of our formation," recalled Weiss. "My belly tank caught fire and I had holes in the fuselage and wings. I released the belly tank and dove to extinguish the fire."

Lt Hayashi stayed long enough to make certain that Oshibuchi's squadron had the lead. His formation stayed above while the 701st dived down into the Hellcats. Confident that everything was well in hand, Hayashi peeled away and took his flock back toward the coast, racing for his airfield.

The *Shiden-Kais* broke off from above, making two-aircraft section attacks, recovering together, joining up, and climbing for altitude advantage, then attacking again. Their training showed. Within the first minute, the Americans realized that their adversaries were skilled and dangerous; this was not going to be another "turkey shoot".

For both the Americans and the Japanese, the only strategy was offense. VBF-17 had a strong team, but the 343 *Kokutai* pilots were determined to win their first battle over their home territory.

The huge, almost surreal, dogfight sprawled across the sky in streaks of black greasy smoke and bursts of orange fireballs; confusion reigned. Because the Hellcats and Georges looked similar in a head-on confrontation, pilots from both sides hesitated and held their fire for fear of hitting their own. It is probable that a few aircraft were shot down by friendly fire.

Lt Ichimura recalled: "I hurried into battle and joined the combat. After that, I did not know what I was doing! The white star markings on one wing and the white lines on the wing for identification came into my sight, persistently one after another. Each time we closed in, tracers from our four 20 mm cannon crossed with those from the Americans' 13 mm guns."

"If a Zero was making an attack forward of 180 degrees, the only defense was to counterattack head-on," explained Charles Weiss. "If the attack came from the rear 180 degrees, the only defense was to dive straight away. The only defenses were crossfire, the armor plate behind the pilot's seat, and the engine firewall ahead of the pilot."

With so many enemy aircraft around him, all Weiss had to do was to point and squeeze the trigger. With his keen shooting eye, he claimed two fighters destroyed and three probables. However, he would never make it back to the *Hornet* to report his successes.

A George came at Weiss from the rear at an angle to place 20 mm shells into his dashboard. He was shocked to see his instrument panel disintegrate before his eyes, but he snapped out of the line of fire in time to save his own life. The assailant was probably Ens Akio Matsuba. He claimed a Hellcat with one burst at the start of combat, then another while covering the lead aircraft and his wingman.

One George pilot spotted a Hellcat down below and executed a diving attack. Weiss saw him coming,

RIGHT: Weiss Division: Left to right: Lt Charles F. Weiss, Lt(jg) Robert A. Clark, Ens Harold Yeremian, and Lt Robert Junghans.

pulled up, and opened fire, but his ailerons felt dead. "I was getting hits on him and he was getting hits on me!" he recalled.

The two fighters clipped wings and that was the last thing the Grumman pilot remembered. He was thrown out of the cockpit and blacked out. When he regained consciousness, he was floating down in his parachute. For Lt Charles Franklin Weiss, his troubles were just beginning.

Back at Matsuyama, the airfield lookout shouted a warning and Genda and his staff ran outside to scan the sky. "I sighted a formation of aircraft heading for Kure almost straight above the airfield, slightly north-northeast," recalled Genda. "At first I thought they were friendlies, but at the next moment, judged that they were F6Fs as their wingtips were squared off."

The enemy aircraft were the *Cabot* group and VF-9 passing overhead, with VF-83 about five minutes behind.

Ens Clark found himself in a real predicament. He was sucked into the giant mêlée without his division and section leader, Lt Weiss. When the Japanese plowed through their division, "Chuck" Weiss dived to the right and his wingman lost him from view. Clark was unable to drop his belly tank. He flamed a *Shiden* while chasing it, only to find another enemy pilot firing at him from behind. He took his aircraft nearly to the ground and evaded his pursuer, then rejoined the fight. His radio was cluttered with desperate calls for help and responses. He saw four or five parachutes floating downward.

"Whilst chasing an enemy aircraft, I looked back and found an F6F adopting an attack position on me," stated Ens Matsuba. "I judged that he was still distant. I fired a burst at the plane before me when I was shot and injured in my foot from sideways, almost at the same time as I fired."

Ens Robert A. Clark had spotted Matsuba and gave his opponent a 90-degree deflection shot, which tore out the canopy above the pilot's head. A .50 caliber tracer or incendiary bullet creased Matsuba's cheek. The Japanese pilot, momentarily stunned, rolled his fighter over and dived in an inverted position toward the deck. Although Clark believed he had sent his opponent down to a watery grave, Matsuba finally regained control of his aircraft and escaped. Clark was credited with two victories during the action. Although Matsuba was safe in his now 'well-ventilated' cockpit, Clark had avenged his leader.

BELOW: Ens Akio Matsuba experienced a 'close shave' during an encounter with Ens Robert Clark, who claimed him as destroyed. (A. Matsuba)

Lt Ryoichi Yamada's division was behind and above the friendlies engaged in deadly combat against the Grummans. In the swirling dogfight, he worried that he may not be able to distinguish friend and foe. His duty was to provide top cover for Squadron 701 below. His two comrades, CPO Sugitaki and LA Yamada, clung closely behind.

"I hurried to a position straight above the combat," explained Yamada. "My wingman tended to stray out of formation. It could not be helped as he was a young Leading Airman and this was his first combat. I squeezed the throttle to put my formation in order."

"The combat had turned into a mêlée. I watched and found *Shiden-Kais* hunting down the Grummans. As the zone of combat expanded, some of the Grummans were forced to get out of it. Once getting out of the zone, they could possibly regroup and come back with full advantage. I immediately knew that my duty was to hit such aircraft like swatting flies with a swatter, and began maneuvering to attack a four-aircraft Grumman formation. My wingman was straying again. So I tried a practice attack. He recovered his usual state of mind and followed me as he did in training."

The Hellcats had been spotted from above and there is strong evidence that Lt Yamada's three-aircraft *Shiden* division bounced Lt(jg) Charles Beall, Jr.'s division, consisting of Beall, Ens Raymond Evans, Lt(jg) Byron A. Eberts, and Ens Noel L. Hannah.

"I concentrated my energy and started an actual attack run, approaching 100 meters just behind a Grumman," recalled Yamada. "He had not noticed me yet. I closed so near that I needed no gunsight and pressed the firing button on the throttle lever. Within half a second, the Grumman caught fire. I sent him another burst for a quarter second. His wing flew off. The stout-looking Grumman banked. The power of the 20 mm cannon was great. My wingman shot down another Grumman, too. It was a key to victory in formation fighting to prevent enemy attacks while another section engaged."

LEFT: Beall Division. From left to right: Lt(jg) Charles T. Beall Jr., Lt(jg) Byron A. Eberts, Ens Noel L. Hannah, Ens George Salvaggio (did not participate in the 19 March mission), and Ens Raymond Evans.

As the mid-wing *Shidens* plowed through his division, Ens Evans pulled up and fired a full deflection shot into the Japanese on Beall's tail. Beall then latched onto an opponent's tail and fired a four-second burst. Lt(jg) Eberts claimed three aircraft, shooting down the first in a head-on run, and destroying the rest from behind as they chased after his comrades. Lt Yamada probably hit Ens Noel L. Hannah; he never returned. The odd man in Lt Yamada's three-aircraft division, CPO Sugitaki, became separated in the dogfight while flying cover and had to fend for himself.

While VBF-17 was heavily engaged in combat, there were simultaneous battles elsewhere. It seemed as though the morning would never end.

Air raid over Matsuyama Airfield

Lt Yoshishige Hayashi had heard an urgent warning that enemy aircraft were nearing Matsuyama Airfield and brought his Squadron 407 to its aid. Placing his aircraft east of Matsuyama, he could probably see the enemy making preparations for their attack. Arriving over the airfield from the east, Hayashi intended to swing his divisions to the rear of the Americans and gave the signal to dive, but the hurried ambush was not successful. The dive resulted in a head-on encounter at almost the same altitude. Six Hellcats eagerly took up the challenge and a tremendous dogfight ensued.

According to the VF-83 mission report, six pilots fought between 20 and 30 enemy fighters over the airfield (misidentified as Jacks and Oscars). The Americans were both surprised and irritated time and time again when the Japanese popped their flaps, forcing the Hellcats to overrun.

Down on the ground, CPO Mitsuo Hori's heart swelled with pride when he saw 16 Japanese fighters racing back to the airfield. The double ace was envious of his comrades and regretted not being in the air. No sooner had both formations clashed when an aircraft flamed and fell. It was a stumpy-looking machine.

"We got it! A Grumman was going down! *Victory*! Excited shouts of joy whooped around the squadron's command hut," Hori recounted. "The combat was going to be won by our *Shiden-Kais*! A total of more than 40 friendly and enemy fighters had engaged. Again another Grumman went down, trailing black smoke."

Capt Genda and some of his staff also watched the action, but Genda was not swayed by blind optimism.

While the air battle between the *Shiden-Kais* led by Lt Hayashi and the Hellcats from VF-83 waged over Matsuyama Airfield, the dogfight between Oshibuchi's squadron and the *Hornet* F6Fs ran to the southwest of the airfield. The two dogfights overlapped over the coast of Matsuyama, and, for a while, a large gathering of aircraft developed over the airfield.

LEFT: Author Henry Sakaida, flanked by Isamu Miyazaki (left) and Chitoshi Isozaki (right), in front of a concrete hangar on the site of the former Matsuyama Airfield, November 1980. Many such hangars are located in the midst of rice fields and today are used for storage purposes. In 1945, such reinforced hangars provided protection from strafing attacks.

"While we were absorbed in the movements of the air battle," recounted Capt Genda, "suddenly we heard sounds of *shu, bon, shu, bon* (perhaps rockets) together with the piercing reports of our AA guns. Looking up at the eastern sky, we saw several formations of about 30 aircraft heading in a column for Kure. Some of them had started bombing and strafing our airfield."

Diving down, the F6Fs went for the eight to ten decoy aircraft parked north of the airfield and launched their HVARs. The decoys were *Shiden-Kais* – misidentified as Oscars by VF-83 pilots – which had been wrecked in training accidents. After they had been stripped of parts, the hulks were positioned to invite strafing. The Grummans fired a total of 57 rockets, all in the first run. Several batteries of 20 mm guns were positioned at the end of the decoy line. The Americans came under intense ground fire as they made their first pass. They burned and destroyed three parked aircraft and damaged eight to ten others.

Lt Hayashi could not direct the fighting for long. His landing gear lowered unexpectedly and he could not retract it. To continue fighting under the circumstances would have been suicidal. He wisely disengaged and fled toward Iwakuni Airfield.

Despite the Japanese claims, none of the VF-83 Hellcats went down in the dogfighting or were shot down by anti-aircraft fire. Three aircraft had 7.7 mm holes in them, which could have come from the cover *Shidens* armed with two 7.7 mm machine guns in the nose. Only one Hellcat was lost when Ens W.J. Morton ditched his aircraft after he ran out of fuel; he was rescued by a destroyer. The squadron put in claims for five Jacks and an Oscar, plus an Oscar probable. It is likely that their claims were accurate, although the identification of the Japanese aircraft types was not.

BELOW: The 20 mm Type 99 Cannon (Model 2, Mark 4) was the standard armament in the Shiden-Kai. However, Japanese pilots would have preferred the firepower and blast effect of six .50 caliber machine guns. (Photo credit: Maru)

As the dogfight between VF-83 and Squadron 407 subsided, some of the *Shiden-Kais* joined with Squadron 701 in their struggles against VBF-17. In these two combined dogfights, Squadron 407 lost Lt Kozo Shima, CPO Shiro Endo, PO1/c Haruhiko Takeshima, PO2/c Tokimasa Shinoda, LA Mihiro Nakatani, and LA Yoichi Saiki. Most of these pilots fell victim to the Hellcats of VF-83.

While the dogfight with the Hellcats of VBF-17 continued to rage, one *Shiden-Kai* roared back to Matsuyama Airfield amidst a swarm of enemy aircraft. The pilot of this aircraft, Ens Akio Matsuba, jumped out and hurried to the command center by truck where he limped forward to attention, bathed in sweat with a red, blistering burn across his cheek.

Saluting his senior officers crisply, he reported directly to Capt Genda, advising, in his opinion that the power of the 20 mm cannon was "enormous". "If our range is right," the veteran pilot told Genda, "… the enemy can be knocked out with one burst!"

After Matsuba had checked in, Squadron 701 assembled its maintenance-complete *Shiden-Kais* and put up a division led by CPO Sanpei Shiono, a veteran pilot, with PO1/c Tsutomu Murayama, LA Itsuo Waki, and an unidentified pilot.

The fight moves inland

Two divisions of VBF-17 Hellcats, including that of Conant, moved inland from the sea, east past Nakayama, in a running gunfight, which took them toward Hirota Village. To the Americans, it seemed that the Japanese were up to their old tricks. Using decoys to simulate inexperienced flyers, they would entice a novice Hellcat pilot to give chase. The unlucky flyer would then quickly find a George on his tail.

Lt(jg) McCormick had just completed four passes on the enemy when his radio was knocked out. He then found himself climbing up toward the tail of a Japanese aircraft. His adversary never realized that there was a Hellcat tailgating him and continued to fly straight and level. However, "Mac" McCormick was equally unaware that a George was positioning itself on his own tail.

At point-blank range, McCormick pressed the firing button on his six .50-caliber machine guns. The silence was deafening, then panic gripped him. He was so close, he could shear off his opponent's tail! Suddenly, cannon shells peppered his Grumman. The armor plate behind his seat absorbed the impact of the explosive 20 mm rounds, which felt like sledgehammer blows; shrapnel ricocheted in the cockpit, biting into his left forearm and elbow as he shielded his face.

Instinctively, McCormick hauled his fighter around to escape, but his opponent had him boresighted. An incendiary shell ignited his fuel tank and an inferno roared up from beneath the cockpit floor. Thrashing about wildly while engulfed in flames, McCormick quickly unbuckled his straps, slid back the canopy, and went over the side. Villagers below, who had been watching the dogfight, saw two parachutes blossom and drift slowly down.

Lt Strother T. Kipp and his division, consisting of Ens Edwin Matthews, Lt(jg) Monte Harouff, and Ens Ervin Holley, fought with fury. The 'hunting' had been good so far with every member scoring. Matthews downed one enemy aircraft while Harouff got another and damaged five others. Kipp dispatched two fighters and Holley accounted for two plus a probable. Miraculously, the team was still intact. Their luck was about to change.

ABOVE: Kipp Division. Left to right: Ens Ervin Holley, Ens Edwin "Pappy" Matthews, Lt Strother T. Kipp, and Lt(jg) Monte Harouff.

About ten minutes before the fighting ended, a *Shiden-Kai* plowed straight down through Kipp's division. Ens "Pappy" Matthews went after the audacious Japanese, who lured him far from his comrades. He chased his opponent down to the deck where the George was seen to crash into the sea. However, a lone Hellcat was an inviting target and the Japanese formated on him. Matthews failed to return from this mission and was declared missing in action.

By the time the dogfights subsided, there were approximately ten Hellcats remaining in the area at 9,000 ft. Kipp's four-aircraft division, reduced to three aircraft, had started to leave for home when they ran into more trouble.

Lt Kunio Matsuzaki, the 2nd *Shotai* leader of S701, and his division, consisting of CPO Toshio Komabashiri, PO1/c Jiro Funakoshi and LA Hisao Sugita, looked down from among the clouds and found a three-aircraft division of Hellcats heading southward. The enemy's altitude was some 2,800 m over the coast in the vicinity of Matsuyama Airfield. The four *Shiden-Kais* dived into the Hellcat formation with their cannon firing.

"*The Japs began diving down from above, two planes at a time, attacking on both sides simultaneously,*" recorded the VBF-17 mission report.

Lt Matsuzaki came down tail-on between Kipp and Lt(jg) Harouff, with his four cannon blazing. Matsuzaki put a hole in Kipp's right wing. LA Sugita saw fires on the three enemy fighters in his dive and the four Japanese aircraft pulled up, looking for another target. Though the young Japanese pilot saw flames emitting from the Grummans, Kipp's wingmen had escaped apparently unscathed.

Ens Ervin D. Holley was above and to the right of the section leader and turned into the attacking aircraft, firing a full deflection shot which found its mark. Flames burst from the cowling and wing root. Sugita was stunned to see his division leader's aircraft suddenly hit from above. Matsuzaki pulled to the left, on fire, and gradually lost altitude, leaving a long plume of thick smoke.

Lt Kipp struggled to stay airborne. Holley and Harouff joined up on their division leader and headed southwest across Shikoku toward the sea. The fire continued to burn in the wing, but Kipp was still able to fly straight and level at 200 knots. Suddenly, five miles southwest of Matsuyama Airfield, the right wing broke off and the fuselage aft of the cockpit separated. Lt Strother T. Kipp never escaped his aircraft.

LA Sugita split-Sed away and was separated from his other two comrades. With several .50 caliber holes in both wings, his *Shiden-Kai* vibrated violently and one of his landing gears came down. He decided to make haste for home. Coming in for a landing, his other wheel would not deploy. When the one wheel made contact with the ground, his aircraft turned and was damaged, but he was able to walk away.

Lt Richard W. Moore's division claimed five victories. In their initial conflict, he latched onto a Japanese fighter with his wingman, Ens Gaylord Edling, behind him. Two *Shidens* suddenly dived between the two Americans. Edling flipped his Hellcat to the left behind his opponent, who also went left, and opened up with a barrage from 900 ft into his opponent's tail. Raising his nose, Edling pulled up his lead and fired into the cockpit area. The enemy aircraft abruptly turned to the left and went into the sea. His victim may have been LA Yoichi Saiki.

ABOVE: LA Yoichi Saiki of Squadron 407 poses for a photograph next to 343-B-03 with one of the squadron's fuel trucks visible in the background. Note the wooden brace on the rudder to prevent wind damage. The name of the aircraft's chief mechanic, Suzuki, is written under the large 03.

BELOW: Lt(jg) Yukihiro Watabe of S701 had predicted his own death three days prior. His ghostly image appeared to his mother on the morning he was killed.

Typically in a large and furious dogfight, friendly aircraft become separated. Ens Edling became lost from his division and could not locate it. He wisely decided to make for home. Along the way, he was attacked twice, but opened full throttle and evaded his pursuers. He sped back to the *Hornet.*

Lt Moore tailed in behind a *Shiden* and started knocking pieces from it. Flames erupted from the wing root and the Japanese aircraft started down. As Moore prepared to deliver the final blow, his Hellcat received hits in the engine from out of nowhere. Oil covered his windshield; his engine began vibrating violently. Not wanting to push his luck, Moore opted out and headed for home. VBF-17 had been in combat for 25 minutes.

If Lt Moore thought his journey home was going to be easy, he was wrong. He was attacked twice along the way, probably by aircraft from the Ichimura division. One opponent set his right wing on fire, but it flickered out. Within sight of the Task Force screen, Moore's engine lost power and he ditched his mount on the sea. He made a no-flap landing and sprained an ankle, but he was safely aboard a submarine in less than 15 minutes.

The Japanese pilots were astonished to see numerous spots of red, yellow, and other colors on the blue of the sea, as if paint had been dropped into the water. They appeared to be firm evidence of Hellcat crashes. The vivid colors may have come from aviation fuel or dye markers which the American pilots carried.

The VBF-17 pilots claimed 25 victories and nine probables, with the top scorer being Ens Ervin Holley with three kills and a probable. However, it had been a costly mission. The following pilots did not return: Lt Charles F. Weiss (POW); Lt(jg) Forrest McCormick (POW); Ens Edwin W. Matthews (POW); Lt Strother T. Kipp (KIA); Lt(jg) Roger W. Karr (KIA); and Ens Noel L. Hannah (KIA).

The Japanese losses were equally disastrous. Lt(jg) Yukihiro Watabe, a division leader of S701, shot down one Hellcat in the vicinity of Nakayama Town. His victim was either Lt(jg) Forrest McCormick or Ens Edwin Matthews. Watabe had predicted his own demise just three days prior. He crashed in a valley between Mt. Kaijo and Mt. Shoji. His mother, Mrs. Fusae Watabe, who had been sleeping that morning, was awakened by her son calling out "I'm home now!" She sat upright and saw the upper figure of her son clearly on the bedroom wall. Then his image faded away forever. Joining the 21-year-old Watabe were Lt Kunio Matsuzaki, CPO Kikuichi Ishikawa, and LA Itsuo Waki. Waki was hit over his airfield and crashed in the mountains.

As testimony to the high caliber of their opposition, the VBF-17 mission report summarized: "*It was the opinion of the more experienced pilots of this squadron, who participated in this mélée, that the Jap pilots encountered here were superior to those met in the Tokyo area. They handled their aircraft well, were exceedingly aggressive and exhibited good organization, discipline, and tactics. Their tactics were similar to those of the US Navy. They appeared to be well trained and experienced in combat flying.*"

Likewise, the Japanese paid respect to their counterparts. Lt Ichimura offered these words: "The US pilots were mostly veterans from the task force under the command of Halsey and Spruance. Moreover, our adversaries had been fighter pilots. This was surmised later from the victory markings painted on their fighters which had crashed on the ground."

Helldivers and Avengers over Kure

Over Shikoku Cdr George "Bunky" Ottinger cruised amongst the 45 fighters and bombers of his Air Group 84 from the *Bunker Hill.* The Group was on a "morning special strike" to the enemy naval bastion at Kure. Their primary task was to inflict punishing blows to the Kure Naval Arsenal and the many installations clustered around the harbor area. The ACI officer's report of over 160 heavy anti-aircraft guns around the target worried him.

AG-84 was the first formation to arrive over Kure and it was warmly greeted by the Japanese, who threw up a colorful and deadly display of flak. Red, yellow, green, purple, and black puffs dotted the clear sky. When the presence of three flattops and a battleship materialized in the harbor, Ottinger reveled in his great luck and directed his eager men to the new targets.

LEFT: Herbert L. Long photographed at Guadalcanal in June 1943. He fired rockets into the carrier Ryuho and it never sailed again. (R.F. Toliver)

The SB2Cs went into their high-speed dives from 13,000 ft, led by LCdr J.P. Conn. Picking out what he believed was a *Kongo* class battleship, he plowed through the intense flak. Halfway through his dive, he was hit in the electrical system and could not release his bomb. Veering away from his intended victim, Lt (jg) J.D. Welsh took the lead and released one 1,000 lb bomb and a pair of 250 lb bombs. The results were not observed, and Welsh and his backseater failed to return from the mission.

No enemy fighters were observed by AG-84 during its wild rampage over Kure Harbor. VBF-17's Hellcats had been scheduled to be over Kure before the arrival of AG-84, but they had been intercepted by Japanese fighters. With the enemy flyers tied up in heavy dogfights to the south, the Avengers, Helldivers, and Corsairs could work without aerial interference.

Maj Herbert L. Long of VMF-451, a veteran from the Solomons Campaign with seven victories at the time, wanted to sink a carrier with his rockets. If successful, it would have been a first. The gutsy New Yorker lined up the *Ryuho* in his sights and salvoed all of his HVARs. The shotgun blast of rockets hit the hangar deck and the bulkhead, killing 20 crewmen and wounding another 30. Although not sunk, this ship never sailed again.

Once AG-84 had expended its ordnance, it vacated the area for the next wave of assault aircraft which were on their way in. In obtaining four hits on an *Unryu* class carrier and one on a battleship, six out of the 11 SB2Cs were hit. One Helldiver was shot down and another ditched on the way back.

"March 19th will remain a nightmare of multi-colored flak as long as the squadron lives," stated the VT-17 narrative report. "*Piercing this lethal pyrotechnic display with a torpedo bomber was akin to sticking one's fingers into a bowl of hot colored marbles.*" VT-17 sortied a dozen TBM Avengers on this mission and claimed 14 500 lb bomb hits on eight ships.

Of the 85 Avengers which flew on this mission, only one aircraft was shot down over Kure. Lt Talmadge Westmoreland ordered his two crewmen to jump after they were hit. He had made the classic mistake of going back to check their handiwork. He was the last to bail out and his flaming TBM-3 plowed into residential buildings. Angry civilians beat him almost to death. Westmoreland, AOM3/c Harold W. West, and ARM3/c Robert H. Williams were all captured. All three would survive the war.

Bob Williams would spend the next five months thinking about a decision he had made prior to take-off. "Harold West tapped me on the shoulder when we were in the plane and asked me if I wanted his Spam sandwich. I said '*No thanks.*' How I regretted that. I thought about that Spam sandwich until the day I was liberated!"

LEFT: Though often inaccurate, High Velocity Aerial Rockets (HVARs) were nevertheless destructive when they struck their targets. Here a crewman on the wing of a Hellcat loads the aircraft's three .50 caliber machine guns while an armorer attends to the warhead of an HVAR. (NMNA)

Kanno looks for a fight

Lt Kanno assembled his men at 3,000 m, spotted a large enemy formation and announced the fact over his radio. As the Japanese formation closed the distance, the tiny specks took the shape of Grumman Hellcats and single-engined bombers – probably the combined force of *Bunker Hill* and *Cabot* aircraft on their way toward Kure Harbor.

"I looked at the lead aircraft of the squadron which was directing its nose toward the enemy formation," remembered Tsuneharu Tamura. "Kanno test-fired his 20 mm cannon to

RIGHT: The Curtiss SB2C Helldiver replaced the Douglas SBD Dauntless as the Navy's standard divebomber. Despite its ungainly appearance, it proved itself a real workhorse. This Yorktown aircraft was photographed during the Battle of the Philippine Sea in June 1944. (USN)

LEFT: Japanese warships and shore batteries in Kure Harbor under intense bombardment. (NMNA)

BELOW: A Japanese battleship endeavours to evade American bombardment during the Kure attack. (NMNA)

ABOVE: 19 March 1945 and the sky above Kure Harbor is dotted with the deadly bursts of anti-aircraft fire. (USN)

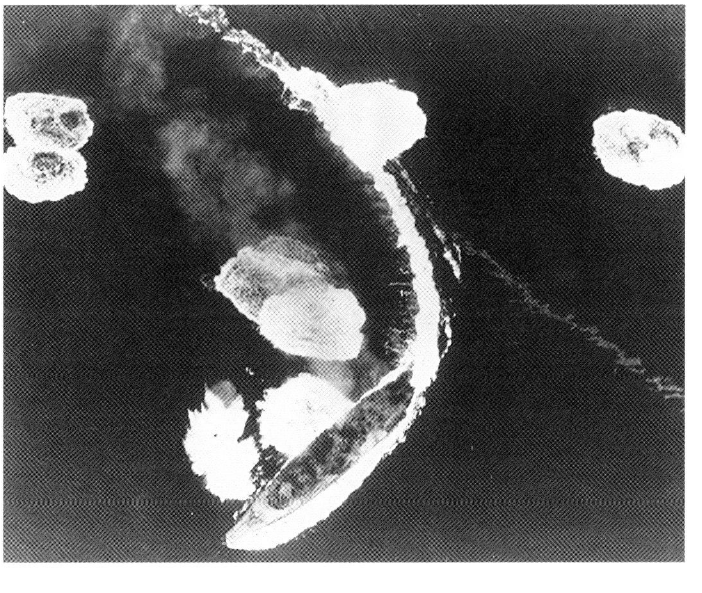

ABOVE: A very rare photograph showing two TBM crewman of VT-17 in Japanese custody after parachuting out on 19 March 1945. Lt(jg) Talmadge Westmoreland (left) was beaten by angry civilians when his aircraft crashed into a village. (Torao Masui)

ABOVE: Fifty-five years later, former B-29 POW Raymond 'Hap' Halloran and Torao Masui, visit the spot where the two VT-17 crewmen were photographed. Seima Masui, Torao's father, heard that American flyers had been captured nearby and would be passing by his house. He had his photographic studio employee snap the picture to above left. The photograph was mislaid for half a century, until it was finally rediscovered by Torao Masui, who was rebuilding his house. (Hap Halloran)

RIGHT: A captured American Navy flyer and his one-man dinghy. The Japanese were amazed at the amount of survival gear carried by American flight crewmen. It was customary to blindfold all prisoners during transport for security purposes. (Y. Kumoi)

let us know the enemy's direction. The enemy aircraft still looked small. Our No.1 aircraft, Sugita, hand-signaled to us to open cowl flaps fully, turn on our gunsight, change the flaps to automatic for combat, and test-fire 20 mm cannon. All at once, the four pilots gripped their shooting levers on the throttle. Shells flew forward with booming sounds through the cloth caps covering the muzzles." Squadron 301 positioned themselves 460 m over and behind the large American formation. CPO Sugita signaled his division to stay tight and follow him.

According to the mission report of Air Group 29, the *Cabot* group made a slow climb to 13,000 ft over the south shore of Shikoku as they advanced toward Kure. They reported 20 enemy fighters in the distance. "*Several times they resorted to mild aerobatics in an effort to entice the strike group fighters to one side,*" read the report. "*These attempts were fruitless and the enemy never came within gunnery range...*"

It is most likely that the Japanese fighters which attempted to entice the Hellcats into a dogfight were those led by Lt Kanno. The Americans did not take the bait and forced Lt Kanno to look elsewhere for an encounter.

US Navy Corsairs vs. the Georges

Ten Navy Corsairs from VBF-10 were en route to attack the Kure Naval Base and Air Depot. This strike was led by Cdr John J. Hyland, the air group commander. He assumed the responsibility of coordinating the attacks of other carrier aircraft working along with the *Intrepid*. Of the ten F4U-1Ds, which sortied at 05.50 hrs, four were armed with rockets and another four carried bombs.

A few minutes before the assault, the VBF-10 pilots could see a group of eight Japanese aircraft circling at 6,000 ft above them at eleven o'clock, tail-chasing each other in a circle. It is believed that they were a group of the 2nd *Chutai* of Lt Kanno's 301st, led by Lt(jg) Isaburo Inoue, a graduate from Kyoto University. Apparently as a result of their diving attack on a group of Hellcats and bombers, Kanno's divisions became separated. The eight *Shiden-Kais* were trying to attract attention, but Hyland ignored them for the more important task at hand.

"The commander led us across the bay, making a 180-degree turn, and we started our approach toward the oil-storage tanks," recalled Roy "Eric" Erickson. At about 12,000 ft, we released our belly tanks and then armed our bombs in the dive. Our division began the attack followed closely by the other six aircraft. We were met with an absolutely ferocious barrage of anti-aircraft fire. Dropping our bombs, firing our rockets and strafing the target, we pulled out over the bay at 3,000 ft to avoid small arms ground fire, which was as capable of killing us as anything else. Then a peculiar thing happened.

"I informed Cdr Hyland that I'd sighted a Rufe floatplane taking off in front of us. Hyland went after the Rufe and Hill and I started to climb. I was confused – should I join Hyland or continue with Hill? In a split second I realized there really wasn't a decision to make. My duty was to fly wing, and I stuck right with him."

RIGHT: Ens Roy D. 'Eric' Erickson (R. Erickson)

Hyland swooped down on the enemy floatplane and 'splashed' it. His first victory was also the first for VBF-10.

Lt Robert "Windy" Hill and his wingman, Ens Erickson, climbed for altitude, not realizing that they had become separated from their comrades. To make matters worse, Hill could not jettison his belly tank. Erickson kept a watchful eye on the enemy aircraft above them as the pair continued to climb. Sensing an advantageous situation, the Japanese decided to strike.

Two of the Georges did a snap roll and flew straight down toward the Corsairs. They were most likely Lt(jg) Inoue and his wingman. As they came within range, Ens Erickson climbed up into them, pulled back on the stick and blacked out for a few seconds. "Thank God, I had my anti-blackout suit on, for I could still see the oncoming aircraft. Without the suit I would have blacked out completely" recalled Erickson. "My vision was clear as I put my gunsight directly on the lead plane and fired."

Erickson did not miss his target. The gray underbellies of the two diving Georges, flying tightly, flashed past and the lead aircraft passed over the Corsair belching black greasy smoke and flames. Erickson explained: "Windy was below me and wasn't able to confirm my kill, but now, out of formation, I wisely decided to form on Windy and not follow the plane that I'd lit up."

The Japanese were infuriated and threw themselves at the pair of F4Us. The two aviators attempted to weave with another section for mutual protection, but to Lt Hill's dismay, their comrades deserted them, diving full power toward the deck. The F4U group headed for Saijo Airfield, leaving Hill and Erickson to fend for themselves.

The two Americans were now in danger of being overwhelmed. Hill still could not drop his belly tank and the excess load only served to lessen his chances of survival. Erickson made a turn to join up and was horrified to see a George sitting on his flight leader's tail, guns firing. It was only a matter of seconds before a lucky hit would spell his end.

"*Shoot the son-of-a-bitch, shoot the son-of-a-bitch!*" screamed Hill into the radio. Going back into their first weave, Erickson retorted in exasperation: "*What the hell do you think I'm trying to do?*"

"In a matter of seconds I had the unwary Jap pilot perfectly bracketed in my sights and then, with all my six guns blazing, the George blew apart!" recalled Erickson. "The front part of his plane flew on straight and level, but the tail section, sheared off behind the cockpit, spun crazily away. Gaining on him fast, I flew through all kinds of flaming debris, instinctively ducking to avoid getting hit by all the fragments." Erickson's victim may have been PO1/c Noriyoshi Kubo.

BELOW: Lt Robert "Windy" Hill (R. Erickson)

Lt(jg) Tatsutoshi Hashimoto and his wingman, LA Hideo Izawa, both from Kanno's squadron, had taken off late and were behind the main formation. The neophyte tried to cling to his flight leader. The two pilots then joined friendly aircraft over Ao Shima, and in the ensuing dogfight became separated. Hashimoto was seen diving into a formation of enemy aircraft. Izawa had just joined up with a group of friendly aircraft when he crossed paths with the two retreating Corsairs.

Erickson and Hill were racing toward Shikoku from Kure. When Ens Erickson made a turn to join up with Hill, his partner was already shooting at another

ABOVE: LA Hideo Izawa.

George. Hill recollected: "As I started my attack run, I kept telling myself, '*Get in closer before you open fire. Get the proper lead in your gunsight!*' My trigger finger was twitching, the blood was pounding through my veins, and I was trembling with excitement as I closed on the enemy. '*Get closer, closer, he hasn't seen you!*' Finally, I was within range. '*Now!*'"

A thunderous hail of .50 caliber shells slammed into the George. LA Hideo Izawa was shocked when bullets came out of nowhere and peppered his fuselage. His mount had taken the full blast from the Corsair, but it continued to fly straight and level. Knowing that he had taken damaging hits, Izawa quickly unbuckled himself and went over the side of his burning aircraft. He swept past Ens Erickson, who could see the enemy pilot's brand new flight suit and an astonished expression on his face!

At around 245 m above the ground, Izawa pulled his ripcord and his parachute blossomed. Ens Erickson tried to get his sights on the dangling pilot, but to no avail. He later decided that it was not the proper thing to do. Izawa hit the ground and was badly injured, but he felt lucky to be alive. He landed near the coast and would spend months in hospital.

The two F4Us continued on their way. However, the pair seemed to attract the enemy like a magnet. "Cruising back over Shikoku Island," Erickson recalled, "I looked over to Windy, who had positioned me a hundred yards to the side of him, and I couldn't believe my eyes. There sat a Tojo on his tail and all four of the Jap's cannon were blinking his way! I shouted a warning to Windy and we immediately broke to weave, but I couldn't get my sights on the enemy with the proper lead. Not wanting to waste ammo, I didn't fire. Having learned my lessons well the first time around I went further out than seemed necessary. Coming back on the second weave, I had a straight 90-degree deflection shot. I put my sights directly on Windy's head to get the proper lead on the enemy aircraft, and it took real nerves of steel to pull the trigger. True to my training, the tracers seemed to bend directly into the Tojo! He went ablaze and slid to earth as if he was on a greased wire. In my mind, this was my third victory of the day!"

Hill and Erickson resumed their homeward trek to the *Intrepid* while their eight comrades strafed various airfields. However, their ordeal was not yet over. "Looking into the rear-view mirror, I saw eight dots in the distance, and they were definitely chasing us!" remembered Erickson. "I told Windy, '*Get it in gear!*' He was

LEFT: American aviators who participated in the attack on Kure never forgot the sight of the carrier Franklin engulfed in flames. The heroic crew saved the ship, but over 800 men died. (USN)

giving it all it he had but we were still losing ground. His belly tank was jammed and wouldn't release, and only a few of his guns were operable." Erickson cursed and thought, "*Here we go again!*"

The pair of Corsairs was down to 500 ft, heading for a picket destroyer. Lt(jg) Hill radioed to the destroyer: "Two friendly aircraft are heading your way! Let them go by, but shoot those eight bastards chasing us!" Nervously looking over his shoulder, Hill suddenly realized that the pursuers were Corsairs! Switching to the emergency frequency to contact the Corsairs, he yelled out "Quit shooting at us, you SOBs! We're friendly!" The eight Corsairs got the message and turned away.

Nearing their carrier, the two tired pilots were horrified to see one of their carriers bellowing black smoke, which spiralled nearly a mile high! "What do we do now?" wondered Erickson, thinking that the burning vessel was their carrier. However, they were picking up a strong homing signal. Every carrier had a homer to help returning pilots find their own carriers. As it happened, the burning aircraft carrier was the *Franklin.*

Far out at sea, some of the carriers came under attack by aircraft from the Japanese 5th Air Fleet. It was 07.08 hrs on the carrier *Franklin* when a radial-engined Judy placed two bomb hits on the flight deck as it was launching aircraft toward Kobe. The resulting explosions turned the carrier into a raging inferno, killing over 800 men. With the flagship of TG58.2 badly crippled and out of the war, R/Adm Ralph E. Davison took his staff aboard the *Hancock* to continue air operations and fighter direction.

Hill and Erickson landed on the *Intrepid* without further incident. All in all, it certainly had not been a dull day. Each pilot received credit for two enemy aircraft destroyed, but more importantly, they had survived to enjoy their well-deserved bragging rights.

"Two-fisted sky duel" over Kure

Fifteen Marine F4Us from VMF-123 cruised toward Hiroshima Airfield at 12,000 ft. It was impossible for the formation to stay at 20,000 ft due to dense cloud layers and the inability to maintain contact with the ground. Maj Mobley and his men could see a formation of Hellcats orbiting south of Shikoku Island. They also saw another group of what they thought were Hellcats orbiting some distance away. The two groups that Mobley saw were probably VF-83 and VF-9 Hellcats.

As Mobley's formation continued onward, they heard a radio message from a formation of F6Fs that "bogeys" had been spotted. Mobley quickly realized that they were talking about his Corsairs. "We're friendly bogeys!" he shouted over the radio. This brief exchange was probably heard by LA Tsuneharu Tamura sitting in the cockpit of his George, as well as the radiomen in the Matsuyama command center. The two formations, Corsairs and Hellcats, moved on a parallel course toward Kure.

Looking back over their shoulders, Capt George C. DeFabio and 2/Lt Edward Milhem spotted a formation of about 20 aircraft. Their position was six o'clock, up 2,000 ft, and about five miles back. With so many assault aircraft in the air, and as a precaution, Mobley identified himself and asked the unknown formation to identify itself. Silence. Mobley turned to observe the unknowns and felt somewhat reassured that they were friendlies because they appeared to be American in make-up and flying formation.

The Marines crossed over the Inland Sea and made landfall east of Kure Harbor. The plan of attack called for a north-to-south assault on Hiroshima Airfield, an army facility. Little did the Americans know that the enemy airfield offered no opposition. Its few fighters had been grounded by lack of fuel and were carefully hidden from possible enemy attack. As the American formation droned on, anti-aircraft fire opened up and flak bursts appeared at its altitude and about three to five miles ahead. Maj Mobley was suspicious and believed that the black and white bursts were going to be used to spot the predicted course and altitude of his formation as an aid to enemy aircraft. Mobley's assessment would prove to be surprisingly accurate.

The Corsairs crossed the coast when they heard chatter over their radios to the effect that another carrier group was under attack over Kure Harbor and was requesting immediate assistance. As the Leathernecks turned their formation to assist, Lt Kanno gave orders to attack.

The Japanese dived in sections, from left to right, and started to bracket the Corsairs. Capt DeFabio's division turned to meet the enemy. Lt Claude Barnhill and his wingman, 2/Lt Ralph A. Russell, received the brunt of the incoming Japanese attack.

"*Barnhill dropped back to cover him and an unnoticed Jap peppered his plane,*" stated the mission report. "*Though he was damaged and his plane was hard to control, Barnhill managed to snap a full deflection shot at a Jap who was attacking Russell – and the burst frightened the Jap away. Then Barnhill was hit again. He never saw Russell after that. The second attack on Barnhill's aircraft left 20 mm holes in both wings, a wrecked instrument panel, and a cockpit full of smoke. Barnhill figured the aircraft was on fire but when he attempted to open the hood, it refused to budge. Later the smoke cleared. With his left aileron out of commission, Barnhill had considerable difficulty in keeping the Corsair from rolling to the left. Captain DeFabio covered his retirement.*"

In the initial Japanese pass, two F4Us were hit. 2/Lt Ralph A. Russell parachuted over Kure Naval Base. The second pilot, 1/Lt Gordon K. Wooster, was last seen going down over Hiroshima.

The Marine Corsair pilots were immediately impressed with the skilled pilots manning the new aircraft with a low mid-wing, diamond-shaped wings and a radial engine. These enemy aircraft were fast and packed a tremendous punch with their 20 mm cannon; 1/Lt Archie J. Clapp could attest to that: "When the enemy aircraft pounced on us, they were able to give me special attention."

His wingman, 2/Lt Vincent Byerlein, had returned to the carrier shortly after launch due to engine trouble. His absence left Clapp exposed on his six o'clock position as well as making him "Tail End Charlie."

"On the very first pass, I saw large holes opened in both wings," continued Clapp. "As it turned out, they also punched large holes in the fuselage and tail section that I couldn't see. Accordingly, it was all I could do to control the aircraft well enough to stay in the fight and take potshots at the enemy firing at my squadron mates while they were doing the same for me. I certainly didn't have time to confirm any damage I might have done to the enemy, and lacked a wingman who might have done that. According to my old logbook, I was flying Bureau Number 82339 that day, which was one of the three aircraft jettisoned as soon as they got aboard the ship."

As the dogfight intensified, the Marines claimed six victories within minutes. Capt William A. Cantrell saved a number of his comrades who had Georges on their tails. The combat report gave this narrative: "*As soon as he turned, Cantrell spotted three Japs pumping lead into three Corsairs. The nearest Jap was 300 yards ahead. As soon as the Jap came into his sights, he opened fire. Holding down on the trigger until the gun barrels smoked, Cantrell watched burst after burst hitting the wings and the fuselage. Flames poured from the right side of the engine. In a little more than two minutes, he had destroyed two enemy aircraft.*"

Cantrell's mount was violently jolted; explosions rocked his aircraft. One of Kanno's men scored with a high side run, putting a large hole in his left wing and blowing off his cowl flaps. Hot hydraulic fluid sprayed into the cockpit, clouding his goggles. To make matters worse, his right foot was numb; he had been hit by shrapnel.

Cantrell evaded the attacker and rejoined his men. With disregard for his own safety, he once again went to the aid of another comrade who had radioed for help. A George was making a high side run on a Corsair. A hail of cannon fire raked the Japanese aircraft, forcing it to abruptly abandon its prey.

BELOW: Capt William A. Cantrell claimed two 343 Kokutai aircraft destroyed and two damaged, and rescued a number of his comrades from George fighters of Squadron 301. (USMC)

Capt William Roques worked himself out of the center of the dogfight maelstrom and ran head-on into a George. "*Roques evidently killed the pilot with the first burst,*" detailed the mission report. "*Flames were seen to suddenly pop from the right side of the fuselage and the Nip went straight down. Before Roques realized it, there was a Jap on his tail and 20 mm bursts damaged his right elevator and stabilizer.*" The Corsair pilot got away.

Lt Dwight Mayo had all he could handle, and more. In chasing off a George from a comrade's tail, he had his hydraulic system shot out, which popped his tail wheel down. Now, his guns and rockets failed to fire and he knew he was in a sorry state. This did not prevent him for staying in the fight and scaring off the enemy with aggressive tactics. But after a while, he decided it was time to go home.

When Mayo pulled out, 2/Lts Franklin Kurchinski and Robert McInnis went to escort him. But a pair of Georges beat the two Marines to Mayo and went in for the kill. Mayo saw what was coming, and in a desperate move, pulled around sharply. With his 'flipper' gone, his Corsair went into an inverted spin, with the two Georges diving down, and the two Corsairs right behind. All were so intent on seeing what would happen to Mayo that no shots were fired.

Mayo was not about to ditch. He wrapped his arms and legs about the control stick, and with superhuman strength, pulled his battered crate out of the dive. Seeing that he was coping, Kurchinski and McInnis then went after the Georges. They damaged one jointly, and McInnis claimed the other as destroyed. Mayo made it back to the carrier and made a perfect wheels-up landing. His rockets were hurled forward on the deck but did not explode. A gang of men relegated his aircraft to the deep.

The wild dogfight moved closer to Kure Harbor where many warships were anchored. It seemed every vessel in the harbor, including shore batteries, threw up a colorful wall of deadly flak. Blue, red, green, and black bursts dotted the sky. If the Japanese were sending a threatening message, the Marines received it.

Maj Mobley fired at a George with a couple of bursts, hitting the wing root and engine with a 20-degree deflection shot. A George made a pass at Mobley's section. The Marine simply blew his opponent's tail away. While going to the aid of a comrade, a George pounced on him. Mobley was hit in the tail section and wings. The riddled Corsair ducked under a formation but his opponent was tenacious. The persistent Japanese regained his position, and sitting behind the F4U, continued where he had left off, pumping 20 mm projectiles into the battered American aircraft.

Mobley thought he was doomed. Smoke filled his cockpit. He dropped his belly tank and prepared to open the canopy. Suddenly the smoke dissipated. The urge to bail out no longer seemed attractive. Miraculously, the Japanese assailant had disappeared.

After a good half-hour of combat, the Marines and Kanno's formation disengaged in the vicinity of Osaki-Kami-Shima. Maj Mobley's compass had been destroyed, so he had no course to offer for the retirement back to the carrier. He suggested a southerly route and all of his men followed. Not realizing that Capt Cantrell had been wounded, Mobley turned the lead to him. The captain allowed the more seriously damaged fighters to land first.

Suddenly, the *Bennington* turned to go downwind; something was wrong. Enemy aircraft appeared in the vicinity and Capt Cantrell took the remaining fighters around and out of range as the carrier opened up with anti-aircraft fire. When the danger had passed a half-hour later, Cantrell landed. He was bleeding profusely from his foot wound. His aircraft had landed with only 20 gallons of fuel remaining.

Maj Mobley flew his wreck back to the carrier, holding the stick in both hands. Upon landing, an air officer took one look at his fighter and ordered it shoved over the side.

When the Marines gave their reports, they accounted for ten enemy aircraft destroyed and four damaged. Two pilots were missing in action and one wounded. However, their aircraft had been terribly mauled. In addition to two Corsairs shot down, three were shoved overboard into the sea when they returned, and one was lost when 2/Lt Barnhill bailed out of his crippled fighter by the destroyer picket line; he was rescued without mishap. Five other aircraft sustained various levels of damage but repaired. Many men would sing the praises for the sturdy Vought F4U-1D which had brought them back safely.

BELOW: An F4U-1D of VMF-123 on the Bennington, April 1945. This scene is reminiscent of the one on 19 March 1945 when three battered Corsairs were shoved over the side upon their return from operations. (NMNA)

Despite the claims of horrific carnage inflicted by VMF-123, the Marines may not have scored a single victory, although their mission report stated that "*... they fought for nearly 30 minutes in a blazing two-fisted sky duel in the skies north of the great Japanese naval base of Kure.*"

Three pilots of Squadron 301 did not return from the morning mission and it is more probable that they were downed in fights with other US Navy groups (VBF-10 and VF-9). Some of their aircraft must have sustained damage over Kure. Lt Kanno lost one-third of his aileron.

Kanno's men had impressed the hard-nosed Leathernecks. Their report stated: "*The Jap tactics*

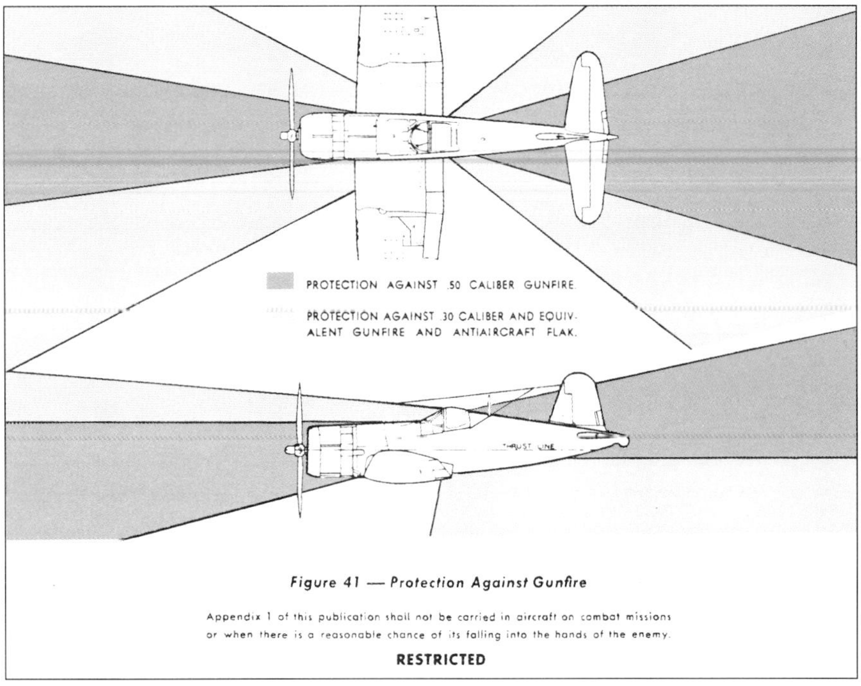

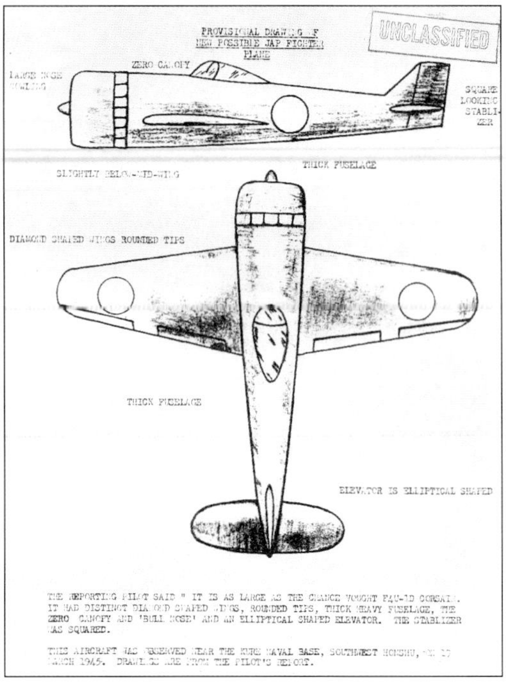

closely resembled standard American tactics. The enemy attacked in divisions and sections. All attacks were well coordinated and nearly always by two-aircraft sections. They struck at our aircraft when we were on the outside of our turns. Their marksmanship and airmanship were as fine as our pilots have ever seen. The pilots were apparently the 'cream' of the Jap air forces."

ABOVE LEFT: As seen on this page from a contemporary manual, the Corsair's front engine compartment and rear armor plate afforded considerable pilot protection, even against the impact of 20 mm cannon fire.

ABOVE: The Shiden-Kai was reported by pilots of VMF-123 as a "...new possible Jap fighter plane". This drawing was attached to the unit's mission reports.

Iwakuni Naval Airfield

Iwakuni Naval Airfield was on TF58's hit list for 19 March. Although it was a major airbase, it was largely deserted. On this day, the 332nd Air Group had only three Zero fighters there and they were carefully hidden from view. Almost all of the unit's aircraft had been transferred to Naruo Airfield to help protect the Osaka and Kobe areas from B-29s. Two Hellcats overflew the airbase on the 18th, probably on a reconnaissance mission.

There were morning fireworks over Kure Harbor when carrier aircraft from Air Group 84 arrived to do battle. Rockets and bombs hit ships, slammed into airfield installations, parked aircraft, locomotives, radio and radar sites, and refineries. Two of three fighter squadrons from the 343 *Kokutai* were tied up temporarily in vicious dogfights to the south. Without aerial interference, the TBMs and SB2Cs concentrated on their objectives.

Lt Takeshi Umemura, a division leader in the 332nd, was in the radio room down in the underground command shelter when the raid began. During the lull in bombardment, a *Shiden-Kai* circled the airfield and landed. The pilot, Lt Yoshishige Hayashi, CO of Squadron 407, went straight to the command center. When the two classmates from the Naval Academy met, Hayashi thought he was seeing a ghost!

Rumors had been circulated that Umemura had been killed in action. "Hayashi looked at me in amazement," remembered Umemura years later. "I had him sit down and served him a cup of coffee." His friend suffered a harrowing experience and explained that in the midst of a combat with Hellcats over Matsuyama, his landing gear had popped out and he was not able to retract it. He was forced to flee at low altitude.

Although the *Shiden-Kai* was rated as an excellent fighter, undercarriage problems had plagued it from its inception. Lt Hayashi was the first pilot in the group to have his landing gear come down during combat. When asked during his post-war interrogation about this vexing problem, Capt Genda answered: "It was very poor. Much worse than on the average Japanese aircraft. In dive-bombing or strafing, if the aircraft reached 420 mph, the gear was apt to fall apart in the air. About one-fourth of the pilots we lost in combat were lost for this reason. It got to be a better plane toward the end."

During the lull in fighting, Lt Hayashi jumped into his fighter and took off for Matsuyama, approximately a ten-minute flight. It is not known if the mechanics were able to correct the landing gear problem.

Snakes!

Lt Ryoichi Yamada's three-aircraft *Shiden* flight landed at Matsuyama toward 08.00 hrs; his fellow pilots' engines were not in good running order and it would have been irresponsible to commit them to further combat. However, the young lieutenant was still in a fighting mood. After checking his machine guns and fuel, he munched on some rice balls and then took off alone on his second sortie. He crawled along the surface of the water to avoid detection and went across the Inland Sea, gaining altitude over the Chugoku Mountains in Honshu. He then headed south over the sea to look for enemy aircraft over Matsuyama. Soon, he joined up with WO Narunobu Ibusuki along the way.

A strange sight greeted Lt Yamada far in the distance. It was a formation of eight Corsairs in a type of a "snake" formation of four two-aircraft sections formed stepwise. This presented a tactical problem for Yamada. Hitting the head would cause the tail to rise in retaliation. Likewise, hitting the tail would cause the head to rise. Attacking the center would cause both the head and tail to rise. This peculiar formation seemed very flexible, difficult to attack, and dangerous. The Corsair formation was VBF-10, with Cdr John J. Hyland leading. Having strafed Saijo, Niihama, and Kannonji, they were on their way to Matsuyama Airfield.

Lt Yamada and his wingman closed from behind. WO Ibusuki fell back to cover his flight leader. Yamada picked out a "Tail End Charlie" who was below the rest and commenced his attack. The two F4U-1Ds at the head apparently saw the two Japanese stalkers and started to turn right. The fast-approaching *Shiden* opened up from 50 m, but the two 20 mm cannon remained silent. The two nose-mounted 7.7 mm guns chattered, sending a stream of bullets into the American aircraft. The target seemed unaffected. Yamada overshot his intended victim and turned right; he should have turned left. Cdr Hyland and his wingman were soon on Yamada's tail. The six other Corsairs went in pursuit of Ibusuki.

The end seemed imminent as the two Corsairs closed the distance behind the *Shiden*. "As I was chased by two aircraft, I flew low to Matsuyama where there was a battery of anti-aircraft guns," related Yamada. "I was going to have them shoot down the two behind me. However, they didn't notice me as I came in so fast and so low. The enemy followed me there, too. So I ended up having a serious dogfight with the two!"

For seven minutes, airfield personnel looked up to see the spectacle of a duel to the death. The outnumbered 'underdog' weaved his aircraft repeatedly, denying Cdr Hyland and his wingman an easy victory, but the odds were against him. The Corsairs would not let up and the Japanese pilot was running out of time.

Explaining his predicament, Yamada recalled: "In air combat training, the higher one attacks the lower. The lower one is in an inferior position. Over the airfield, I was in such a position about five or six times. At last, I did what's called 'the last-ditch maneuver' in American terminology. You take them down to the ocean surface to crash. But they must have been very good pilots because they didn't go down! Also, I was impressed that the two never came in at the same time.

"If you glide, you cannot get hit. I continued to glide until finally, I was about to go up by twisting. Then some bullets hit. One of them came and stopped under the sole of my boot. I sensed that I was hit, but I couldn't be certain because I couldn't see. So then we passed each other and I waved my scarf. The other guy was on his way back to the carrier and he also waved. That was our goodbye. I wanted to meet this guy if I ever had the opportunity."

The *Shiden* had been hit numerous times and its oil pipe was severed. The landing gear of the mid-wing *Shiden* had a

BELOW: Cdr John J. Hyland, CAG-10. As Adm Hyland, he went on to command the Pacific Fleet during the 1960s. Pencil sketch courtesy of R.D. "Eric" Erickson.

BELOW RIGHT: Lt Ryoichi Yamada of Squadron 701 dueled with Cdr John J. Hyland and his wingman over Matsuyama Airfield. Yamada later retired as a General in the Japan Self Defense Air Force.

two-step retracting and lowering system and, as a result of damage, the landing gear of Yamada's aircraft was not fully lowered.

Lt Yamada bolted away and sought refuge behind Kogo Shima, a small island about four and a half miles northwest of the airfield. When the Americans left the area, Lt Yamada slid down to the airfield and landed near some decoy aircraft. He figured that his old *Shiden* would be disposed of anyway. The propeller blades contacted the ground and were buckled. With his parachute pack on his shoulder, the sweat-drenched pilot proceeded to headquarters to make a personal report to Capt Genda.

WO Ibusuki could not shake off his pursuers; they were determined to nail his fighter. After Lt Yamada had concluded his dogfight with the pair of Corsairs, it was now Ibusuki's turn to demonstrate his skills. He brought two Corsairs over the airfield, hoping that ground gunners would oblige him by knocking them down. The anti-aircraft guns remained strangely silent as a dogfight commenced at 500 m. The veteran warrant officer slid his aircraft out of the line of fire, time and time again. The two American pilots, probably worrying about their fuel supply, climbed steeply and flew off. Ibusuki circled the airfield to make sure there were no enemy aircraft still loitering and then landed.

BELOW: Lt Naoshi Kanno.

Lt Kanno's unfortunate encounter

Lt Kanno and some of his men disengaged from the wild encounter with the Marines over Kure. It had cost Kanno a third of his aileron. The *Shiden-Kais* were now low on fuel and ammunition, and a withdrawal to the south, back toward Shikoku, was ordered.

Unknown to Kanno, he was on a collision course with Fighting Nine from the *Yorktown.* The Hellcat pilots had just finished strafing nearby Saijo Airfield and were on their way to Matsuyama when six to ten Japanese aircraft were spotted at 08.10 hrs. Lt Leslie DeCew, the six-victory ace and flight leader of the sweep, took three of his four divisions to strafe Matsuyama Airfield, leaving Lt Bert Eckard's division to tangle with Kanno's men.

ABOVE: The first of Lt Bert Eckard's seven victories was Lt Naoshi Kanno, commander of Squadron 301. (AFAA)

Despite the damaged condition of his fighter, Kanno took the lead and made a run on a Hellcat flown by Eckard. However, Kanno overshot the division leader and the American did not waste time in firing a 15-degree deflection shot into the George. The tracers ignited the fuel tank under Kanno's feet and sent a sheet of flame roaring upward into the cockpit.

CPO Yasuharu Nikko made a stern run on a Hellcat, but Ens Joseph Kaelin caught him with a full deflection shot. The George burst into flames, smoked, burst into flames again and was last seen at 1,220 m going straight down. Nikko crashed in the sea off Hojo City on the northeast coast of Shikoku, about nine miles north of Matsuyama Airfield.

Lt Kanno fought to control his aircraft as he headed downward. It was a rule amongst pilots that they were to avoid crashing their aircraft into civilian areas if at all possible. He steered his aircraft out over the water. The Americans were dueling with his men and did not follow. With flames licking his face, the young lieutenant bailed out of his burning aircraft. He had forgotten to attach a static line from his parachute to his seat, thus his parachute did not automatically deploy as he left his aircraft. He struggled frantically to grasp the ripcord while dropping like a rock. With just seconds to live, his parachute blossomed.

ABOVE:
CPO Yasuharu Nikko.

In the skirmish with Kanno's 301st, the Japanese pilots forced the Americans to weave defensively, but their opponents claimed the match. Lt Bert Eckard and Ens Joe Kaelin received one victory apiece. Lt(jg) Emmett Lawrence received credit for two destroyed. This was a classic example of several pilots shooting at the same aircraft and receiving credit; only two Georges went down.

Lt Kanno landed on an unplanted rice field near the outskirts of Imabari City, about 32 km northeast of Matsuyama Airfield. A farmer advanced cautiously with a pitchfork, mistaking the 23-year-old Japanese squadron leader for an American. With burns to his face and hands, and a foul mood to match, the pilot warned the old man to back off. Relieved to discover that the airman was a fellow countryman, the farmer offered apologies while others crowded around him. Eager to report back to base, Kanno quickly located a bicycle and pedaled furiously toward a village. His frantic efforts to reach base were slowed by crowds of well-wishers who mobbed him, offering cups of sake.

The plight of Ens Takada

ABOVE: Ens Joseph Kaelin's first victory was over CPO Yasuharu Nikko. On 11 May 1945, he became an ace with his fifth and final victory. (AFAA)

Ens Mitsuru Takada's Myrt raced away from the American Task Force. With a top speed of 609 km/h at 6,100 m the reconnaissance aircraft was just about equal in speed to the Grumman F6F-5. However, the aircraft was having problems.

CPO Hiroshi Kageura frantically transmitted the coded messages to headquarters about the movements of enemy aircraft: "*...returning now due to engine trouble ...large group of enemy aircraft heading north over the south coast of Shikoku ...Another enemy formation sighted over Kochi ...A combined force of fighters and bombers, headed north...*"

The messages came in, one after another. Another Myrt was ordered into the air to guide some *Shiden-Kais* to meet the incoming Americans. Kageura continued to tap out the messages: "*Further, three enemy formations sighted. One hundred bombers heading north, 32 kilometers west of Kochi!*"

WO Mitsuya Tanaka, a member of the recon squadron, listened to the messages in the command center. He strained his ears to hear the signals which now indicated that at least one Myrt was in action against the enemy.

With a solitary shout of "*Keep it up!*" the radio room remained otherwise silent. Everyone stared at the radio equipment, imagining the plight of their helpless comrades. There was no hope. There came a long dash tone. After about three seconds, the dash ceased. Then silence.

"Airborne aircraft should land on other airfields!"

During the lull in the strafing, Lt Hayashi returned from Iwakuni without his faithful wingman, CPO Shiro Endo. Radiomen on the ground heard the lieutenant shouting on the radio "*Keep it up, Endo!*" as they fought over the airfield. When Hayashi's landing gear came down during the dogfight, he was forced to flee. The young

Declaration
17 August 1945

The Commander-in-Chief of the Imperial Navy Combined Fleet hereby declares to the entire Navy and acknowledges the distinguished service of:

Navy Ens Mitsuru Takada, 4th Reconnaissance Squadron, 343 Air Group
Navy Flight Petty Officer Hiroshi Kageura, 4th Reconnaissance Squadron, 343 Air Group
Navy Flight Petty Officer Minoru Endo, 4th Reconnaissance Squadron, 343 Air Group

While flying a reconnaissance mission at the forefront of a fighter interception force at a time when a strong enemy task force was at sea south of Shikoku on 19 March 1945, the aforementioned *Saiun* crew members caught sight of the enemy fleet off Tosa and successfully reported its position by maintaining contact with it in the face of intensive anti-aircraft fire, thus greatly contributing to our operations from that point.

On return to base, because of incidental engine trouble, however, they encountered a large group of hundreds of enemy carrier aircraft consisting of fighters and bombers. They fought bravely. Finally, the crew and the aircraft were hit. They flew resolutely at an enemy formation as soon as they knew that they would not be able to return to base, and dared to deliver a ramming attack on it, successfully destroying two of the enemy aircraft. They sacrificed themselves in the eternal great cause and displayed the spirit of the Imperial Navy.

Jisaburo Ozawa
Vice Admiral, Japanese Imperial Navy

Citation
24 March 1945

The Commander-in-Chief of the Imperial Navy Combined Fleet takes pleasure in presenting this citation to the 343 *Kokutai*, for services as set forth in the following:

For outstanding service in connection with the interception operations against the enemy, while serving as a fighter air group based at Matsuyama Airfield during the day of 19 March 1945, when the enemy task force had attempted to assault bases in the western part of the Inland Sea with its main strike force of carrier-based aircraft. The 343 *Kokutai* intercepted from Matsuyama Base and destroyed more than 60 enemy aircraft in almost no time by resourceful tactics and conducting combat professionally and bravely, thus highly raising the morale of the entire Navy.

Soemu Toyoda
Admiral, Japanese Imperial Navy

subordinate was probably trying to protect his leader when he was hit. Hayashi was anxious to get back into the air to make up for leaving combat and also to avenge his wingman's death. The squadron leader was waiting with his division members CPOs Miyuki Shimotsuru and Mitsuo Nakashima on the airfield.

Between 08.15 and 08.20 hrs, aircraft from VF-9 and VBF-10 began their assault on Matsuyama Airfield. Lt Goro Ichimura and his two surviving division members, LA Magoshichi Koma (Ichimura's wingman) and CPO Tsuneo Ozeki arrived over the base. They had lost Ozeki's wingman, LA Yoichi Saiki. The lieutenant could see the .50-caliber shells from the Hellcats "...bouncing off the roof of the hangar as if beans were popping on a fry pan." Matsuyama Airfield had a number of protected sheet metal hangars with iron plates installed in the roofing, as well as thick all-concrete hangars.

Cdr Tadashi Nakashima, deputy commander of the 343rd, had worried that his fighters would be coming back to base for refueling, right in the midst of strafing attacks. He shouted over the radio: "Airborne aircraft should land on other airfields!" Naval airfields within flying distance included Naruo, Omura, Oita, and Iwakuni.

The groundcrews were scrambling to fuel and rearm the fighters when suddenly, someone shouted to take cover. Everyone complied except for two. There was an instant staccato of anti-aircraft fire and a high pitched sound of approaching aircraft. CPO Mitsuo Nakashima was in the cockpit revving up his engine when a VBF-10 Corsair came in low over the airfield. Maintenance PO2/c Hikoshiro Kitagawa was under the wing, about to pull away the wheel chock. A trail of .50 caliber shells ripped up the earth in its path toward the target and shredded up both aircraft and men. The F4U-1D streaked away. Five fighters were destroyed by Cdr John J. Hyland's VBF-10 pilots in their strafing attacks. One Corsair was hit by anti-aircraft fire and was forced to ditch when it ran out of fuel; the pilot was rescued.

When the Corsairs left the airfield, Lt Ichimura decided to take a chance and land with his wingmen even though there were VF-9 Hellcats lurking overhead. The *Shidens* were almost out of ammunition and virtually flying on vapor. "After taxiing my plane into the revetment and asking for servicing," recalled Ichimura, "I returned to the command hut, watching the aircraft landing and I saw how the combat was going. Aircraft landed one by one and they were guided into their revetments."

Hayashi was infuriated at losing CPO Nakashima and also his own aircraft; he asked if there was a serviceable fighter left on the airfield. There was, and it was Lt Goro Ichimura's *Shiden* that had just come in. As a graduate of the 69th Class of the Naval Academy, Hayashi was about to pull seniority over Ichimura, Class of 71.

"I went back to my plane and mounted it for my second sortie," remembered Ichimura. "I was about to go out to the taxiway when Lt Hayashi ran to my aircraft, saying '*Only your aircraft is airworthy now. Let me use it!*'"

Lt Hayashi jumped into Ichimura's fighter and flew off to the north, looking for trouble. He did not have long to wait. One division of VF-29 Hellcats was approaching Matsuyama Airfield after a successful photographic run over Kure Harbor and nearby airfields. Midway between Kure and Matsuyama, two Georges attacked them. One of the Japanese was most likely Lt Hayashi.

Turning toward the Japanese fighter that had glued itself to Lt Bruce Jaques's tail, Lt(jg) John Herb fired a long burst at the tenacious enemy. Herb closed in from nine o'clock to six o'clock to a range of 300 ft. Tracers were seen to hit the enemy aircraft's cockpit. The Japanese pilot rolled into a steep dive toward the sea. Herb received credit for its destruction.

The remaining Japanese made a head-on run against Lt(jg) James Van Fleet. Both fired and Van Fleet passed below his opponent. When the enemy pilot pulled up, Lt Jaques got a head-on shot, but the damage claim was awarded to Van Fleet. The intruder dived and got away.

Capt Genda recalled: "The aircraft that made a second sortie engaged enemy aircraft sporadically during their retreat from Kure Harbor. We were not as successful as the first sortie because we were not able to hit enemy aircraft in such a large formation."

BELOW: Lt Bruce Jaques may have exchanged fire with Lt Yoshishige Hayashi, but his opponent escaped. Jaques finished the war as an ace with 5.5 victories. (AFAA)

Slaughter over Hiroshima

ABOVE: The Nakajima Ki-43 Hayabusa (Peregrine Falcon) had served as the standard Japanese Army fighter since the beginning of the war. It was given the Allied codename 'Oscar'. With just two nose-mounted 12.7mm guns and no armor protection, it could not compete against the Hellcat.

When word reached Ozuki Army Airfield that American carrier aircraft were raiding Kure Harbor, four pilots ran from the command tent to their fighters. It was an odd mixture of pilots and aircraft. Three were instructors from the Akeno Fighter School flying Oscars and one was a one-eyed transport pilot in a Frank. In the lead was Maj Junzo Yamamoto, the head instructor for the branch of the Akeno Fighter School at Miyakonojo. His wingman was Sgt Matsui, an assistant instructor. Capt Kaoru Sakuragi was paired with M/Sgt Yukio Shimokawa of the 9th Aero Transportation Unit.

One-eyed Shimokawa had earned the moniker of "Shimokawa, The Courageous Man" as a fighter pilot with the 50th *Sentai* in Burma during 1942-43. He had downed over 14 British and American aircraft and had been severely wounded twice. While attacking a B-24 Liberator, he was hit in the eye but struggled back to base, and finally sent home to Japan. After recovery, he worked ferrying fighters to the frontline, hating his assignment. He wanted to experience action once more.

The Ki-43 *Hayabusa* (codenamed Oscar) was the workhorse of the Japanese Army Air Force during the early days of the war. It was now a 'lightweight' compared to the Imperial Navy's A6M5 Zero (codenamed Zeke). A top speed of 530 km/h at 4,000 meters, two nose-mounted 12.7 mm guns, and no armor protection made it obsolete against the Hellcat. In the hands of inexperienced pilots, it stood almost no chance of survival. The Oscars at Ozuki Army Airfield were trainers belonging to the Akeno Fighter School. On the morning of the 18th, they were ordered to advance to Ozuki when TF58 was detected to be approaching.

The Nakajima Ki-84 *Hayate* (codenamed Frank) was considered one of the most capable JAAF fighters of the war. It was frequently mistaken for the Navy's George. It was armed with a pair of 12.7 mm guns in the nose and a 20 mm cannon mounted in each wing. With a maximum speed of 631 km/h at 6,120 m, the Frank was a deft middleweight and a formidable opponent for the Grumman F6F-5.

On the way to Kure, the four Akeno fighters were probably joined by eight Oscars and Franks which had taken off from other airfields. It was a desperate mission by desperate pilots to protect their homeland.

Four divisions of VF-17 Hellcats, launched from the *Hornet* at 08.00 hrs, were in the process of hitting warships in Kure Harbor when the Japanese Army fighters from Ozuki were spotted. Two divisions of Hellcats split off to go after the intruders. James L. Pearce, then a lieutenant, remembered the fateful encounter. "*I can still vividly recall Tommy Harris tally-hoing a group of Franks off to our ten o'clock position*," he wrote. "*We were trailing in behind Ned Langden's two divisions as they dived down into a hell of a lot of vari-colored anti-aircraft bursts coming up from the ships they were attacking. I was happy to divert away from that horrendous AA barrage to take off after the Franks. I don't think they saw us until just before we jumped them.*"

Lt Takeshi Umemura at Iwakuni Naval Airfield looked up to see the Army fighters flying at full speed on an easterly course at about 2,000 m north of the airfield. He focused his binoculars and then saw their opposition. There were at least ten Grummans waiting for them. "No sooner had I shouted to my men when a fight started," Umemura recalled. "Before both groups finished one turn, four or five columns of black smoke fell straight to the ground. I suspected that the Grummans would not easily belch smoke."

RIGHT: Pearce Division, which was heavily involved in the action of 19 March 1945. From left to right: Lt(jg) George C. Johnson, Lt(jg) Murray Winfield, Lt James L. Pearce, and Lt(jg) Jack Crawford.

The Japanese fighters were at a great disadvantage. One by one, they were blown out of the sky. Lt Pearce, in the lead, started the massacre by exploding a fighter with a full deflection shot when he caught it doing a split-S maneuver to escape. He hit another Japanese broadside, which caused it to smoke and go down. A comrade then entered the fray, flaming the same fighter; the pilot bailed out.

When Lt(jg) George C. Johnson, Jr. dived down behind an enemy fighter, firing all the way and scoring damaging hits, he, in turn, was hit by a Frank. The Grumman was holed in the empennage with a 20 mm round and in the cockpit by a 12.7 mm bullet. The damage was very minor and Johnson broke off his attack and rejoined his division.

Lt(jg) Thomas S. Harris was on the verge of becoming an ace. He had scored his first victory (a Val divebomber) on 11 November 1943 and then ran into a very long dry spell. A second victory was obtained on 16 February 1945 when he destroyed a Zeke over Tokyo. On 18 March, he downed two more Zeke fighters and claimed a 'probable' and a 'damaged'. The division leader and original "eagle eyes" of the squadron was not prepared to go home empty-handed.

A pair of enemy aircraft was sighted at eleven o'clock at a slightly higher altitude, fleeing from the dogfight. Harris and his wingman, Lt(jg) Harry J. Sundberg, pulled up into a head-on attack and flamed one each. At that point Harris became an ace with his fifth victory, but he wanted more.

ABOVE: A Grumman F6F-5 Hellcat of VF-17 is prepared for a sortie. Japanese pilots considered the Hellcat to be their most dangerous opponent. Unlike the F4U, the Hellcat was much more maneuverable. (NMNA)

As Harris and "Swede" Sundberg came out of their runs, another pair of enemy aircraft was seen above and behind them. Not all of the Japanese pilots proved to be such easy prey. Of the toughest one-to-one dogfight of his career, Tommy Harris gave this account: "I have never seen an airplane that was so difficult to shoot down. We dogfought all the way down to the water and I was behind him, and he was constantly turning. I had damaged him somewhere in the engine and there was some oily smoke coming out. The only way I could hit him was when he would reverse in a sort of a snap roll-type action. As he reversed, I would make a very short burst, only a fraction of a second, and get a few rounds into him, and we kept this up until I was wringing with sweat, and we kept losing altitude. We had flaps down part of the time, throttle idle and then full forward. This fellow just would not let go until I think he was finally losing power. We gradually got down, and the last thing he did was hit the water."

The tough Japanese pilot that Harris opposed was Capt Kaoru Sakuragi. Eyewitnesses on the ground reported two friendly aircraft flying into a formation of four Grummans around 09.00 hrs. A vicious dogfight ensued, with the aircraft diving steeply and climbing, firing the entire time. Spent shell cases fell like rain on the playground at Jigozen Primary School.

Asano Sakaida, the grandmother of the co-author, witnessed the low-level dogfight between Harris and Sakuragi. She had been visiting relatives at Aojima that morning and went outside to see what the commotion was. Hundreds of people were watching the air battle, their faces turned to the sky. Then one aircraft streamed smoke, howled downwards, and crashed into the sea.

Young schoolboys and girls clapped with joy, thinking that the stricken aircraft was American. Fishermen went out to the crash site, but later returned with Sakuragi's body wrapped in his parachute. Soldiers and civilians dedicated prayers to their vanquished hero and spoke respectfully of the Americans' skill.

The fighting was far from over. An Oscar came in to attack Sundberg's aircraft. He pulled inside of the Japanese aircraft and 'hosed' him. Hit in the engine and cockpit, the enemy aircraft went down burning. This was his third victory.

The determined Japanese Army pilots were being viciously mauled. One managed to evade the Americans and land at Iwakuni. The pale-faced pilot was called before Lt Umemura to give his report. "They're all gone!" sighed the dazed corporal. "Our flight leader was shot down, too! I was the only one left and I had to crawl between the mountains to escape!"

After the short briefing, the demoralized pilot, in a breach of military etiquette, turned and left without saluting.

M/Sgt Shimokawa fought desperately and claimed two Grummans. Now, he was the only Japanese left in the air fighting. Four Hellcats began chasing the lone Japanese like hounds after a fox, forcing him down to 300 m. They fired on the Frank but they could not hit him.

Shimokawa had opened up his engine to full throttle, but the Hellcats seemed to be gaining on him. He headed into a canyon, a fatal trap. When he pulled up to clear the mountain, Lt(jg) Tillman E. Pool fired a long burst from eight o'clock. The streams of bullets converged into the Japanese fighter, igniting the fuel tanks. Shimokawa bailed out. "Just as my parachute opened, it hooked up on a tree, but I was OK," he recalled. Shimokawa would receive a rare citation for his actions and survived the war. Tillman Pool had scored the third of his six eventual victories.

Some 40 minutes after they had taken off, Maj Yamamoto returned. He looked very pale and asked: "Has any other pilot returned?" Five minutes later, Sgt Matsui landed his battered aircraft with a large bullet hole in one wing. Later that night, M/Sgt Shimokawa showed up with scratches on his face and hands.

FAR LEFT: Capt Kaoru Sakuragi was a fighter instructor from the Akeno Fighter School who put Tommy Harris through some of the toughest moments of his life. (Sakuragi Family)

LEFT: Lt(jg) Tommy S. Harris scored his fifth and sixth victories over Hiroshima. He ended the war with a tally of nine. (T. Harris)

BELOW: Entries in Tommy Harris' flight logbook show that he downed four Japanese fighters in two days. (T. Harris)

RIGHT: Asano Sakaida, the co-author's grandmother, was one of hundreds of people in Miyajima who witnessed the dogfight between Lt Harris and Capt Sakuragi. She had never seen anything like it and often recounted the story in later years.

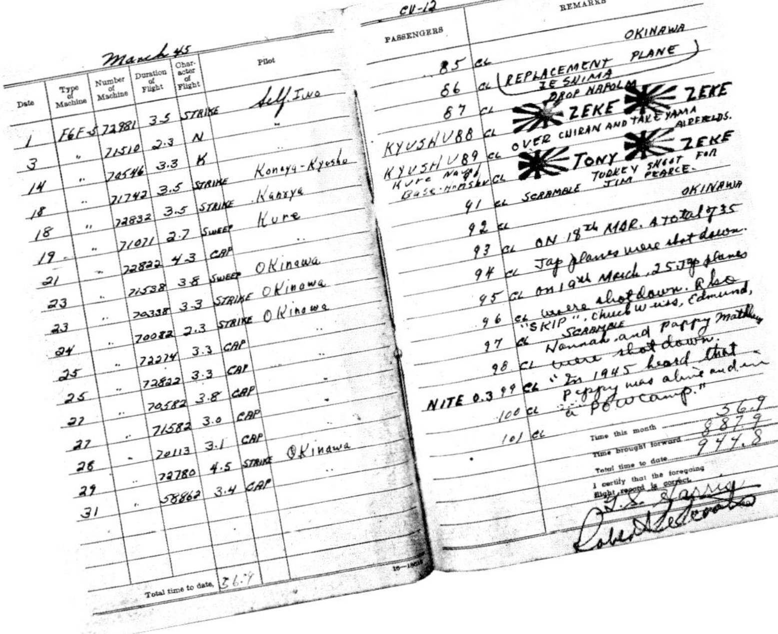

March 45

Date	Type of Machine	Number of Machine	Duration of Flight	Character of Flight	Pilot
1	F6F-5	72981	3.5	STRIKE	Self. Iwo
3	"	71510	2.3	N	"
14	"	70546	3.3	K	"
18	"	71742	3.5	STRIKE	Konoya-Kyushu
18	"	72832	3.5	STRIKE	Kanoya
19	"	71071	2.7	SWEEP	Kure
21	"	72822	4.3	CAP	"
23	"	71538	3.8	SWEEP	Okinawa
23	"	70338	3.3	STRIKE	Okinawa
24	"	70082	2.3	STRIKE	Okinawa
25	"	72274	3.3	CAP	"
25	"	72822	3.3	CAP	"
27	"	70582	3.8	CAP	"
27	"	71582	3.0	CAP	"
28	"	70113	3.1	CAP	"
29	"	72780	4.5	STRIKE	Okinawa
31	"	58862	3.4	CAP	"

Total time to date, 56.9

CV-12

Passengers	Remarks
85 CL	OKINAWA (REPLACEMENT PLANE)
86 CL	IE SHIMA DROP NAPOLM
87 CL	ZEKE ZEKE
KYUSHU 88 CL	OVER CHIRAN AND TAKEYAMA AIRFIELDS.
KYUSHU 89 CL	TONY ZEKE
Kure Naval Base-Honshu CL	TURKEY SHOOT FOR JIM PEARCE.
91 CL	SCRAMBLE OKINAWA
92 CL	ON 18th MAR. A total of 35
93 CL	Jap planes were shot down.
94 CL	On 19th March, 25 Jap planes
95 CL	were shot down. Also
96 CL	"SKIP", Chuck Weiss, Edmund,
97 CL	SCRAMBLE Hannah and Pappy Matthews
98 CL	were shot down.
NITE 0.3 99 CL	"In 1945 heard that Pappy was alive and in
100 CL	a POW camp."
101 CL	

Time this month 56.9
Time brought forward 887.9
Total time to date 944.8
I certify that the foregoing flight record is correct.
T. S. Harris

When VF-17 returned to the carrier, they were shocked. James Pearce remembered: "...it was the VBF-17 group led by a guy named Ed Conant that encountered the Georges during their earlier fighter sweep into the Kure area. The Georges really cut the earlier VBF group badly, quite a few F6Fs were shot down. We, of course, were over the target about the time the survivors of that fracas were returning to the ship so we didn't know of their disastrous encounter with the Georges until we returned to the ship later in the morning. As a general rule, the Japanese fighter pilots we were encountering at that stage of the war were mediocre at best, so the Georges really caught the VBF boys with their pants down."

Thanks to the sacrifice by the Japanese Army pilots, Iwakuni was spared. The airfield had been on the sweep list for both VF-17 and VBF-17, but both were intercepted and thwarted. Iwakuni Naval Airfield had a charmed life and remained unscathed throughout the war.

The day's accounting

Singly and in pairs, the *Shiden-Kais* returned to base even though orders had been given to land at other airfields. "We inquired with the airfields around the West Island Sea area about our aircraft making emergency landings," recounted Genda. "A few of them sent word that our aircraft had landed."

By 15.00 hrs, all survivors were accounted for. Lt Kanno arrived on base in a car in the afternoon. His face was red from burns but he was still full of fighting spirit and touted the superiority of the *Shiden-Kai* and the power of the four 20 mm cannon. Lt Yoshishige Hayashi, CO of Squadron 407, was the second to return, jubilant about his victory over a Hellcat near Kure on his last mission. The last to arrive was Lt Takashi Oshibuchi, the squadron leader of Squadron 701, who was almost given up for dead. He had landed at Naruo Airfield, with which he was very familiar, in distant Hyogo Prefecture. Everyone was relieved when Oshibuchi returned to Matsuyama Airfield around 14.30 hrs.

Surprisingly, there was very little damage to the airfield. Furthermore, all of the pilots seemed very satisfied with their new fighter. Some had taken punishing damage and had survived. There were no particular comments about the quality of the American pilots, but the Japanese pilots admired the Hellcats and Corsairs for their rugged performance. Most of them had fought these American fighters in the Marianas and the Philippines. They had been eager to know before the combat if the *Shiden-Kai* could take them on. Now, they had their answer. The pilots assembled to discuss their experiences and combat tactics, and the fate of their missing comrades.

CPO Shoichi Sugita was proud of his division. They had received the divisional "Victory Prize." In their hut, he congratulated LA Tamura. This came as a mild surprise because Tamura had anticipated being scolded for leaving his leader during combat. "You did a good job!" shouted Sugita and he gave his bewildered subordinate a pack of cigarettes.

BELOW: CPO Katsue Kato became top scorer in the 343 Kokutai during the combat of 19 March 1945. Though he claimed nine victories, they were never officially recognized. He flew in Lt Kanno's division and was killed in action on 16 April 1945. (K. Osuo)

Amongst the star performers that morning was CPO Katsue Kato who flew in the No.3 position of Lt Kanno's division in S301. He claimed nine victories and was the top scorer in the day's action.

Task Force 58's strike against Kure and surrounding areas was flown by over 300 aircraft with at least 12 fighter squadrons totaling 111 F6Fs and 52 F4Us. There were 15 bomber squadrons comprising 73 SB2Cs and 85 TBMs.

The US Navy's raid on Kure failed to sink the *Yamato*. It did, however, damage three carriers, *Amagi, Katsuragi*, and *Ryuho*. The carrier *Katsuragi* was well camouflaged and escaped detection by recon aircraft prior to the assault on Kure. It had docked alongside Mitsugojima Island and was disguised as a bridge connecting the island to the shore; trees were even set up on the deck. The battleships *Haruna* and *Hyuga,* and the light cruiser *Oyodo* were also damaged, but no major warships went to the bottom. The battleship *Ise* survived without a scratch even though it was not camouflaged with netting. R/Adm Kakuro Mutaguchi, the commander, placed first priority on anti-aircraft fire, and netting would have prevented his gunners from firing freely.

Claims by both the 343 *Kokutai* and TF58 aviators were somewhat similar, with the 343rd claiming 52 aerial victories. TF58 fighter pilots recorded 63 enemy aircraft destroyed during the Kure strike. There was the usual overclaiming on both sides due to confusion in the heat of battle.

ABOVE: A crowd gathers around a TBM-3 of VT-84 on the Bunker Hill after its return from Kure. Each Avenger carried four 500 lb bombs. The squadron claimed six hits on two carriers. All 14 TBMs returned and none were damaged. (NMNA)

The true scope of the carnage can be determined by the losses sustained.

TF58 lost 14 fighters (including aircraft which returned damaged but which were disposed of at sea) and eight fighter pilots missing. In addition to the fighter losses, 11 SB2C Helldivers and two TBM Avengers were hit by intensive anti-aircraft fire and failed to return.

In the aerial dogfight arena, 343 *Kokutai* lost 15 fighters and 13 pilots. Another notable loss for the 343rd was the three-man crew of the Myrt reconnaissance aircraft. Residents of Higashi Tsuno Village saw one aircraft belching white smoke overhead. It crashed on a steep slope in a mountainous forested area. The wreckage of Ens Mitsuru Takada's Myrt was scattered over three mountain peaks. The bodies of Takada and CPOs Hiroshi Kageura and Minoru Endo were recovered with the help of the village elder, civilian guards, and the Women's Defense Association. The Myrt crews' efforts prevented the great disaster that Capt Genda had dreaded and allowed him to unsheathe his blade. Everyone hoped that their deaths were swift and merciful.

Wounded pilots included Lt Kanno, senior veteran pilot Ens Akio Matsuba, LA Hideo Izawa, and PO1/c Tsutomu Murayama. Six groundcrew were also wounded in the bombing and strafing attacks. Kanno and Matsuba were sent together by aircraft to the Kamegawa Naval Hospital in Oita Prefecture. Both would return in a few weeks to continue fighting.

Capt Genda was pleased to hear favorable comments made by the American aviators shot down and captured during the strike. "I heard that some prisoners said '*We had thought the Japanese fighters were not well coordinated in combat flying and were not good in marksmanship either, but the fighter group we encountered was formidable beyond our expectation. They were well coordinated and excellent in marksmanship.*' It was somehow odd to be pleased with those words from enemy pilots, but it may prove that the fighter squadrons of the *Tsurugi Butai* were unsurpassed at the time!"

The 343 *Kokutai* claimed a magnificent victory, but it was tempered by the loss of its personnel. A personal tragedy was magnified by the visit of a woman to Matsuyama Airfield. CPO Yasuharu Nikko's death was shocking enough, but the misery was compounded by the arrival of his elder sister on the day he was killed. She had traveled down from the remote Hokkaido Prefecture in Northern Japan. Nikko's comrades debated the question of whether to inform the sister of his death. After much discussion, it was decided.

RIGHT: 20 March 1945: a group of Squadron 301 CPO pilots host a visit from three women who had come to visit CPO Yasuharu Nikko's sister to inform her of her brother's death. Front row, from left to right: Nobuya Komeda, Koichi Niizato, and Ichiro Muraki. Second row: Kumiko Yasutaka, Kiyo Matsuura (Nikko's sister), Teruko Yasutaka, Yoshikazu Miyamoto (half standing), and Tesshin Kamei. Standing: Hiroshi Tanaka, Eiichiro Sakurai, Yoshio Aoyama, Seiichiro Sato, and Tomokazu Kasai.

The small group of pilots disobeyed strict orders and left base for the grim task of delivering the bad news. They were hoping that if officers caught them, their CO, Lt Kanno, would come to their defense. The men visited the inn where Nikko's sister, Kiyo Matsuura, was staying, but no one could work up the courage to say anything. Then she realized what they were trying to say. Without shedding a tear, she sighed: "I understand."

Another grim reminder added to the sadness. "A few days after the combat, a civilian found a life jacket with Lt Kozo Shima's name on it on the beach and brought it to the air group," recounted Goro Ichimura. "There was a 13 mm hole in it. We then knew that he had been hit in the cockpit and died a heroic death."

US Aircraft Losses and Pilot Casualties on the Kure Strike, 19 March 1945

Carrier Squadron	Pilot/Crew	Cause of Loss
BENNINGTON		
VMF-123		
F4U-1D BuNo.57903	Maj Thomas Mobley Jr.	Attacked by 343 *Kokutai*, aircraft dumped overboard
F4U-1D BuNo.82473	Lt Dwight Mayo	Attacked by 343 *Kokutai*, aircraft dumped overboard
F4U-1D BuNo.82350	Lt Claude O. Barnhill	Attacked by 343 *Kokutai*, aircraft dumped overboard
F4U-1D BuNo.82339	1/Lt Archie J. Clapp	Attacked by 343 *Kokutai*, parachuted from aircraft, rescued
F4U-1D BuNo.82362	1/Lt Gordon K. Wooster	KIA by 343 *Kokutai*
F4U-1D BuNo.82516	2/Lt Ralph A. Russell	KIA by 343 *Kokutai*
VF-82		
F6F-5 BuNo.72209	Lt(jg) Earl A. McAllister	KIA, crashed while strafing
VB-82		
SB2C-4 BuNo. 20593	Lt Donald D. Worden	Anti-aircraft, POW, survived
	ARM3c Clifford A. Brown	Anti-aircraft, POW, survived
BUNKER HILL		
VB-84		
SB2C-4 BuNo.20790	Lt Charles L. Carper	Anti-aircraft, ditched, rescued
	ARM2c Millard Wilcox	Anti-aircraft, ditched, drowned
SB2C-4 BuNo.20803	Lt(jg) John D. Welsh	Anti-aircraft, POW, survived
	ARM2c Donald M. McKown	Anti-aircraft, KIA
ESSEX		
VBF-83		
F4U-1D BuNo.57601	?	Operational, ditched, rescued
F4U-1D BuNo.57546	?	Anti-aircraft, crash-landed on carrier, pilot saved
F4U-1D BuNo.57414	?	Operational, pilot saved
VF-83		
F6F-5 BuNo.72873	Ens William J. Morton	Out of gas, ditched, rescued
VB-83		
SB2C-4 BuNo.20711	Lt(jg) Wilston J. Lay	KIA anti-aircraft
	ARM1c Arne W. Lorentzen	KIA anti-aircraft
SB2C-4 BuNo.20788	Lt(jg) Donald L. Willis	KIA anti-aircraft
	ARM3c Carl C. Roberson	KIA anti-aircraft
SB2C-4 BuNo.20702	Lt(jg) Eugene J. Tougas	Anti-aircraft, POW, survived
	ARM2c Charles J. Richardson	KIA anti-aircraft
VT-83		
TBM-3 BuNo.68641	(Crew unknown)	Anti-aircraft, ditched, rescued

Carrier Squadron	Pilot/Crew	Cause of Loss
HORNET		
VBF-17		
F6F-5 BuNo.70859	Lt Strother T. Kipp	KIA by 343 *Kokutai*
F6F-5 BuNo.71754	Lt(jg) Forrest McCormick	Shot down by 343 *Kokutai*, POW, survived
F6F-5 BuNo.70991	Lt Edwin Matthews	Shot down by 343 *Kokutai*, POW, survived
F6F-5 BuNo.72372	Lt Charles F. Weiss	Collided with aircraft of 343 *Kokutai*, POW, survived
F6F-5 BuNo.58857	Lt Richard W. Moore	Ditched, rescued
F6F-5 BuNo.71781	Lt Frederick J. Prinz	Returned, aircraft dumped overboard
F6F-5 BuNo.70675	Lt(jg) Roger W. Karr	KIA by 343 *Kokutai*
F6F-5 BuNo.72569	Ens Noel L. Hannah	KIA by 343 *Kokutai*
VB-17		
SB2C-3 BuNo.19561	Lt Douglas J. Yerka	KIA anti-aircraft
	ARM2c Edward Peternell	KIA anti-aircraft
SB2C-3 BuNo.19658	Lt(jg) William A. Laws	KIA anti-aircraft
	ARM2c Eugene S. Burrow	KIA anti-aircraft
VT-17		
TBM-3 BuNo.68064	Lt Talmadge Westmoreland	Anti-aircraft, POW, survived
	ARM2c Robert H. Williams	Anti-aircraft, POW, survived
	AOM3c Harold W. West	Anti-aircraft, POW, survived
INTREPID		
VBF-10		
F4U1-D BuNo.82834	?	Anti-aircraft, ditched, rescued
VB-10		
SB2C-4 BuNo.20971	Ens Robert Brinick	Anti-aircraft, POW, survived
	ARM3c Crawford H. Burnette	Anti-aircraft, POW, survived
SB2C-4 BuNo.20120	Ens Rumsey	Anti-aircraft, ditched, rescued
WASP		
VB-86		
SB2C-4 BuNo.20772	Ens Harold R. Eyer	Anti-aircraft, POW, survived
	ARM3c William McReynolds	Anti-aircraft KIA
YORKTOWN		
VF-9		
F6F-5 BuNo.70630	Lt(jg) Richard R. Prior	KIA strafing Saijo

Endnote

1. *Sakura* was the code for the *Shiden-Kais* on standby. Mt. Niitaka in Taiwan is higher than Mt. Fuji and was the highest mountain in the Japanese Empire. Ironically, "Climb Mt. Niitaka" was the code that Adm Yamamoto used for the Pearl Harbor raid when he ordered Adm Nagumo to proceed to Hawaii.

"How miserable war was"
Captivity

The American pilots who fell to the guns of the 343 *Kokutai* on 19 March 1945 never believed that they would be shot down. Confidence and youthful bravado are the traits of fighter pilots. In the half century following the cessation of hostilities, through Japanese records and eyewitness accounts, it has been possible to ascertain the fates and experiences of some of these Americans. The following accounts illustrate the cruel hardships that followed their capture.

1/Lt Gordon Keith Wooster of VMF-123 was last seen by his Marine comrades going down northeast of Kure Naval Base after being hit by *Shiden-Kais*. He crashed his F4U-1D in a mountainous area near the Yasuura Naval Barracks. Naval personnel pulled him out of the wreckage and gave him first aid; he was badly burned and his teeth were broken. Wooster could barely speak to the interpreter because of his injuries and asked for water. He was able to say that he was married.

Capt Fumio Matsuda, the chief Naval doctor at the base, gave Wooster careful and courteous treatment. However, the extent of his injuries was so great that the slender American with the reddish-brown hair died the following day. He was cremated and buried at Kanadachi Crematory at Kure. Wooster's remains were delivered to the US Occupation Force at Kure on 27 December 1945.

Trouble awaited any unlucky American airman who parachuted onto Japanese soil. In general, civilians were not educated as to the correct treatment of downed flyers and prisoners. Due to intensive American bombing and enormous civilian casualties, it was only natural that hostility prevailed. Civilians were more a threat to the welfare of downed American aircrew than Japanese soldiers and military policemen. Veterans who fought in the Russo-Japanese War of 1904-05 and also during the First World War knew that prisoners should be treated in a humane way.

Japan's failure to ratify the provisions of the Geneva Convention, which dictated humane treatment of all prisoners, would have a very detrimental effect. Many of the prisoners were medically neglected and left in poor health. Completely out of the war effort, they faced a new enemy – death by illness and/or starvation.

Of the six pilots of VBF-17 missing in action from 19 March, three became prisoners. Lt Charles Franklin Weiss had collided with a George in a head-on attack and was knocked out of his cockpit. "I never remembered getting out," he recalled. "I don't remember pulling the ripcord. When I came to, I was floating down and I landed right in the middle of the Inland Sea."

After spending the night bobbing up and down in his one-man dinghy, a small fishing boat with three colorfully costumed men came out to him. They circled him for more than an hour.

Lt Charles F. Weiss

Weiss pleaded with the men to take him in. He was white as a sheet and freezing to death. The Japanese finally hauled him aboard and relieved him of his .38 revolver and his watch.

He was eventually taken to Kure and then on to Ofuna Prison Camp. He was periodically beaten for simply asking for medical treatment and suffered from malnourishment. He found solace only when he lapsed into deep sleep. His dreams were always vivid and had the same theme: escape. He dreamt that he escaped prison with the aid of a beautiful Japanese woman who helped him make his way to Iwakuni Airfield, which was the primary target of his squadron's mission on the day of the strike, and from where he stole a George and flew away to freedom.

"I was free and on my way to a carrier or a friendly island," Weiss wrote. "It was the most real nightmare I ever had, but in the excitement and whooping, I was awakened by several guards in my solitary confinement and it took a few blows to make me realize I was still captive. I looked forward to every night when I went to sleep. From there, it was all downhill but it was a most real experience and one that I would never forget. I was living a dream, a beautiful dream."

Weiss spent the last week in the death cell, praying and waiting to die. He was dying from dysentery and malnutrition and weighed a mere 70 pounds.

"After several days, out of nowhere, a fellow prisoner risked his life and fed me through the bars, with some of his own rations. I had never met him and he didn't know who I was. I really didn't know how he was able to reach me because we were so isolated and confined. I later learned that he was an air force officer named Bob Orr. There is no question in my mind that the food and water that he deprived himself of for me kept me alive another day or two until my prayers were answered..."

At liberation, the prison camp doctor who had regularly beat Weiss during his incarceration escorted him to the POW Hospital at Shinagawa on Tokyo Bay. Blood transfusions, vitamin injections, and tender loving care stabilized his condition.

Lt Weiss became the first POW hoisted aboard the new hospital ship *Benevolence.* On his stretcher was a box of Japanese yen. Surprisingly, upon surrender, the Japanese paid him money which was owed to him as a POW.

"I crawled over to a poker game and won even more money!"

On the day the surrender ceremony took place aboard the battleship *Missouri,* Adms Chester Nimitz and William F. 'Bull' Halsey visited Weiss on the hospital ship. The 27-year-old Washingtonian had the distinction of being the first POW released from Japan after the Second World War. He spent much time eating his way back to the United States. Returning home, he traded in the Japanese money for a substantial amount of American dollars.

Lt(jg) Forrest McCormick, a 24-year-old redhead from Medicine Lodge, Kansas, was blasted out of the sky by a George. He parachuted from 10,000 ft into a pine forest near Hirota Village. A group of local men organized a search party and headed into Sarudani (Valley of the Monkeys). They armed themselves with hoes, rakes, pitchforks, and an ancient shotgun.

McCormick landed heavily in a clearing and unbuckled himself. He accepted the fact that fate had dealt him a cruel hand and he proceeded to

Lt(jg) Forrest McCormick in the cockpit of his Hellcat. He would remember his last breakfast of cold scrambled eggs for the rest of his life. (F. McCormick)

dress his wounds with his first aid kit. Within half an hour, he heard voices further down the slope. He made no effort to hide himself and was soon spotted by some farmers.

"One of them approached cautiously and gestured for me to surrender my weapon, but I was unarmed," recalled McCormick. "In my haste to take off that morning, I had forgotten my revolver. I had my left arm in a sling and gestured back that I was unarmed."

When the farmers were finally convinced that the American posed no threat, they rushed in and quickly tied him up. He was pushed down the slope and in the process, a few of the men used the opportunity to deliver a few blows. The pilot was thrown roughly onto the back of an old truck for the short ride into the village.

When the residents saw their men coming back, word spread quickly. The locals came out to see the captured American being paraded through their village. Standing over six feet, the lanky prisoner towered over the people and looked strangely alien; many believed that the Americans were a race of giants. Frightened children clung to their mothers. For many of them, he was the first Caucasian they had ever seen. Suddenly, a man rushed up, pencil and paper in hand, and asked him to write his name; McCormick obliged.

Haruko Shirakata (left) at age 12, with her grammar school teacher and a friend. She watched her parents treat Forrest McCormick and admired him from afar. (H. Mihara)

As the crowd grew around him, the mood became ugly.

"*Tatake!*" ("Hit him!") came the shouts.

"*Korose!*" ("Kill him!") someone else screamed. The collective anger swelled. The impoverished villagers, many of whom had lost loved ones in the war, began to vent their hostilities. McCormick knew he was in serious trouble.

Mayumi Shirakata, a 48-year-old nurse and wife of the local doctor, came out to the road to see what the commotion was. She was surprised to see the American and shocked at the uncivil behavior of her neighbors. The diminutive woman pushed her way through the crowd and placed herself between McCormick and the mob.

"Stop!" she screamed. "Stop, all of you! This young man may have parents. We also have sons in this war. Imagine your sons in this situation. He is just fighting for his country. He is blameless. He is hurt and needs help. I am going to take him to our clinic. Don't you dare lay a hand on him!"

The angry villagers, stung into momentary silence by the courageous woman, slowly backed away. She escorted the pilot to her husband's clinic,

Dr Sadao Shirakata ran a small clinic in the impoverished village of Hirota on Shikoku Island. (H. Mihara)

Dr Shirakata and his wife/nurse Mayumi were respected members of their small community. He passed away at age 94, she at 100 in 1998. (H. Mihara)

In 1980, co-author Henry Sakaida traveled to the home of Mayumi Shirakata in Japan to deliver personal thanks from Forrest McCormick. Left to right: Haruko Shirakata Mihara, NHK-TV reporter, Mayumi Shirakata, and the author.

followed by the crowd. When they entered the clinic, the mob became unruly again. They peered through the windows and shouted in anger while a few threw stones. Extremely agitated, Dr Sadao Shirakata roared at the crowd and they melted away.

"On that day, my husband had many patients," Mrs. Shirakata remembered. "But he made them wait and took care of the pilot."

McCormick could not remove his flight suit, so a village policeman assisted. Nurse Shirakata was amazed to see zippers, which were still strange to the Japanese people. *"America has made progress!"* she thought.

While Dr Shirakata cleansed the wounds and applied bandages, their 12-year-old daughter Haruko stood by and gazed at the foreigner. She remembered his torn flight suit, which was stained with blood.

"I cannot forget the time when I watched him and our eyes met," she reflected in 1979. "His eyes were light green or blue and very beautiful, sad, and weak. I thought he was 19 years old or so. I felt pity at the sight of his neck wound. How miserable war was."

After treatment, Lt(jg) McCormick was taken to a nearby house where he was kept under tight security. Late in the afternoon, military police arrived and loaded him onto the back of a flatbed truck. For security purposes, they forced him to wear welder's goggles.

Mrs. Shirakata was very worried for the American. She knew how harsh the military police were toward captured airmen.

"He was such a handsome young man," she recalled. "There was nothing more I could do for him, so I said a little prayer."

"From the military installation the next morning, I was moved with about a half dozen other prisoners to a train where I passed out due to the pain," wrote McCormick. "Later, I was forced to walk off the train, then boarded a boat which took us across the strait to where I believe was Kure."

Ens Edwin 'Pappy' Matthews fared even worse than McCormick. He too was shot down by a George and parachuted into an apple orchard not far from Hirota.

"I was shot up pretty bad," he reported after the war. "I had shrapnel wounds in my left leg plus bullet wounds inflicted by a civilian who captured me. I also had burns on my left leg, face, and both hands."

The civilians beat him to a pulp and dragged him to the same village where McCormick was being held. Because his injuries were so severe, Dr Shirakata ordered the villagers to take Matthews on a stretcher to a hospital, which could provide better treatment. It was learned after the war that the wounded pilot received no such care.

One by one, American prisoners were transported to Kure. One of the interrogators who came into contact with the prisoners was Lt(jg) Hiroshi Goto of the Kure Guard Unit: "I was not good at speaking English, but my senior officer ordered me to interrogate the American pilot, saying '*You can do it as you are a graduate of the Navy Academy.*'"

While most schools and colleges stopped English lessons during the war, the Navy Academy at nearby Etajima offered English, German, and French courses. All midshipmen were required to master one of these three languages.

Goto recalled that one of the POWs had been a college student who had played football. The American would give only his name, the name of his carrier, and the location where he had been launched. The Japanese officers were surprised at how well-prepared the American flyers were. Each pilot had an "escape map" of Kure and surrounding areas printed on their silk mufflers. Their one-man dinghy contained survival gear, which included fishing hooks, food, signal mirror, and a flare gun. The only piece of survival gear that the Japanese pilots had was their life jackets.

Lt(jg) McCormick was surprised to see Lt Charles Weiss and Ens 'Pappy' Matthews in custody and was happy that they were alive. Kure Naval Station did not have a POW prison, so the newly captured flyers were accommodated in the naval brig. They were thrown into a cell with several young Japanese.

McCormick remembered: "I didn't get the feeling that this was a POW prison. It seemed like a regular jail. I guess they had problems of their own. I was shoved around, but there was no torture. The guards did, however, use their bamboo sticks quite freely at the slightest irritation."

All of the Americans captured on the 19 March raid on Kure, including seven SB2C and three TBM crewmen, survived captivity. Forrest McCormick worked for the US Post Office and died in Enid, Oklahoma, in November 1989. Edwin 'Pappy' Matthews resided in San Jose, California, for the last years of his life. According to his comrades, he was never quite the same after his POW ordeal. He passed away in the mid-1990s. Charles Weiss retired from the Navy in 1961 when he failed his physical for the rank of Captain due to his POW internment. He spent the next 24 years at the New York Stock Exchange. At the time of this writing, the healthy 83-year-old gentleman resides in the San Francisco Bay area.

This photograph brought happiness to Mayumi Shirakata, who had prayed for the young man for over 35 years. She had had no idea if he had survived captivity. Left to right: Anita, and sons Jon, Bradley, and Jeffrey, and Forrest McCormick. (F. McCormick)

Chapter 3

'Violet Lightning'

The Kawanishi N1K2-J *Shiden-Kai*

「紫の電光」

The story of the 343 *Kokutai* would not be complete without an examination of the Kawanishi N1K2-J *Shiden-Kai.* In the annals of Second World War Japanese military aviation history, the name *Shiden-Kai* is closely associated with Capt Minoru Genda's élite air group. The root word '*Shiden'* means 'violet lightning' in Japanese, and it was under this name that the aircraft became known. The term can also mean 'gleam of a sword (or blade)'.

Both the *Shiden* and its lineal descendant, the *Shiden-Kai* ('*kai'* meaning 'improved'), owed their existence to the Kawanishi N1K1 *Kyofu* ('Mighty Wind') floatplane, known to the Allies by its codename 'Rex'. In December 1941, the Kawanishi Aircraft Company submitted to Naval GHQ a proposal to design a fast land-based fighter variant centered on the floatplane design. It was a bold endeavor because the company had never produced a land-based fighter before. V/Adm Rikizo Tada was impressed enough to give his blessing, but no official specifications were issued. The initial engineering and construction of the *Shiden* would be a private venture. To its way of thinking, Kawanishi believed that if it built a better fighter, the Navy would buy it. The company appointed Shizuo Kikuhara to head the engineering team. The Navy assigned Engineering Cdr Junjiro Suzuki to offer guidance.

The *Shiden* prototype, designated X-1 – Experimental Interceptor No.1 – was completed in December 1942. It was fitted with an 18-cylinder Homare 11 radial engine rated at 1,820 hp. The oversized, four-bladed propeller dictated a mid-wing design and lengthy and complex undercarriage legs. During take-off, the legs would hydraulically contract in a two-step process and fold into the wheel wells. For landings, this process was reversed.

Lt Takumi Hoashi became the first pilot to fly the new fighter. It was rolled out at Itami Airfield on 31 December. Company test pilot Wasuke Otokuni was scheduled to fly it, but Hoashi proposed that he fly it first, a breach of company regulations. An agreement was worked out whereby Hoashi would simply roll it onto the runway. However, the temptation to actually fly the machine proved too much and Hoashi flew off.

The next day, 1 January 1943, Hoashi flew the *Shiden* officially and put it through its paces. It was exceptional to work at the end of the year and on New Year's Day in Japan, yet this is indicative that the Navy was very serious about the new fighter and Kawanishi was ready to work for the Navy's needs. The prototype was found to be underpowered, and Hoashi complained about poor visibility during taxiing due to the length of the undercarriage. Furthermore, the landing gears were unsuitable for unpaved runways. There was also the problem of excessive propeller torque as the aircraft took off. Despite the test pilot's negative views, the *Shiden* handled beautifully in flight and showed great potential for improvement. Although the Navy was not very impressed with the initial results, it decided not to abandon the project.

The origins of the land-based Shiden fighter types lay in the Kawanishi N1K1 Kyofu (Mighty Wind), known also under its Allied codeword as 'Rex'.

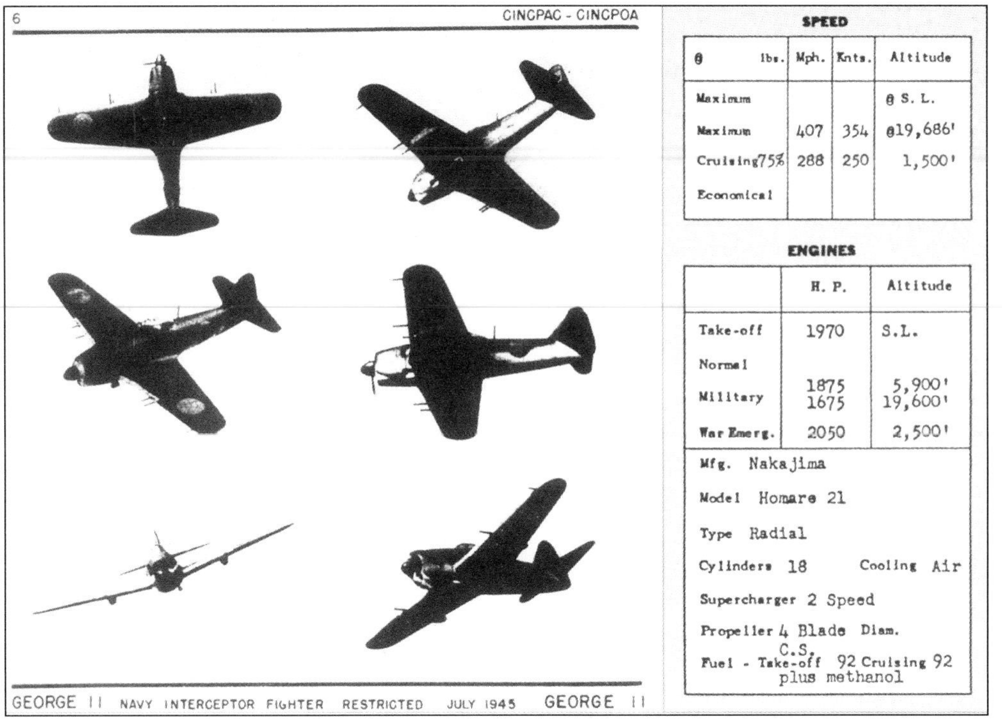

SPEED

@ lbs.	Mph.	Knts.	Altitude
Maximum			@ S. L.
Maximum	407	354	@19,686'
Cruising75%	288	250	1,500'
Economical			

ENGINES

	H. P.	Altitude
Take-off	1970	S.L.
Normal		
Military	1875 1675	5,900' 19,600'
War Emerg.	2050	2,500'

Mfg. Nakajima
Model Homare 21
Type Radial
Cylinders 18 Cooling Air
Supercharger 2 Speed
Propeller 4 Blade Diam.
Fuel - C.S. Take-off 92 Cruising 92 plus methanol

LEFT: American pilots often confused the George for the JNAF Raiden (Jack) and the JAAF Hayate (Frank) and Shoki (Tojo) because of its large radial engine. Likewise, the George was sometimes mistaken by the Japanese, at a distance, for the Grumman Hellcat and the P-47 Thunderbolt!

BELOW: The Americans learned of the Shiden through captured examples in the Philippines and gave the aircraft the Allied codename of George. This Technical Air Intelligence bulletin is dated October 1944. Note the tail number 201-53, which indicates that the drawing represents a 201 Kokutai aircraft.

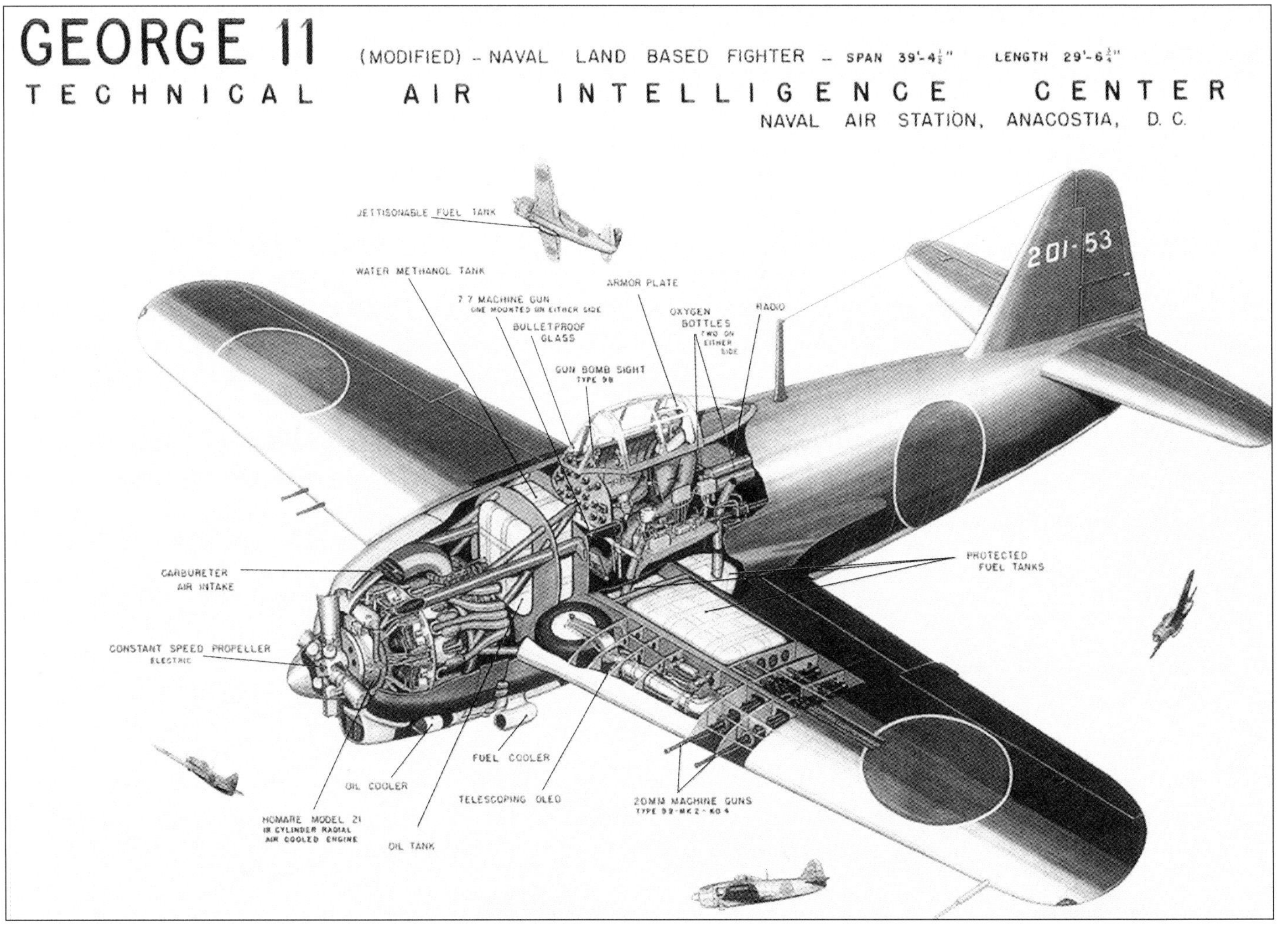

ABOVE : N1K1-J Shiden Model 11B carried two wing-mounted 20 mm cannon but did not have the two 7.7 mm nose guns of the earlier models. It is seen here with a 400 liter drop tank. (Maru)

The *Shiden's* major advantage was its unique combat flap system, which no other aircraft possessed; it was a purely Japanese innovation. In the Zero fighter, the pilot had to exert himself physically to manually engage the flaps. During a dogfight, if the pilot banked his aircraft radically to the left to make a turn, the aircraft would bank abruptly to the right. This was due to stall in the right wing, in which the angle of incidence became greater. With combat flaps, they would automatically lower at a steady rate, enabling the pilot to make a sharp turn without forcing the stick. The smoothness in making a turn added considerably to the pilot's comfort. The combat flap system can be compared to the power steering in today's large automobiles.

On 5 June 1943, Kawanishi's test pilot, Daisaku Okamoto, flew the new and improved *Shiden* prototype with the combat flap system. Shortly afterwards, Lt Yoshio Shiga, who had replaced Lt Hoashi as the Navy's test pilot assigned to the project, took her up. A mock dogfight took place over Naruo Airfield between Shiga and Wasuke Otokuni, pitting the newer version against the original without the automatic flaps. The results were remarkable. Shiga wound up on Otokuni's tail every time.

The Navy received the first *Shiden* on 24 July 1943.

As the standard Zero was losing its sting to a new generation of American fighters, Kawanishi was given approval to proceed with development of the N1K1-J *Shiden* Model 11, and 300 aircraft of this type were manufactured and accepted by the Navy in October 1944. It was armed with two 7.7 mm machine guns in the nose and a pair of 20 mm cannon mounted to each wing – one in the wing and another installed in a gondola.

But it was out in the field where the real testing came. Many *Shidens* were sidelined with engine and undercarriage problems stemming from poor machine tooling and inferior quality of materials. It achieved a maximum speed of 583 km/h at 5,900 m. The Americans gave it the codename of "George."

Only four days after the *Shiden's* initial test flight, Kawanishi's design team, again headed by Kikuhara, started work on the *Shiden-Kai*. The designers lowered the wing from the mid-wing position and then redesigned the fuselage, undercarriage and tail surfaces. This new model, designated the N1K2-J *Shiden-Kai*, had fewer parts than its predecessor and was easier and faster to manufacture.

The weapons system was changed when the 7.7 mm nose guns were eliminated and replaced by two 20 mm cannon in each wing. The firing speed of the cannon was 500 rounds per minute with a velocity of 700 m per second. The trajectory of the shells was synchronized to converge at a distance of 200 m. For comparison, the American Browning .50 caliber machine gun had a cyclic rate of 800 rounds per minute at a velocity of 850 m per second.

The knockdown power of the 20 mm cannon was impressive. However, its slow rate of fire demanded high levels of marksmanship which many younger pilots did not possess.

It was widely believed in post-war circles that the *Shiden-Kai* had armor protection for the pilot. There is no evidence to support this. Some surviving veterans state that there was no armor plate behind the pilot's seat, only a wooden panel. Two *Shiden-Kais* refurbished in the United States (343-A-19 held by the National Museum of Naval Aviation and the National Air and Space Museum) had no armor plates. An RAAF Intelligence Summary published after the surrender, stated: "*Internal tankage is protected with laminated rubber over metal. Bullet-proof windshield is provided but no other pilot armor.*"

The *Shiden-Kai* was first flown by Kawanishi's test pilots on 1 January 1944 at Naruo Airfield, newly constructed next to the production lines. Munekichi Okayasu took her up for 15 minutes and Wasuke Otokuni for 35. About a week later, Lt Yoshio Shiga and Lt(jg) Hajime Koga of the Naval *Kugisho* (Air Technical Department) flew it and delighted in its breathtaking performance.

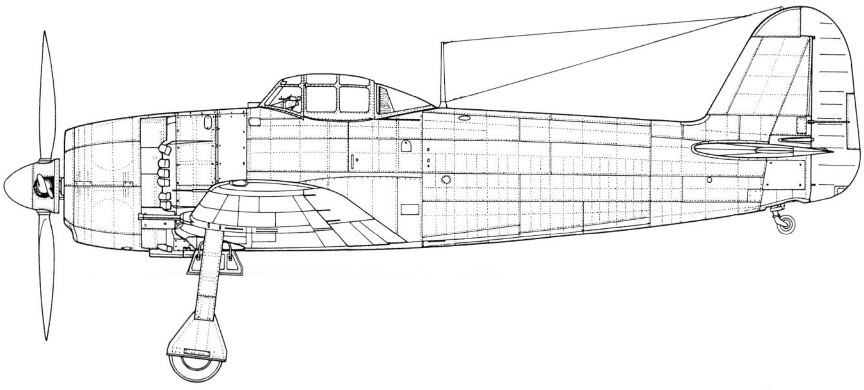

The Kawanishi N1K2-J *Shiden* (*above*)

Model 21 (*first 100 production aircraft*)

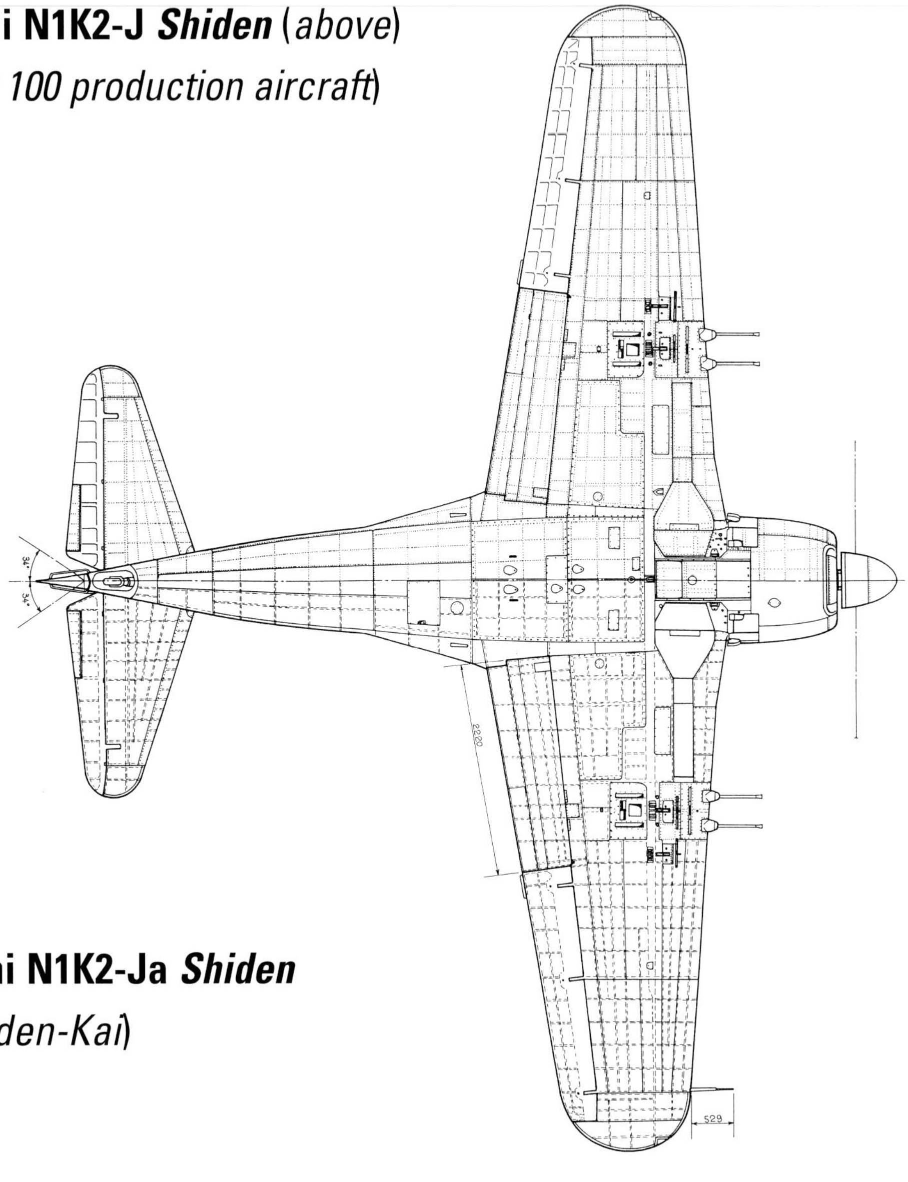

The Kawanishi N1K2-Ja *Shiden*

Model 21 (*Shiden-Kai*)

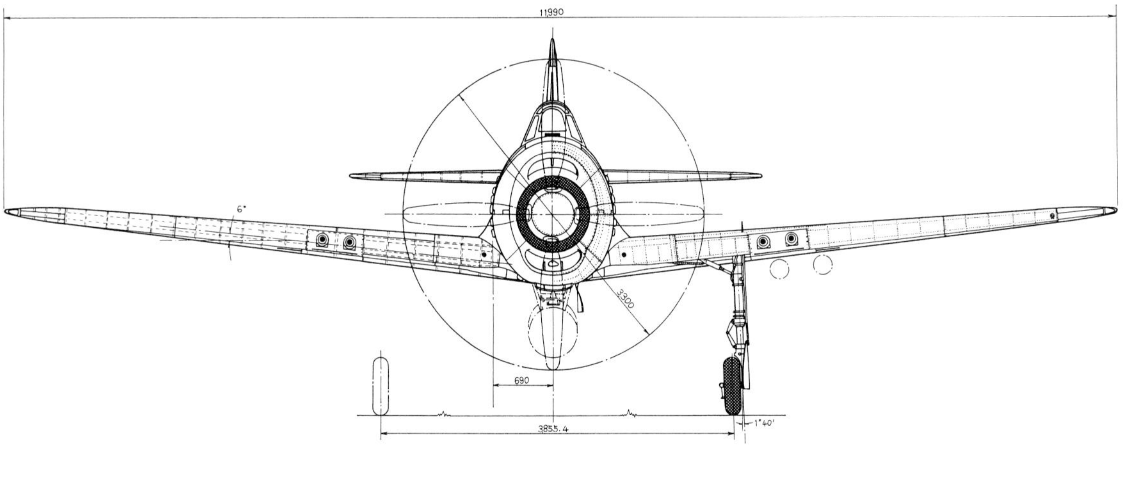

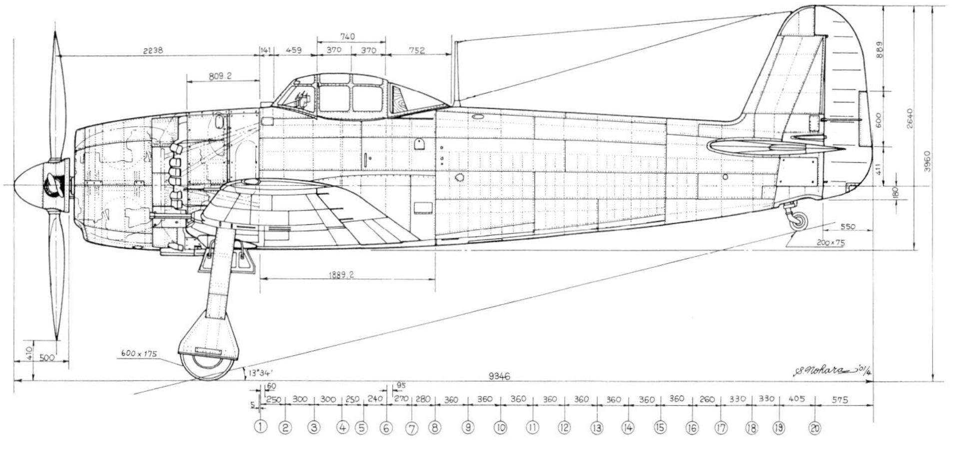

The Kawanishi N1K2-Ja *Shiden*

Model 21 (*Shiden-Kai*)

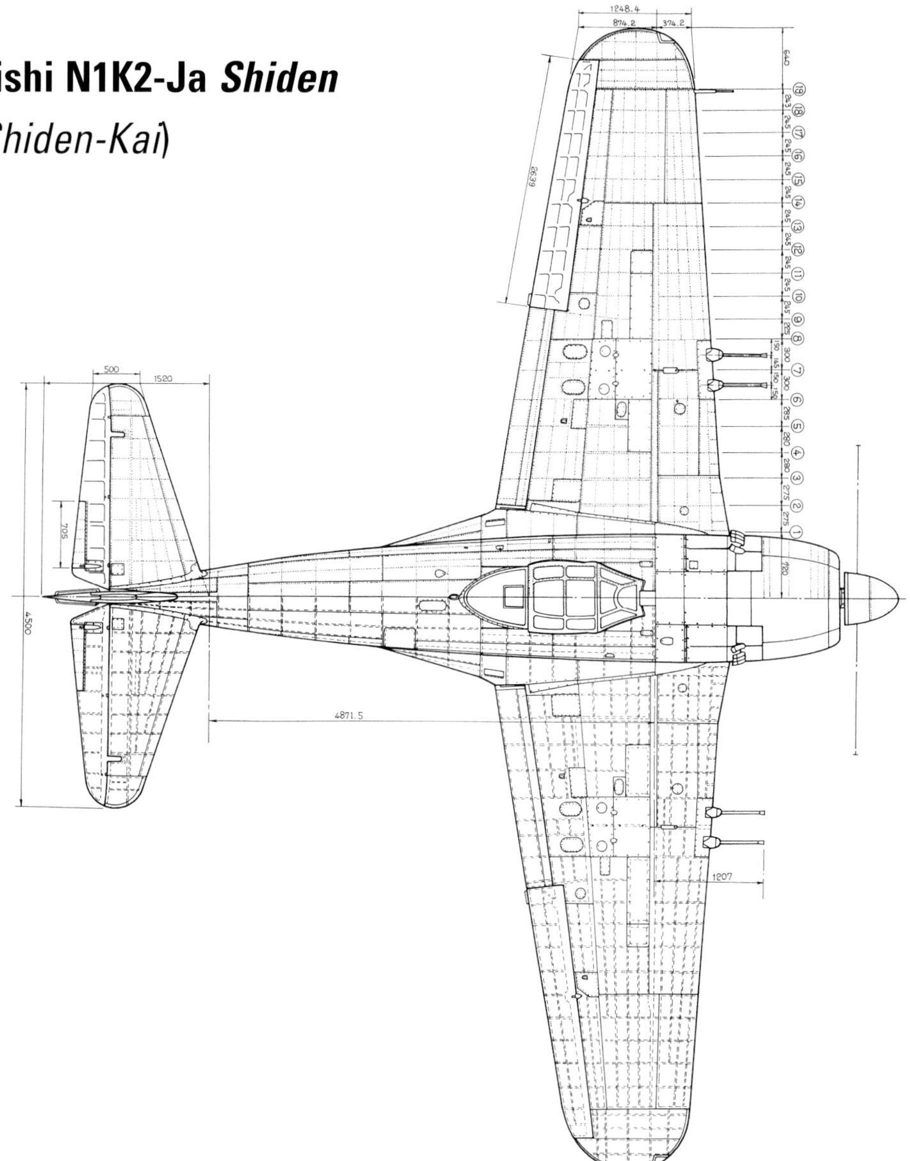

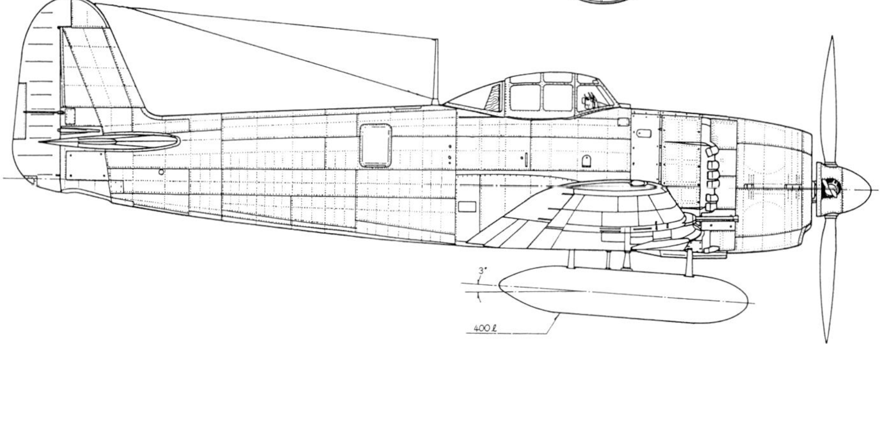

The N1K2-J Shiden-Kai was considered to be a superior aircraft in comparison to the Zero and the Shiden. LCdr Marshall Beebe, a VF-17 ace and former test pilot who had once fought the George opined that: "... it was equal to, or better than the F6F." Yoshio Shiga, who test flew it, compared it to a "...horny downtown girl." (Maru)

"It was a test flight under limited conditions," recalled Shiga. "There were still problems with the engine and propeller. At an altitude of 3,000 meters there was no problem with stability in either direction, including landing gears and flaps down. Stall came abruptly; this was different from the Zero fighter. Visibility was OK. There was no problem with the three rudders. The *Shiden-Kai* was totally different from the *Shiden*. I sent a signal 'OK' by making a loose dive towards the runway and a short-turn landing after pulling up at a low altitude.

"The *Shiden-Kai* had a peculiar tendency near stalling, but it was an excellent heavy fighter that could exceed the light Zero fighter. If the Zero was a proper young lady, then the *Shiden-Kai* was a horny downtown girl! It was not sophisticated, but the 20 mm cannon were effective."

Those who had mastered the Zero transitioned easily. It was the beginners who had the most difficulty. The control was very sensitive. Being rough on the stick could lead to autorotation whereby the aircraft would spin out of control. This "horny downtown girl" demanded respect. Novices who handled her wrongly usually wound up dead or injured. There were many training accidents. Then there was the problem with brakes: sometimes they would lock up and cause the aircraft to nose over. Despite these negative factors, those machines which entered combat flew well against Hellcats and Corsairs.

When the 343 *Kokutai* was formed at Matsuyama on 25 December 1944, the goal was to have all the unit's pilots fly the *Shiden-Kai*. Three squadrons were established and the allowance for each unit was 48 aircraft. It was after the unit's baptism of fire on 19 March 1945 that all squadrons made the complete conversion to the *Shiden-Kai*. The older *Shidens* were relegated to training duties with Squadron 401.

In combat performance, the *Shiden-Kai* was quite impressive when compared to the A6M5 Zero Model 52. It had a maximum speed of about 596 km/h at 5,600 m. According to VF-17 skipper, LCdr Marshall Beebe, an accomplished test pilot and fighter ace: "The *Shiden-Kai* was equal to or better than the F6F."

LCdr Yoshio Shiga agreed, but added: "The performance of the Homare engine was lower than expected. Fuel quality was down to 85 octane because the gas was mixed with oil extracted from pine tree roots which was a volatile oil. However, even under these conditions, pilots such as Oshibuchi and Kanno said that the *Shiden-Kai* was good up to around 30,000 feet. In my opinion, the *Shiden-Kai* could deal with the F6Fs, but had a hard time against the P-51s."

In combating B-29s over Kyushu, the *Shiden-Kai* was not very effective. This was because the aircraft had poor climbing speed and lost engine performance at high altitude. The horsepower of the engine decreased in a steep curve at altitudes over 6,400 m, while B-29s could maintain their maximum speed. Additionally, *Shiden-Kai* pilots had been trained in the fighter-to-fighter role and were not accustomed to attacking B-29s.

The *Shiden-Kai* was a rugged fighter in comparison to the standard Zero fighter.

The B-29 can be credited with the destruction of *Shiden-Kai* production. After they first bombed the Kawanishi Konan Factory at Kobe, where Emily flying boats were manufactured, they began raiding the *Shiden-Kai* Naruo Factory on 9 June 1945. To increase production, the company started operations at Himeji in March. It had completed 20 aircraft in May before Superfortresses bombed the factory on 22 June.

The N1K2-J *Shiden-Kai* was considered the best fighter of the Pacific War by JNAF pilots and is fondly remembered by the veterans who flew it.

Chapter 4

Okinawa – Baptism of Fire

12 April 1945

沖縄特攻進撃路啓開

銃火の洗礼

The Japanese continued to lose ground as the Allied noose tightened around the home islands. The Battle of the Philippine Sea in June 1944 sealed the fate of the Japanese Navy's carrier force. With the destruction of hundreds of aircraft and the loss of three carriers with experienced crews (*Shokaku*, *Hiyo,* and *Taiho*), carrier operations almost ceased. The Battle of Leyte Gulf (23-25 October) spearheaded the invasion of the Philippines and ultimately neutralized the once invincible Imperial Naval Fleet. The carriers *Zuikaku* and *Zuiho* were sunk. All remaining aircraft subsequently became land-based.

The desperate situation in the Philippines gave rise to the *Kamikaze* (Divine Wind) suicide squadrons, named after the great typhoon which twice destroyed the Mongol invasion fleets which made for Japan in 1274 and 1281. Six hundred and sixty four years later, Japan faced another enemy armada, and its people prayed for another Divine Wind to save their sacred homeland.

With few aircraft and little hope, the idea of a manned bomb gained acceptance. Surviving pilots reasoned that it was better to die a glorious death in the sky than perish helplessly on the ground. It was a last chance to strike back and make sacrifice count. The enticing theory was that one pilot could sink an aircraft carrier.

The first *Kamikazes* were not volunteers. They were selected mainly from a group of *Kohi* 10 (Aviation Group A, 10th Class) NCO pilots by Cdr Asaichi Tamai, vice commander of the 201 *Kokutai*. It is said that those first *Kamikaze* pilots were chosen from two squadrons whose commanders were either dead or missing.

V/Adm Takijiro Onishi, Commander of the 1st Air Fleet, was driven into a tight corner. His task was to support V/Adm Kurita's fleet, utilizing the giant battleships *Yamato* and *Musashi* by rushing into Leyte Gulf to batter MacArthur's troops. Onishi's *Kamikazes* had to damage the flight decks of the American carriers and make them inoperative for a week.

The first successful – and much publicized – sortie occurred on 25 October 1944 when Lt Yukio Seki of the 201 *Kokutai* led five Zero fighters on a one-way flight. The formation managed to sink the escort carrier *St. Lo* and damaged three others. The theory of 'one aircraft, one carrier' was now proven. Word quickly spread and *Kamikaze* operations took on increased significance. Men volunteered in droves for the so-called 'Special Attack Corps'. Those who perished would receive a posthumous double rank promotion. Their deaths would bring honor to their families.

In April 1945, an all-out coordinated effort by both Japanese Navy and Army *Kamikaze* units was approved by GHQ, with V/Adm Matome Ugaki, commander of the 5th Air Fleet at Kanoya, taking charge of operations under Adm Soemu Toyoda, Commander-in-Chief of the Combined Fleet. The main targets were the enemy carriers in Okinawan waters, followed by battleships, and support vessels. Even the remaining ships of the battered fleet were to be sacrificed. The giant battleship *Yamato* would also set sail with only enough fuel to reach Okinawa. It was to beach itself and inflict as much damage as possible. The Imperial Navy GHQ was determined to make the *Kamikazes* successful; the fate of the nation depended on it. The 343 *Kokutai*, operating under the 3rd Air Fleet, was attached to the 5th Air Fleet on 1 April to cooperate with the *Kamikaze*.

When Cdr Tadashi Nakashima was brought into the 343 *Kokutai* as its deputy commander, the unit's NCO pilots were afraid that he would requisition them as suicide pilots. The men were willing to die in combat, but not in desperate suicide attacks. It is thought that Capt Genda did not initially realize that Cdr Nakashima's arrival would affect the morale of the NCO pilots.

BELOW: V/Adm Takijiro Onishi, commander of the First Air Fleet, conceived the formation of the Kamikaze Special Attack Corps and initially directed its operations.

LEFT: V/Adm Matome Ugaki, commander of the 5th Air Fleet, took charge of Kikusui Operations, which ran from 6 April to 22 June 1945. Following the announcement of the Emperor's surrender and to atone for the many deaths under his command, Ugaki stripped his uniform of its gold braid and took off from Oita with a flight of Judy bombers on the afternoon of 15 August. A radio transmission was heard, saying that the flight was diving on an enemy ship. The following day, Ugaki's aircraft and body were discovered on a sandbar at Iheya Jima, about 30 km northwest of the northern tip of Okinawa.

ABOVE: V/Adm Shigeru Fukudome, commander of land-based naval air forces in the Philippines, toasts his Kamikaze volunteers. Hopelessly isolated from Japan and with few aircraft and no hope for survival, there was never any shortage of volunteers.

ABOVE: A determined suicide pilot dives on the carrier Suwannee on 26 October 1944 in the Philippines. The most important target for the Kamikazes was the aircraft carrier against which they were instructed to crash into the flight elevators. (USN)

ABOVE: Direct hit! The Suwannee explodes under the impact of a Kamikaze. More than 100 men were killed and 170 wounded in the attack. However, the carrier remained afloat. Initial attacks were successful because the Americans were caught off guard by such desperate tactics. (USN)

LEFT: Destroyers such as the Morris, on picket duty, bore the brunt of the Kamikaze fury. (USN)

RIGHT: The scene of destruction inside the hangar of the escort carrier Sangamon following a suicide attack. (USN)

However, Nakashima was transferred to the 721st *Kokutai* in late March. A medium bomber group, the 721st was formed specifically as a suicide unit in October 1944. The unit's main 'ordnance' was the rocket-propelled *Ohka* ('Cherry Blossom'), a bullet-shaped, rocket-propelled manned bomb with a 1,200 kg warhead which was carried under the belly of a Betty bomber and released a few miles from its designated target. Nakashima's new assignment as the unit's CO was to train the pilots of the escort fighter squadrons which were attached to the air group.

Nakashima's departure from the 343 *Kokutai* brought welcome relief to some of the men in the unit. There was 'bad blood' between this officer and the enlisted pilots because of the deputy commander's involvement in *Kamikaze* operations in the Philippines. Lt Takeshi Ninomiya, a weapons officer, recalled that in his conversations with Lt Naoshi Kanno, the young squadron leader told him: "I drove him away!"

According to an account given to Japanese aviation writer Naoki Kodachi for his book *Reisen Saigo No Shogen II*, CPO Tomokazu Kasai recalled that the animosity started during the opening days of the *Kamikaze* attacks. On 27 October 1944, Lt Kanno and his men escorted *Suisei* (Judy) bombers over Leyte at dusk and landed at Cebu in the dark. This was probably the first time Kanno made a night landing. There was no light along the runway, but the pilots landed safely. Kanno reported to the then LCdr Nakashima. For reasons unknown, an argument ensued between the two officers. Kanno became furious and pulled out his 7.65 mm Browning pistol which accidentally discharged, with the bullet hitting the tip of his heavy leather flight boot. The injury to his big toe was only minor. CPO Kasai stepped in between the feuding men and quickly hustled his commander away.

With relations off to a bad start, things became even worse. Two days after this incident, Kanno and his men had their aircraft commandeered by Nakashima for *Kamikaze* use. The young squadron leader was enraged. In order to return to base, he and his pilots were forced to fly as passengers in an old transport aircraft. En route, they were attacked by a lone P-38. The transport pilot, unable to deal with such a situation, virtually gave up hope. But Lt Kanno was not so willing to accept the situation and took control of the aircraft even though he had no experience in piloting a large machine. Taking evasive action, he made an emergency landing on a small airfield on Lubang Island. Even before the transport came to a stop, he ordered everyone to bail out and take cover. The P-38 came down and strafed the aircraft until it flamed. Kanno's group returned to Mabalacat by ship.

Soon a rumor spread that pilots who landed on Cebu would be forced to become *Kamikazes*. Pilots warned their comrades not to land there.

With the departure of Nakashima from the 343rd, Capt Genda and LCdr Shiga took on the task of handling combat operations. Cdr Takahide Aioi, who would arrive at his post in May, would handle other air group matters from Matsuyama Airfield, the unit's official base, with a total of 3,000 personnel.

Under the codename *Kikusui* (Floating Chrysanthemum), *Kamikaze* missions had been mounted against Task Force 58 since 18 March 1945 when the massive fleet had approached Kyushu and Shikoku. The name *Kikusui* was now applied to the *Kamikaze* attacks planned from 6 April at the start of the American landings on Okinawa. The *Kikusui* symbolized loyalty to the Emperor, in the context of fighting with one's back to the wall.

Since TF-58 had approached Okinawa in the latter part of March, Zero fighters from the 203 *Kokutai* based at Kagoshima, Izumi, and Kasanohara were sortied on escort and intercept missions. At the end of March, Zero fighters and *Shidens* from the 601 *Kokutai* at Katori in Chiba Prefecture, were based at Kokubu Airfield No.1 and took part in the intercept missions over the Amami Islands. The 252 *Kokutai* based at Mobara and Tateyama (Chiba Prefecture) also relocated to Kokubu at the same time and was placed under the command of the 601 *Kokutai*. When *Kikusui* No.1 commenced, Zero fighters from the 302 *Kokutai* at Atsugi were based at Kasanohara and shared facilities with the 210th. Thus, the Navy's main fighter groups were mobilized to cooperate with the *Kamikaze* operations.

Capt Genda received orders to move his unit to Kanoya on Kyushu Island and, on 4 April, he flew in a *Shiden-Kai* to the new base, escorted by LCdr Yoshio Shiga and another pilot. Key staff officers were to follow. The new location brought the unit closer to the frontline. Their assignment was to open the corridor for the flow of *Kamikaze* aircraft heading toward Okinawa. Due to the fast pace of events, training was abandoned.

The new assignment meant also that, for the first time, the 343 *Kokutai* was to embark on a very long-range mission. It would have to fly from Kanoya Airfield to Amami Oshima and back again. The one-way

LEFT: Leading Airman Ikuji Matsumura photographed in March 1945 at Matsuyama with 343-B-20 in the background. Note the large number '20' applied within the 'meatball'.

RIGHT: Another photograph of LA Matsumura, this time with 343-B-03. Matsumura lost an eye in an accident during a fuel/distance test flight on 28 March 1945 and never flew again.

BELOW: Shiden-Kai 343-A-15 at Matsuyama Airfield on 10 April 1945 just before Squadron 301 departed for its new base at Kanoya. Lt Naoshi Kanno frequently flew this aircraft. As with other units, pilots of the 343 Kokutai flew aircraft on an availability basis. Poor-quality fuel caused maintenance problems, making the assignment of individual aircraft to units impossible. However, aircraft marked such as this were reserved for flight leaders. (K. Osuo)

distance was roughly 375 km and would take about two hours. With an allowance for combat at full power for a maximum of 15 mins factored into the equation, fuel consumption looked daunting.

A test flight comprising 12 *Shiden-Kais* took place on 28 March to ascertain fuel consumption for the mission. Each aircraft was fully armed and carried a single 400 ltr drop tank. Lt Yoshishige Hayashi led three divisions and flew for two and a half hours over the Inland Sea between Okayama and Oita. The data was extrapolated and the mission was deemed feasible, in theory.

Lt Kanno's Squadron 301 landed at its new base on 10 April and prepared for the upcoming *Kikusui* No.2 mission slated for 12 April. He had appropriated aircraft 343-A-15 as his own and directed his crew-chief to paint two large diagonal yellow stripes on the fuselage behind the cockpit. "Enemy aircraft will willingly gather around my plane with the yellow marking," he explained. "Then I will kill them!"

This idea caught the attention of the other squadron leaders. Lt Takashi Oshibuchi's 343-C-45 and Lt Yoshishige Hayashi's 343-B-30 soon had sporty stripes on the fuselage of their aircraft.

12 April 1945

The sudden relocation of the 343 *Kokutai* from Matsuyama to Kanoya was not welcomed by the mechanics. The *Shiden-Kais* needed continuous servicing which was disrupted during the transfer. Additionally, poor-quality aviation fuel continued to cause havoc with the Homare Type 21 engines. To make matters worse, the unit was ordered to sortie on 12 April. It was a case of too much, too soon.

The combat mission was assigned as interception over the Amami Shoto (islands) which consisted of Amami Oshima and the smaller Kikaiga Shima. These islands are located midway between Kyushu and Okinawa along a chain of islands known as the Nansei Shoto (Southwest Islands). The picturesque islands served as stepping stones from Okinawa to the mainland of Japan and, as natural landmarks, as navigation aids for both Japanese and American pilots. Although the Amami Shoto was in Japanese territory, it was also within the patrol zones of American carrier fighters. Any Japanese foray into the area was going to be tremendously costly.

Eleven divisions of *Shiden-Kais*, the entire force of airworthy aircraft – 44 machines, were prepared to sortie from Kanoya. Since Lt Kanno was the first to arrive at the new base, he was selected to lead this mission. Lts Oshibuchi and Hayashi were still back at Matsuyama.

The large formation was composed of Lt Kanno leading the 1st *Chutai* from Squadron 301 consisting of five divisions and Lt Masaji Matsumura (Naval Academy 71st Class), the *buntaicho* of Squadron 301, leading the 2nd *Chutai* from another group of Squadron 301, consisting of four divisions. Lt Ryoichi Yamada, the *buntaicho* of Squadron 701, led the 3rd *Chutai* consisting of two divisions. Two aircraft had to cancel take-off and eight aircraft dropped out along the way, including division leaders CPO Isamu Miyazaki and WO Wataru Tomisugi from the Matsumura *Chutai*.

ABOVE: CPO Takumi Sugitaki stands by his Jack whilst serving with the 302 Kokutai at Atsugi. (T. Takaishi)

Kanno's total strength decreased to 34 aircraft, although effectively his force was short of a further aircraft because CPO Takumi Sugitaki had taken off very late due to engine problems and lagged far behind. More than half of the Squadron 301 pilots flew Squadron 701 and Squadron 407 aircraft. Apparently, there had been a servicing problem and the mission was a makeshift assembly of aircraft from the three squadrons.

"I was too young and so serious!" remembered Sugitaki (now Takaishi). "I took off after all the other aircraft had taken off." On the way to Amami he saw the *Shiden-Kai* formation ahead of him and at higher altitude. He could not climb to join them. Lt Kanno called him on the wireless: "The aircraft flying behind, join up with the formation!"

Sugitaki explained his situation, but received no response. The lone *Shiden-Kai* fell behind and the friendly formation was soon out of sight. He proceeded alone to Amami Oshima.

The Georges pressed on. At 11.10 hrs, the formation set its course to 210 degrees, heading southwest. Tokara Retto appeared to the left. At 12.30 hrs, they made a left turn west of Amami Oshima; at 12.45 hrs, the group turned north, south of Kikaiga Shima.

CPO Sugitaki was nearing his destination when he made a fortuitous discovery. There was a lone Corsair not more than 1,000 m below him. It was just begging to be bounced. As he dived towards the F4U, the American saw him coming and escaped in a very steep turn. Sugitaki saw three or four divisions of fighters and assumed they were his comrades. While chasing his prey, he looked up and recognised the white star markings on the wings over him. The first Corsair had apparently been a decoy! About 15 F4Us started to attack him.

The F4U pilots exhibited good turning tactics and the lone George pilot knew he could never escape, so he decided to take one with him. Sugitaki plowed straight into the formation, quite prepared to die, but his opponents were not. The Corsairs evaded him and dispersed upward. The George then dived toward the sea with a few gull-winged fighters firing at it. It had taken many hits and started to stream smoke.

One American closed in as if he wanted to join up with Sugitaki. The Corsair pilot gazed at the George below him, probably wondering why the enemy aircraft would not fall. Sugitaki saw his opponent and the F4U made a steep right turn in a climb. Having survived a brief encounter with three divisions of F4Us, the bewildered chief petty officer fled the scene and headed for home.

ABOVE: Kikaiga Shima under attack by VBF-17 Hellcats on 12 April 1945, just before they engaged the 343 Kokutai.

CPO Tomokazu Kasai could see Kikaiga Shima in the distance. White smoke was billowing upward as if an air raid was in progress. The formation finally arrived over the island where its pilots caught sight of two divisions of Hellcats strafing Wan Airfield. These were part of a 12-aircraft formation of VBF-17, commanded by Lt Edwin S. Conant. The veteran carrier pilots had fought with the 343 *Kokutai* on 19 March, and now they were about to duel with them again.

Lt Kanno spotted Hellcats at 12.50 hrs, about 20 in number, and heading south at 3,000 m. He reported about 30 F4Us also heading south, cruising along at less than 1,000 m. His group was at 6,000 m with the cover *chutai* at 6,500 m. He gave orders to form battle stations and radioed his division leader to provide cover. Lt Matsumura acknowledged and took his position over Kanno.

While Conant and his second division leader, Lt John V. Delander, were in the midst of completing their strafing run, Lt Kanno started the dive on Conant's third division up above, led by Lt Guye W. Troute. He dived straight down to 2,000 m. CPO Kasai, the 4th man, tried to follow him but in his usual style, the leader did not wait for his men. He would take immediate action and only skilled pilots could follow him. The formation of his division began to collapse.

"Lt Kanno fired a burst at the enemy's first aircraft. Suddenly it drew white smoke and nosed straight down," remembered Kasai. "In succession, the second aircraft fell with white smoke as well."

Troute radioed a warning, but it was too late for Ens Raymond Grosso. He was taken out in the first pass and bailed out, never to be seen again.

Ens Charles Dikoff, Grosso's section leader, saw tracers flashing over his wing and instinctively rolled left in time to see Kanno recovering below. He went after the enemy squadron leader who made a tight right turn. The Hellcat pilot managed to turn inside the George and fired directly at the cockpit, but Kanno rolled over on his back, gunned his engine, and headed for the water. PO2/c Mikio Mitsuishi tried to catch Dikoff as he pursued Kanno, but the crafty American rolled his aircraft and shot his opponent into the sea.

Lt Conant heard the distress call from Troute. The two divisions of Hellcats reacted liked angry hornets and climbed to meet the challengers. Conant's first burst knocked pieces from the wing and tail from one of the Japanese aircraft, but the enemy machine escaped. Turning around, he spotted a George on the tail of a Hellcat. He weaved, pulled up, and stitched the enemy's cockpit from 300 ft, killing the Japanese pilot; the aircraft exploded. Another George flashed across his path. Conant maneuvered on his tail, closed to 75 ft, and dropped him into the water.

BELOW: CPO Tomokazu Kasai (T. Kasai)

For CPO Kasai, it seemed that the Hellcats were all over him. He recalled: "It was too late. Fighters began to fall here and there. I could not see which aircraft they were, but they looked to be friendly aircraft. I won't be defeated! I shouldn't be separated from my formation. Then abruptly, CPO Sugita made a roll and dived down. Immediately, I followed him. Looking ahead, an F6F was seen. He was in my gunsight. Right at that moment, I forgot that I was to maintain formation with my section leader. I fired my four 20 mm guns. The vibration was pleasant. Tracers were absorbed into the enemy aircraft. I did it at last! He began trailing black smoke and went down. Immediately, I followed him. Then I realized that I was alone and looked for CPO Sugita, but he was nowhere to be seen. I went too far! We were not allowed to leave the formation. Formation fighting was Capt Genda's principle, but I had ignored it."

The fighting intensified over and around the airfield. Ens Dennis L. Miner may have crossed paths with CPO Shoichi Sugita, the 343rd's highest-scoring ace. A George flashed by in a tight right turn and 'Denny' opened fire from 100 ft, hitting the cockpit area. The Japanese pulled up

in a high wingover and came out in a vertical dive, which tore off the cockpit canopy. Although Miner was credited with a kill, if his opponent was Sugita, he escaped this ambush and continued fighting.

At this time, LCdr Marshall U. Beebe was on CAP duty with his dozen F6F-5s ten miles north of Kikaiga Shima. When the CO of VF-17 heard the chatter of pilots in action, he rushed his formation south. At 31, he was considered 'old' for an operational fighter pilot, but he was most certainly good. He had won his wings at NAS Pensacola in 1937 and worked at the Dept. of Engineering in BuAer from 1941 to 1943. With his firm grasp of tactics and F6F design, he was a proven leader. His score stood at eight victories.

Arriving on the scene, Beebe was greeted by a fighter which bolted out of the dogfight and headed straight towards his division. The Californian managed to get a 60 to 20 degree deflection shot into the enemy machine's port side, hitting the engine and wing root. Undaunted, the Japanese tried to maneuver behind Beebe's wingman, but Lt(jg) Frank 'Sis' Sistrunk drove him off Lt(jg) Robert Good's tail.

ABOVE: LCdr Marshall U. Beebe was an accomplished test pilot and ace with 10.5 victories to his credit. He led Fighting 17 when the squadron was re-formed at NAS Alameda, California, on 18 April 1944. (M. Beebe)

ABOVE RIGHT: VF-17 Hellcats on the Hornet ready to sortie. Excellent radio communication and radar-directed combat air patrols by the Americans caused considerable Japanese aircraft losses.

The chase was on. The damaged George headed for the deck with a commanding lead, but Beebe and Sistrunk, using water injection, shaved the distance in about eight minutes. The skipper fired a 20 degree deflection shot, corrected his aim, and starting knocking pieces off the fighter. The desperate pilot used violent evasive tactics at zero altitude, but his fate was sealed. Beebe delivered the *coup de grâce* and his opponent went in, scattering debris over an area of 500 ft.

Lt(jg) Sistrunk felt cheated. With two victories under his belt, he needed three more to become an ace. He headed up to 6,000 ft, made a head-on run against a George, scored some hits, then latched onto its tail. His adversary did a split-S, but it did not save him. The fighter burst into flames in the cockpit area and the pilot parachuted. Sistrunk would return to the *Hornet* with his third victory and a birthday celebration; he had just turned 23. He would attain ace status four days later with his fifth kill.

Lt Matsumura had received quite a shock. It appeared to the *chutai* leader that a large group of enemy fighters – originally estimated at approximately 50-strong – had suddenly increased to about 80. It was flying in two layers with another group coming in to join up. Matsumura attempted to switch fuel from his drop tank to his wing tanks and pulled the drop lever, but his belly tank would not release. Quickly, he changed the supply back to his auxiliary and prepared to fight.

RIGHT: Lt(jg) Frank Sistrunk (AFAA)

Meanwhile, Lt(jg) Richard 'Cookie' Cowger, leading the third division of VF-17, was heading north, about ten miles off the west coast of Kikaiga Shima at 10,000 ft. His section leader was Lt(jg) Willis 'Medals' Hardy, an exceptional ace who downed five aircraft in one day on 6 April. Hardy spotted four Georges in two-aircraft sections and watched as the Japanese – probably a formation comprising Lt Matsumura and his wingman, PO2/c Susumu Imai, with CPO Seiichiro Sato leading the 2nd section with PO1/c Makoto Nishimura – dived down from 12,000 ft into a division of Hellcats at 3,000 ft altitude. Observing that the 1st *Chutai* aircraft were in peril below, Matsumura was forced to act.

CPO Sato recalled the harrowing ordeal: "The 2nd section had to follow the 1st section to protect them.

ABOVE LEFT: Maj Herman 'Hap' Hansen Jr. at Guadalcanal in 1943. He scored his first victory, over a Zero, at Rendova on 30 June 1943. He remained scoreless until 17 February 1945 when he shot down an Oscar over Tokyo. In all, he achieved 5.5 victories. (R.F. Toliver)

ABOVE: An F4U-1D of VMF-123 takes off from the carrier Bennington. (NMNA)

Surprised at my leader's unexpected maneuver, I dived down, closely following Lt Matsumura, thinking it was no good to go down so low."

Hardy and his wingman, Ens Ernie 'The Baron' Vonder Heyden, dived down to thwart the enemy ambush. The pair of Hellcats arrived at the scene simultaneously with a group of Corsairs. The 'party crashers' were Marines from the *Bennington.*

Maj Herman Hansen Jr., leading 12 F4U-1Ds of VMF-112, was on CAP near Amami Oshima. A flash of red against the ocean blue caught the CO's attention. A group of between 15 and 20 enemy fighters was spotted heading toward Amami at 10,000 ft altitude. They were most likely Lt Kanno's 1st *Chutai* seen after its engagement with VBF-17 Hellcats.

The Corsairs climbed at full throttle toward the enemy. The Japanese fighters dived down to meet them. At the last moment, they turned, making a grave tactical error. Hansen took his men over the top and came in from the rear. Some *Kamikaze* aircraft strayed into the scene with catastrophic results. The CO of VMF-112, celebrating his 25th birthday, knocked down a George and added a Kate torpedo bomber in under two minutes, scoring the first victories for his new squadron and becoming an ace.

Capt Don Owen turned on the first enemy fighter he saw and chased him into the water. He then teamed up with a Hellcat and shared in another kill. A George dived on Owen, but his wingman, 1/Lt 'Big John' Callahan, cut inside the intruder and destroyed it with three bursts. He then caught sight of a George executing a split-S to escape from a Corsair. Callahan climbed on his opponent's tail and fired all the way down; splash two. His third victory was scored when he caught a fighter trying to land on Wan Airfield.

The Georges were making passes in two-aircraft sections on stragglers below. Lt(jg) Willis Hardy found himself flying with an unidentified Leatherneck pilot on his starboard side. The Hellcat and the Corsair were spotted from above and two *Shiden-Kais* focused their attention on the gull-winged Corsair, but their aerial gunnery was poor and they hit nothing. One of the Georges completed its pass and climbed. Hardy went vertical and opened fire from 500 ft, hitting the enemy aircraft in the engine and cockpit. Hardy pushed over to avoid stalling; his opponent snapped out and fell behind the Grumman, then dived toward the sea.

A Corsair chased its opponent right into Hardy's line of sight and he took advantage of the situation by opening fire from five o'clock. The George flamed and crashed into the sea. During this encounter, Hardy's wingman, Ens Vonder Heyden, experienced engine trouble and was not able to engage. While returning to the carrier, a George attacked them, but Hardy inflicted damage on the attacker and drove him off.

Meanwhile the dogfights continued. Lt Matsumura found himself surrounded by American fighters in what was to be his first dogfight. Lt Kanno radioed for him to assemble at the rendezvous point. Matsumura responded by screaming into the radio that he was surrounded. Suddenly, tracer flew past him and he looked behind to see a Hellcat sitting on his tail. He made two turns and escaped into clouds.

"I looked around but I saw only enemy fighters everywhere!" recalled Kasai. "Then I saw three enemy aircraft below. They were at a position suitable for an attack from behind and above. I dived down right away and closed on the first aircraft to about 50 meters. I squeezed the firing button. I saw the pilot lean backwards and then fall forward in the cockpit. I continued to engage the second aircraft. I caught him in the gunsight and squeezed the button again, but my guns did not fire! I could get killed! I made a dive. When I pulled up, I was almost on the deck and fled. I looked up and saw two F6Fs chasing me. I crept along skimming the sea. The enemy aircraft suddenly became four. They were persistent and I was determined to ram them if they attacked me. One of them started to smoke and two of the others pulled away. The remaining enemy aircraft was at an altitude of 3,000 meters." CPO Kasai raced for home.

The combat concluded around 13.15 hrs. Most of the surviving *Shiden-Kais* assembled and headed back toward Kanoya. Lt Ryoichi Yamada was surprised to hear Lt Matsumura's voice on the radio some five to ten minutes after they left the combat area. Matsumura was still fighting.

Some of the *Shiden-Kai* pilots had left the engagement early because of damage to their aircraft. CPO Sugitaki was surprised to come across a lone fighter with its canopy missing. When he drew closer, he recognized CPO Shoichi Sugita at the controls. For a time, they flew side by side and Sugita could see his comrade's right wing tank burning. Sugitaki was nearly overcome by seeping gasoline fumes inside his cockpit. He decided to bail out over Take Island, less than 20 nautical miles from the southern tip of Kyushu.

Sugita saw his comrade land in the water near the coast. When Sugitaki eventually emerged from the sea and crawled up on the rocky coast, Sugita rocked his wings and headed for base. Sugitaki was taken in by an islander and stayed as his guest for a few days until a steam launch came to retrieve him.

The jubilant American pilots returned to their carriers. 1/Lt Herbert V. Lundin of VMF-112 was the first to touch down. Climbing out of his mount, he was greeted by a horde of crewmen. "You better get out your adding machine!" he shouted to the ACI officer. The Leathernecks claimed 20 destroyed and six damaged and yet suffered not a scratch. Their rival unit on the ship, VF-82, lost one aircraft but its pilot, Ens Terry Mills, was rescued. The unit claimed four fighters over Amami Oshima.

On board the *Hornet,* there were wild celebrations. LCdr Marshall Beebe's VF-17 claimed 13 victories and suffered no losses or damage. VBF-17 lost only one pilot and tied the score with VF-17.

The Marines knew that their kills had been hard earned. 2/Lt James Hamilton, who scored one victory, damaged three others and shared a kill, observed that his opponents had fitted two 20 mm cannon on each wing. "*There were no small caliber guns,*" he reported. "*It was hard to burn. There was plenty of armor plate and self-sealing tanks, too.*"

Though it was realized that the F4Us were unable to stay with the George in a turn, the Marines faulted the Japanese for lack of organized tactics.

The Hellcat pilots also noted that their opponents were well-armored and could turn inside the F6F. The VF-17 pilots believed that the Japanese aircraft had excellent climbing characteristics and superior speed. The 343 *Kokutai* pilots, when confronted with Hellcats on their tails, snap-rolled to the left. Their automatic combat flap system saved them time and again, much to the Americans' chagrin.

Most of the returning *Shiden-Kais* had landed at Kanoya by 15.30 hrs. Their pilots had fought a ferocious battle. CPO Shoichi Sugita claimed two Hellcats, plus a kill and a probable over F4Us. When his wingman, CPO Tomokazu Kasai, also made claims for two Hellcats, Sugita became irritated and questioned the authenticity of Kasai's statements. Kasai duly amended his claims to two 'probables'.

Those claiming single victories included Lts Naoshi Kanno and Masaji Matsumura, WO Shoji Shibata, CPOs Mitsuo Hori, Eiichiro Sakurai, and Hiroshi Tanaka, and PO2/c Toru Otsubo (all from Squadron 301). Members of Squadron 701 claimed 12 kills. The unit reportedly engaged over 80 Hellcats and Corsairs, and claimed 20 F6Fs and three F4Us shot down.

Lt Kanno paid dearly on this mission. He lost nine men from his squadron, including his own wingman, while Squadron 701 suffered one loss. Four aircraft made forced landings but three of them eventually returned to base. Lt Matsumura and three others made emergency landings on Tanega Shima, a large island about ten nautical miles from the southern coast of Kyushu. Their fuel tanks were almost dry and the lush green island was a welcome sight. There was a small airstrip there where they could refuel. When Matsumura landed, an enemy aircraft came down and strafed his aircraft on the ground. His mount was damaged and the lieutenant had to borrow an old Zero to return to base the following day.

What caused such severe losses? Cdr Yoshio Shiga blamed the collapse of the Japanese formations on so many aborts. When division leader WO Tomisugi turned back because of engine trouble, two members of his division failed to return.

The following day, Lt Kanno confronted Lt Matsumura. "All of your aircraft would have been destroyed if you continued fighting!" admonished the squadron leader. "You should have thought of your commander's position!" WO Mitsuo Hori, the second division leader added more tactfully: "Sir, you should not have reduced your altitude so much."

Matsumura thanked Hori for his advice, which was given to him so that he could improve his leadership skills. Matsumura was shocked to learn that one of his division members, PO1/c Makoto Nishimura, had not returned.

The tragedy of this mission was brought home by CPO Eiichiro Sakurai of S301. He wrote a letter to Kazunari Hashimoto, the brother of his division leader:

"I was the second man to Lt(jg) Tatsutoshi Hashimoto. My engine was not in good condition when I took off, so my flight was delayed. We fought with 80 enemy fighters over Kikaiga Shima. I was separated from your brother in combat and returned to Kanoya, nursing my engine. It was after landing that I first knew that your brother and CPO Niizato, the number three man and my classmate, had not returned. If the engine had been in good condition, I would not have separated from my leader. I regret this greatly."

Many of the victories achieved by the American carrier pilots who fought the 343rd were claimed as *Kamikaze* aircraft and/or their escorts. In the confusion of jumping from one encounter to another, it became difficult to differentiate between the various Japanese fighter types. Confusion or not, it was still a 'turkey shoot' by American standards. The key to the American victory was their superb radar-directed combat patrols and excellent radio communication.

On the *Bennington,* Maj Herman Hansen wrote eagerly to his mother in Kansas City about the day's events. *"This has been the most unusual birthday I've ever had..."*

Lt(jg) Tatsutoshi Hashimoto.

CPO Koichi Niizato.

PO1/c Osamu Omori.

PO2/c Takeshi Ishikawa.

Chapter 5

Tragedy at Kanoya

15 April 1945

鹿屋の悲劇

Kikusui No.2 ended on 15 April with a brief lull in the fighting. The Japanese 5th Air Fleet was unable to detect the movements of the American task force that day. Only one *Kamikaze* mission, involving ten Zero fighters, was sent out from Kokubu Airfield. The day would be spent preparing for *Kikusui* No.3, which was to begin on the 16th.

The ever-increasing waves of suicide aircraft were of serious concern to the American and British navies which were tasked with providing air support for the Okinawa Campaign. The attacks originated from Kyushu, Japan's southernmost island. Plans were drawn up to destroy the *Kamikaze* staging areas and three US Task Groups were assigned to do the job.

One of the most important bases for the *Kamikazes* was Kanoya Airfield. When a large concentration of aircraft was detected by a reconnaissance flight, 'Fighting 17' – VF-17, from the *Hornet,* was assigned to sweep the airfield with 32 Hellcats along with VF-30 and VF-82 whilst Grummans from the *Independence* and *Randolph* were to attack nearby Kushira Airfield. The first aircraft of the strike force of nearly 90 aircraft took off at 13.15 hrs.

On Kanoya Airfield, mechanics prepared the *Shiden-Kais* for the next day's mission. Lt Naoshi Kanno's Squadron 301had eight aircraft on standby, composed of the Matsumura and Sugita Divisions.

The approach of the large American formation was detected by the Japanese radar and lookout network around 14.50 hrs. Surprised by the American move but judging that he had ample time before their arrival, Capt Genda ordered an immediate scramble.

The Americans – VF-46 Hellcats from the *Independence* – arrived over Kanoya much sooner than expected. The primary target was Kushira, but LCdr Robert A. 'Doc' Weatherup, who was leading eight aircraft from his unit and 20 more from VF-12, saw activity down below. He decided to make an attack and started his dive from 15,000 ft while the VF-12 Hellcats continued to their primary target.

BELOW: Kanoya Airfield under bombardment by B-29s. This airfield was an important staging area for Kamikaze aircraft. (Josh Curtis)

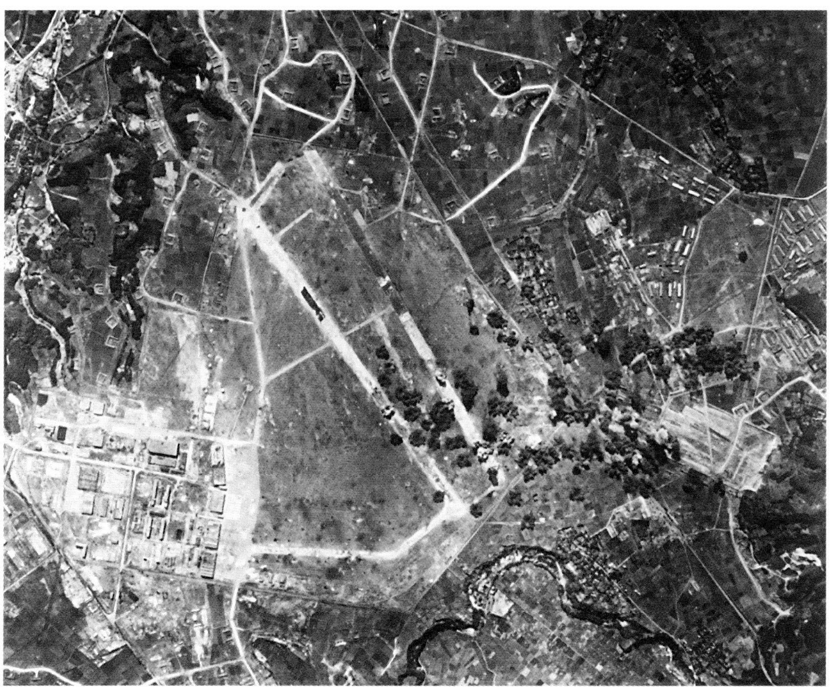

Genda realized that interception was too late. He issued an order to abort the scramble. However, when word reached the field, there was no means to stop it. The signal flag to take off was lowered, but the pilots' attention was focused on the Hellcats coming in.

CPO Shoichi Sugita spotted the advancing Americans and pointed to them as he sprinted toward his aircraft. Right behind him was PO2/c Toyomi Miyazawa. Ens Saburo Sakai, an instructor in Squadron 701, saw the futility of interception and yelled to Sugita to take cover. The single-minded chief petty officer ignored his comrade's warning; he was determined to get airborne. The two pilots jumped into their fighters, started engines, and taxied forward.

Tomokazu Kasai recalled: "At the warning '*Enemy formation sighted, heading north for Kanoya*,' the four aircraft engines were started. There was almost no time for a warm-up. The first and third machines raised violent smoke and dust. CPO Sugita looked back from his cockpit and pointed upward and had begun to take off, when seven or eight Grummans dived in a strafing run at his aircraft. I was stunned and I signaled to remove the wheel chocks when a rocket hit the wing of my aircraft and blew a big hole in it."

CPO Kasai's damaged fighter blocked PO2/c Tsunehara Tamura's aircraft. Tamura recalled in a post-war interview that "...the wheel chocks for my aircraft had not been removed. As I was about to signal to remove them, I found that the groundcrews had run to take cover." All scrambled pilots except two quickly abandoned their fighters. Together with the groundcrews, everyone sought cover as the Hellcats tore across the airfield.

ABOVE LEFT: LCdr Robert A. 'Doc' Weatherup had the best air-to-air gunnery score in his squadron and he attributed his fine shooting eye to the pheasant-hunting days of his youth. However, the only victories he achieved during the war were the two Georges he shot down on 15 April 1945 over Kanoya. (R.A. Weatherup)

ABOVE: CPO Shoichi Sugita, photographed in January 1945. In constant action since mid-1942, he had claimed over 120 victories by the time he joined the 343 Kokutai. However his aggressive nature eventually cost him his life. (Y. Izawa)

"As we approached Kanoya, I could see several aircraft taxiing and apparently preparing to take off," remembered Bob Weatherup. "I decided to attack the airfield. I picked out an aircraft (or decoy) in a revetment and fired my rockets at it and then strafed until I had to start my pull-out."

Sugita rolled down the airfield. He was in the most vulnerable position. Miyazawa, his faithful wingman, was just starting to take off. Rockets and bombs exploded all across the field.

"As I was getting level again, I saw that two aircraft were starting their take-off," continued Weatherup. "I decided to take the lead aircraft, but I was forced to pull pretty hard in a high tight turn of nearly 270 degrees to get on his tail. By this time, the lead aircraft was probably at about 400 ft and starting a slight turn. I decided to take the first aircraft, thinking that it was probably flown by a senior flight leader and that their formation might fall apart without him. I also presumed that some of my other aircraft would take any following aircraft."

LCdr Weatherup used his altitude and speed advantage and pulled a big lead. When Sugita's aircraft was in his gunsight, he started firing. "At first, I had a little too much lead and the target aircraft apparently saw my tracers and relaxed his turn," recalled Weatherup. "Almost immediately, I started to see 'flicks' where my bullets were hitting armor or heavy structure. Soon thereafter, the target aircraft started to nose over. I continued to fire until I had to pull up. The target aircraft did not burn until it crashed, although there may have been a little smoke."

CPO Shoichi Sugita, the renowned 21-year-old "King of Aces," fell out of the sky at low altitude and crashed in a thunderous blast. Capt Genda, LCdr Shiga, Ens Sakai, and others witnessed his demise. CPO Kasai recalled the horror: "An incredible scene came into my eyes. At a burst from a Grumman, Sugita's aircraft banked abruptly and crashed with black smoke at the end of the airfield."

PO2/c Toyomi Miyazawa struggled to gain altitude, but his efforts were also futile. LCdr Weatherup made an overhead run on his opponent, who appeared to be inexperienced. The George took many hits, but did not burn. It dropped from 800 ft and plunged into a pine forest east of the airfield.

LEFT: PO2/c Toyomi Miyazawa became LCdr Weatherup's second victory.

In the late afternoon, WO Isamu Miyazaki went in search of his missing comrades. "I visited the 8th Base Unit, a ground unit at Kanoya," remembered Miyazaki. "There was a tunnel shelter behind the unit and I found two wooden boxes that were respectively large enough to contain a human body. I opened them and found Sugita and Miyazawa. I asked a member of the unit to place them in another appropriate location, explaining that they were important pilots."

A lieutenant commander of the base unit showed up to assess the request. Miyazaki made a call to his headquarters, and LCdr Yoshio Shiga arrived. Shortly afterwards, an officer of the ground unit joined the group. At dusk, the bodies were removed and taken to the medical section of the base unit. Capt Genda and others went to the base hospital to pay their respects. Genda was infuriated when he saw one of the blackened corpses on the floor. "Is this the kind of treatment you give our heroes?" he scowled at the hospital staff. They quickly lifted the body from the floor and placed it on a table.

"*CPO Sugita was Kanno's favorite pilot*," wrote Capt Genda. "*I could imagine his shock. I apologized to him in front of the corpse. I told Lt Kanno that I was to blame for his death because I had delayed the abort order. I had no excuse.*"

The CO made a promise to his young squadron leader to find a replacement pilot of equal skill. Lt Kanno was dazed and said nothing.

Navy GHQ issued a Naval All Units Proclamation on 1 August 1945. CPO Shoichi Sugita was elevated two grades to the rank of ensign for distinguished service. According to the citation, he had shot down 70 enemy aircraft and had 40-plus shared victories.

In the final ignominy of the war, Sugita's ashes were lost in transit to his family in Niigata Prefecture. It was 33 years after the war that his mother, Iyo, first visited her son's crash site at Kanoya Airfield.

RIGHT: Shoichi Sugita's mother, Iyo, and her two sons, pray at the crash site where the great ace fell.

BELOW: Family members pray for the soul of Toyomi Miyazawa at a gravestone erected by the residents of Satozuka-Cho at Kanoya on the site where his aircraft plunged to earth.

ABOVE: Former enemies – now friends. Saburo Sakai and Robert Weatherup at a reunion engineered by co-author Henry Sakaida in California, June 1983. It was 38 years earlier that Sakai saw Weatherup shoot down Shoichi Sugita over Kanoya Airfield.

Chapter 6

Second blooding over the Amami and Kikai Islands

16 April 1945

再び奄美大島
喜界島上空の血戦

April would prove to be a tragic month for S301. CPO Satoshi Yokoshima died in a training accident on the 10th. Nine pilots failed to return from the mission of 12 April and on the 15th, the veteran ace, Shoichi Sugita, and his wingman, Toyomi Miyazawa, went to their deaths.

Despite good leadership, experienced pilots (500 hours average) and a superb fighter aircraft, the 343 *Kokutai* could do little to make a difference in the face of the Allied typhoon of destruction which had been driven by the incredible industrial might of the United States and which continued to gain momentum and ferocity as it swept its way towards Japan. There was no stopping it. One élite unit was simply not enough to turn the tide of war. Winning just one battle would no longer be easy. American radar-directed CAPs, overwhelming numbers, and excellent enemy radio communications cost the Japanese dearly.

On 16 April, the first day of *Kikusui* No.3, Kanoya Airfield was cluttered with fighters and *Kamikaze* aircraft, causing confusion and delays. The 343rd's S407 arrived at the sprawling airfield on the 13th, followed by S701 the following day. The raid on the airfield on the 15th did little to prevent operations.

Kikusui No.3 was an all-out effort involving nearly 400 Navy and Army aircraft. Ninety-four *Kamikazes* would soon sortie. Groundcrews readied five Betty bombers for flight and an *Ohka* rocket-propelled manned bomb was attached to each aircraft's belly. The motley collection of Navy *Kamikazes* included Zero 52s and obsolete Zero 21s with 250 kg bombs, Jill and Kate torpedo aircraft, Val dive bombers, and sleek twin-engined *Ginga* bombers (Frances). Army suicide aircraft comprised Franks, Oscars, Tonys and fixed-landing gear Nates from the China War era. To ensure their success, fighters would seek out American CAPs and engage them in dogfights, thereby allowing the widely dispersed *Kamikazes* to slip through to their targets.

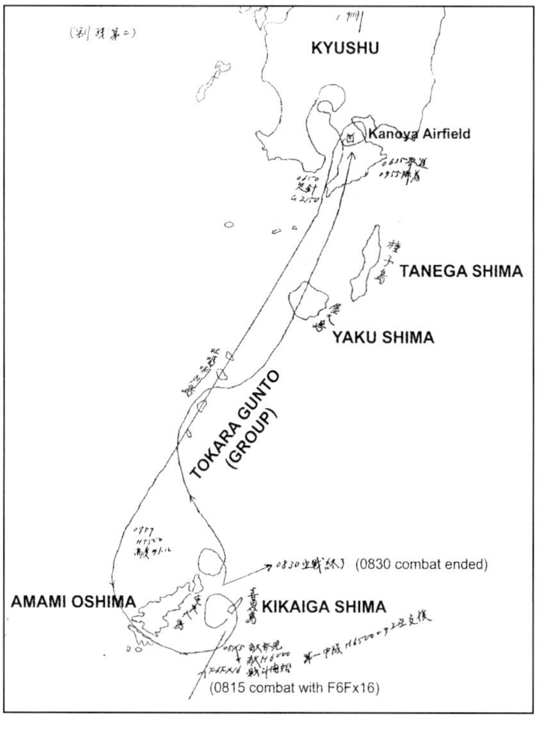

The 343 *Kokutai* was once again called to open up the corridor for the suicide aircraft on their way toward Okinawa. The plan was to fly to Kikaiga Shima and Amami Oshima where contact would be made with the enemy. Lt Takashi Oshibuchi took overall command, ably assisted by squadron leaders Lts Hayashi and Kanno. Of the 40 fighters slated to sortie, four were unable to take off due to mechanical problems.

The *Shiden-Kais* began their long-range mission from Kanoya Airfield at 06.35 hrs. They assembled over Kagoshima Bay and proceeded southwest. Along the way, three aircraft had to abort, leaving Oshibuchi with 33 fighters. The make-up of the formation now consisted of the 1st *Chutai* (Lt Oshibuchi's S701 with eight fighters), 2nd *Chutai* (Lt Yoshishige Hayashi's S407 with 11 fighters), and the 3rd *Chutai* (Lt Naoshi Kanno's S301 with 14). They cruised over the Osumi and Tokara chain of islands at around 19,500 ft, heading for Amami Oshima.

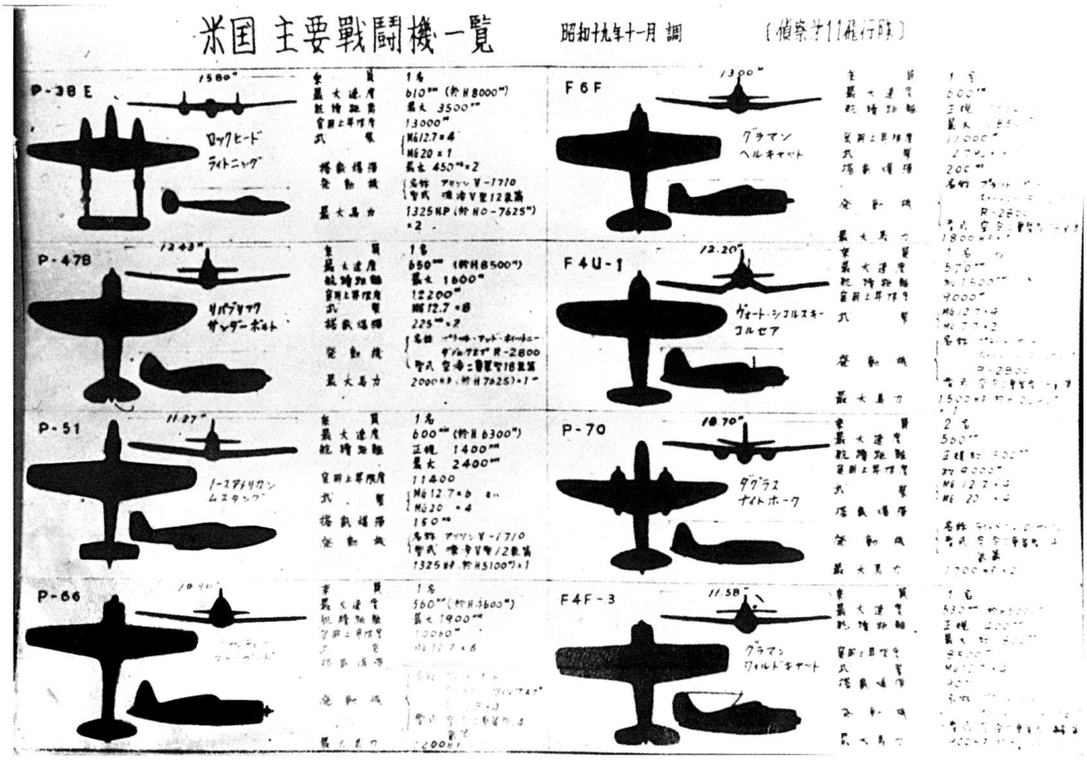

RIGHT: A Japanese chart showing American aircraft, dated November 1944. The Japanese had examined shot-down examples of the Hellcat, Corsair, and other fighter types, but never made improvements to their own fighters based on their findings.

By 07.57 hrs, Lt Oshibuchi could see his destination to the east. The formation would swing around the southern tip of the island and proceed north towards Kikaiga Shima.

However, incoming Japanese aircraft had been detected at 07.30 hrs by American radar and a large number of Hellcats were being vectored to meet them.

Lt James Pearce was leading two divisions of VF-17 Hellcats from the *Hornet*. His group was directed to fly to the southern end of Kikaiga Shima for interception. When he reached that point, he was vectored to 270 degrees to the west towards Amami Oshima. Over the northern tip of Amami Oshima, Pearce spotted the 343 *Kokutai* formation flying at 18,000 ft altitude. He gave orders to climb. It appeared that his group had not been spotted and the element of surprise would now be on his side.

RIGHT: WO Isamu Miyazaki was a survivor. His unit, the 252 Kokutai, was annihilated twice, once in the Marshall Islands and again at Iwo Jima. It was reorganized for a third time and Miyazaki fought with it in the Philippines until he was ordered back to the mainland with two other veterans. He had scored more than 13 victories by the end of the war. (I. Miyazaki)

The 343rd caught a glimpse of a dozen VF-47 Hellcats from the *Bataan* heading north. The American group had an altitude advantage of 1,700 ft. The *Shiden-Kais* then turned north over the southern end of Amami Oshima before turning to the west, crossing Kikaiga Shima, and dropping their belly tanks. With no fuel reserves, their combat endurance would be limited to no more than 15 minutes. Veteran ace WO Isamu Miyazaki of Squadron 301 reported: "*When we spotted 16 enemy aircraft, the 1st* Chutai *separated itself from the 2nd and 3rd* Chutais *to gain altitude higher than the enemy.*"

Lt Hayashi's 11 fighters planned to ambush the VF-47 Hellcats; Kanno took his men up to act as cover.

Lt Oshibuchi's 1st *Chutai* made a left climbing turn in preparation for an attack and then lost sight of the enemy. Since bad radio communication prevented the 1st *Chutai* from finding its comrades, it proceeded to patrol over the northeastern end of Amami Oshima.

If the 343 *Kokutai* pilots thought they had the element of surprise, they were badly mistaken. VF-47 had seen them drop belly tanks. Lt Richard Wilson sounded the 'tally-ho' and then radioed for reinforcements. Lt Hayashi was fixated on his target and was preparing to attack, unaware that his own formation was being targeted by VF-17. To make matters worse for Hayashi, his drop tank would not release and this annoyed him. PO1/c Hisaichi Kubo, flying as 'Tail End Charlie' saw a group of 'bogeys' straight ahead and above, and it appeared to him that his squadron leader had not noticed them. Hayashi's formation proceeded.

The 343rd's first element was caught by surprise when Lt(jg) Charles Watt's division bounced them from behind. "Fearless" Watts, with 5.75 victories to his credit, lived up to his reputation. He singled out one aircraft, which started a shallow left turn and then commenced firing from 800 ft. Streams of .50 caliber shells peppered the cockpit, port wing root and tail. The George began smoking, flipped over and spun all the way down. Another George had the misfortune to cross his path. Watts caught his opponent in a starboard turn and shot it to pieces. The Japanese aircraft smoked heavily and crashed into the sea about three miles west of Kikaiga Shima.

Lt(jg) Werner 'Gary' Gearisch, the squadron's "loverboy of Division Six," attacked a George from astern and directed his fire into the cockpit. The desperate Japanese dived for the deck and executed a Split-S into the water. Gearisch rejoined the dogfight, latched onto another opponent, and placed hits in the cowling. The George began smoking and headed down in a steep spiral.

Lt(jg) George C. Johnson, Jr., Lt Pearce's wingman, was often kidded as "the only pilot to get 200 percent out of his engine in a pinch." This was no joke. 'G.C.' sighted a lone George at 8,000 ft, dropped his belly tank, and went after him like a hawk. He dived from 10,000 ft and coaxed every bit of performance out of the 2,000 hp Pratt and Whitney R-2800 engine. He closed the distance to about 500 ft and then opened up with his six Brownings. The Japanese – flying about 25 ft above the water – took hits in the starboard wing, started burning, and disintegrated over the sea.

BELOW: VF-17 Hellcats on the Hornet prepare to launch. In its dogfights with the 343 Kokutai on 16 April 1945, VF-17 claimed seven victories and returned unscathed. (NMNA)

Lt Hayashi found himself in trouble. The young squadron leader could not drop his belly tank, which curtailed his ability to enter combat. He lost contact with his wingman and was chased and shot at by Hellcats.

Lts(jg) Jack 'Chili' Crawford and Murray 'Windy' Winfield went for two Georges about 200 ft below them. Each selected an opponent, then dived down from their perch to deliver a quick blow. Both Japanese aircraft caught fire and smacked into the sea. Looking around for more game, they spotted a loner and 'hosed' it with their machine guns, but the lucky Japanese somehow escaped. This pilot could have been Lt Hayashi.

Kanno's 3rd *Chutai* initially supported Hayashi's fighters by flying top cover, but when the fighting became intense, they left Miyazaki's division above and dived down into the battle. The 2nd *Chutai* rushed northward to escape the onslaught.

Now, VF-47 entered the fray led by Lt Richard Wilson. The *Shiden-Kais* were misidentified as Army *Tojos* because of their large radial engine.

"*Wilson dove on two Tojos straggling in the rear of the fleeing Japs at about 16,000 feet altitude.*" read the mission report. "*He burned first one and then the other with bursts from the rear. Both went down in flames. The two kills were confirmed by Lt(jg) Charles T. Krug, leading the second division of CAP... Lt(jg) Krug and his wingman, Ensign William Lamborn had gone in on the first attack with Wilson after Lt(jg) Theodore I. Veiock, Jr. had been forced to lag because his aircraft was running hot. Krug flamed a* Tojo *on his initial attack after pumping lead into him from astern. The aircraft went straight in, flaming and smoking, from 17,000 feet. Krug followed the Jap part way down to make sure he was going in. Immediately after the splash, he tally-hoed two more bogeys at about six miles and up.*"

The two Hellcat pilots climbed and closed on the enemy. The enemy aircraft were identified as Tonys, named after the Italian Macchi M.C. 202.[1] They may have belonged to the Army's 59th *Sentai*, based at Chiran on Kyushu. Krug and Lamborn each downed a fighter.

Lts(jg) Howard Elzinga and his wingman, Eugene McSweeney, from Wilson's second section, attacked a formation of eight Georges. Working as a well-drilled team, the two men bracketed one aircraft, then another. Elzinga made a high side run and shot his opponent in the engine and wing roots; the George spun down toward the water. McSweeney accounted for the second fighter.

With all the 'flying lead', accidents were bound to happen. Elzinga had just shot down his second George when his aircraft was raked by an unidentified Hellcat shooting at another aircraft. He was hit in the nose cowl and port wing and he left the dogfight smoking and losing oil. McSweeney and Lt(jg) Theodore Veiock escorted the damaged Hellcat back to the *Bataan.*

When the *Shiden-Kais* started losing their momentum, more Hellcats arrived. This forced WO Isamu Miyazaki to abandon high cover duty and jump into the fray. Lt Wilson had just joined up with VF-17 Hellcats when he saw the enemy diving on them from out of the sun. It was probably Kanno's cover *chutai.* He pulled up and fired about a thousand rounds into one aircraft. Pieces fell from his adversary, but it did not burn, and he claimed it as a probable.

"Lt Kanno and his men dived down to aid the 2nd *Chutai*," recounted Miyazaki. "My division took the cover position 300 meters above him and joined the combat. After a while Grummans started retreating with damage, but those were only the first group. The second and third groups came to their aid by means of radar and good radio communication."

WO Miyazaki grew frustrated at fighting such overwhelming enemy forces: "In this battle, CPO Wataru Tomisugi, 2nd man, and PO1/c Taro Asama, 4th man in my division, became separated and lowered their altitude. Immediately, I headed toward them for assistance, though I was in a dogfight. But I lost sight of them. After the combat ended, I waited for them at the assembly point but they did not come. Both Tomisugi and Asama were nice-looking boys who wore the special violet mufflers. Young as they were, both pilots had good skills and I was looking forward to their further development. I regretted very much that I lost them because my judgement was not good."

At 08.30 hrs (Japan time), Lt Oshibuchi assembled his men and headed back toward Kanoya. Just as they were leaving the area, Zero fighters and *Shidens* from the 201 and 601 *Kokutais* arrived to do battle with the Hellcats. Thirty Zero fighters and 11 *Shiden* fought a raging dogfight with the flyers from the *Bataan.* Very few of the Japanese fighters went down, but Fighting 47 proclaimed it as a "turkey shoot."

ABOVE: Combat postmortem by four VF-17 pilots following the action of 16 April 1945. From left to right: Lt James Pearce, Lt(jg) Jack Crawford, Lt(jg) George C. Johnson and Lt(jg) Murray Winfield. (J. Crawford)

The encounter between the Japanese and the Americans can be summed up by the comparative performance report of VF-47: "*Pilots of VF-47 who participated in this air battle came out of it with a tremendous confidence in the F6F-5. They feel that their aircraft offers them tremendous advantages in firepower, armor protection, and speed over the Japanese* Zeke, Tojo, *or* Tony. *Even the F6Fs that were hit carried their pilots back to base. In contrast, the Jap aircraft burned easily, often broke up when hit, and seemed to lack effective firepower when in a good offensive position. Moreover, the F6F-5 easily matches the speed of these Japanese fighters. It is evident, however, that these Jap aircraft can out-turn an F6F.*"

VF-47 sustained minor bullet damage on two Hellcats and one aircraft made a barrier crash landing. Another aircraft was hit by fire from a Japanese fighter and by friendly anti-aircraft fire from a battleship; it was forced to ditch, but the pilot was rescued. VF-47 claimed 22 enemy fighters destroyed and set a record to crow about.

VF-17's rematch with the 343 *Kokutai* can be best described as stellar; a total of 3,615 rounds of .50 caliber shells were fired in achieving seven victories, one probable, and five damaged. The Hellcats returned without a scratch.

The badly mauled *Shiden-Kais* returned to Kanoya at 09.55 hrs, minus nine pilots. Groundcrews were met by sullen faces and stony silence. Lt Kanno's squadron lost three pilots, but Hayashi was missing six. Three F6Fs went into the record as officially shot down, but it was another disastrous mission for the 343rd and personally regrettable for Lt Oshibuchi. The war of attrition was being lost at an alarming pace. In just four days, 21 pilots were killed in action.

Said WO Miyazaki: "It was a day of defeat. I keenly felt the truth of our principle of formation flying: 'Do not get separated from the division.' Formation fighting is stronger because it is an assembly of aircraft. The more aircraft assemble, the stronger they become."

Kikusui No.3 cost Japan 163 aircraft and sank the American destroyer *Pringle.* The carrier *Intrepid* and the battleship *Missouri* also sustained hits, as did four other ships. Staff officers of the Japanese 5th Fleet silently wondered about the effectiveness of this operation.

But *Kamikazes* would continue to carry out even more sorties.

Endnotes

1. 'Tony' was the Allied codename for the JAAF Kawasaki Ki-61 *Hien.* It had a tapered nose similar to the Italian Macchi, and Allied Intelligence had originally thought that the Japanese were using license-built Italian fighters, hence the Italian name 'Tony'.

Chapter 7

Battling the B-29 Superfortress

B-29 超空の要塞と戦う

The Americans invaded Okinawa on 1 April and the Japanese Army had no plans to reinforce the defenders. The task was thus given to the Imperial Navy and Army Air Forces to inflict as much damage as possible on the Allied invaders. The incessant *Kamikaze* suicide attacks, beginning with *Kikusui* No.1 on 6 April, proved a major threat to the Allied Fleet around Okinawa. The destroyers on picket duty around the carrier task forces became inviting targets for suicide aircraft.

Adm Nimitz was faced with an emergency and demanded that raids by B-29 Superfortresses neutralize the airfields on Kyushu. Gen Curtiss LeMay received orders for his 21st Bomber Command to comply with the new request. However, he was not enthusiastic about using his B-29s to hit tactical targets because they were the wrong machines for the job. The Superfortress was best at bombing strategic targets such as factories, transportation hubs, refineries, and docks. To bomb half-empty airfields at high altitude did not seem cost-effective. Nevertheless, six missions were slated for the first assault, starting on 17 April, and the six targeted airfields included Kokubu.

The 343 *Kokutai* needed more space and maintenance facilities than the crowded Kanoya could offer. Kanoya also became the focus of concentrated B-29 attacks. A decision was made to relocate to Kokubu No.1 Airfield, about 25 miles north of Kanoya, and the unit flew its last mission from Kanoya on 16 April. Capt Genda recalled: "Because the airfield at Kanoya was close to our information and warning network, we knew we would lose the opportunity to make regular, timely intercepts. Following CPO Sugita's incident, we moved to Kokubu which was a bit better than Kanoya at this point, but we would not have enough time to make sorties and conditions on the airfield were not good."

An aerial photograph of Kure Harbor taken by a photo-reconnaissance B-29 on 12 April 1945. Containing ships and cargo facilities, harbors were inviting targets.

In a badly timed move, the 343 *Kokutai* arrived at Kokubu the day the B-29s came over in force. On the 17th, 20 Superfortresses from the 313th Bomb Wing hit the base with fragmentation bombs, but damage was minimal and there were no casualties within the air group. Nor were there were any B-29 losses. This initial assault prompted the three squadron commanders to study and exchange their views on countermeasures.

The *Shiden-Kai* pilots found the B-29s most difficult to oppose. Their aircraft had not been designed for high-altitude interception. According to LCdr Shiga, the horsepower of the Homare 21 engine lowered drastically at more than 6,400 m.

The most effective maneuver against the B-29 was the head-on run. Lt Naoshi Kanno had experimented with this tactic against B-24s at Yap in the Caroline Islands. However, the 20 mm rounds fired in a conventional head-on attack tended to fly over the target as the pilot lowered his aircraft. From this observation, the Navy pilots decided to utilize the principle in a reverse way by shooting while upside down – and it worked. The bullet trajectory did not overfly the target, nor fall short. The pilot would aim for a position

ABOVE: The airfield on Kyushu under bombardment by B-29s. (USN)

ABOVE RIGHT: An industrial area near Iwakuni Airfield after a visit by B-29s. Small, family-run machine shops located in residential areas fed the large factories. This arrangement forced the Americans to bomb civilian areas near the factories, a fact the Japanese public did not understand at the time. (USN)

forward of the bomber. This same method was perfected independently in the CBI by the top JAAF ace of the Second World War, M/Sgt Satoshi Anabuki. The Army pilots called this maneuver the "Anabuki Run."

Fighting the Superfortress demanded a whole new set of skills and the 343rd pilots were forced to learn new tactics. The pilots of the 302 *Kokutai* at Atsugi were veterans at this game. They had trained well for attacks against the four-engined bomber and had fought B-29s on a daily basis over the Tokyo region since November 1944. One of their instructors was Lt Yutaka Morioka, who commanded a Zero squadron at Atsugi. He had learned his techniques from his Naval Academy classmate, Lt Naoshi Kanno.

Lt Kanno visited the 302 *Kokutai* in the autumn of 1944 after returning from the Philippines. When Kanno found out that his classmate was undergoing conversion from Val divebombers to Zero fighters, he offered to teach him his 'tricks of the trade.' "Just lend me a Zero!" Kanno demanded Morioka.

The two classmates climbed to 5,000 m and Kanno instructed. Lt Morioka turned and flew head-on towards his opponent. Kanno made his own diving attack and passed by the nose of Morioka's aircraft. Morioka thought that they would collide but he was also very impressed by the technique. After Kanno departed, Lt Morioka began teaching his subordinates Kanno's bomber-killing method, which later proved effective.

In order to lead by example, Lt Kanno was always the first to attack and his men had to follow. Leading the charge into a formidable B-29 formation was a horrifying experience and required nerves of steel. Many pilots swallowed hard against a dry throat and questioned their sanity when the time came.

The defensive firepower of the Superfortress was tremendous. Ten .50 caliber machine guns and a 20 mm cannon were installed in five remotely-controlled and well-positioned turrets. The computerized gun sighting system allowed gunners to track targets and the speed, distance, and lead angles of enemy fighters were computed automatically. Because of the Superfortress's deadly defense, many inexperienced Japanese pilots were intimidated and were prone to open fire at the American bomber too early and beyond effective range. Because of the extraordinary size of the Superfortress, the pilots miscalculated the distance.

The 313th Bomb Wing came back to bomb Kokubu on 18 April 1945. This time, the flight line of Squadron 407 was damaged and although the 343 *Kokutai* did sortie fighters, its claims are unknown. Three groundcrew were killed and ten were wounded. The Americans reported no anti-aircraft fire from the airfield during the two raids.

Stormy weather canceled bombing raids planned over Kyushu for 19 and 20 April. During these two days, the 343rd readied their aircraft for further B-29 interception missions. They knew that as soon as the weather cleared, the Americans would be back.

Lt Yoshishige Hayashi was in a foul mood. He had lost six good pilots from his squadron just days earlier in their duel with VF-17 Hellcats and this haunted him. He was also engaged in a professional dispute with Lt Kanno – with whom he usually got on very well – about the method of 'killing' B-29s. Kanno insisted that his upside-down diving attack from the front was the best tactic. However, Hayashi believed that the vertical dive was better. It was not like Hayashi to be so argumentative. "I won't come back if I can't shoot down a B-29!" Hayashi bellowed at Kanno. Kanno replied: "If you can't shoot it down when luck is not with you, then it can't be helped! You'll just have to try again!"

Hayashi continued to vent: "No! I simply won't come back if I can't destroy a bomber!"

Kanno grew weary and agitated. Finally, the irate Squadron 301 leader exclaimed: "Then please do as you like! I won't come back either if I can't destroy a bomber!"

On 21 April, LeMay's B-29s returned to finish their job and they came in force with 217 aircraft. They rained 1,071 tons of high explosives across nine airfields. Included in the package were time-delayed bombs, fused to explode up to 36 hours after delivery. Two groundcrew were killed and two wounded.

The combat diary of the 343 *Kokutai* shows that on this morning, 23 *Shiden-Kais* were airborne from Kokubu under the overall command of Lt Takashi Oshibuchi. His 1st and 2nd divisions were short by one aircraft each. The 1st division consisted of Lt Oshibuchi in aircraft 343-C-45, PO1/c Toshio Komabashiri (C41) and CPO Takamoto Kashima (C26); 2nd division consisted of WO Narunobu Ibusuki (C53), CPO Takeo Yamada (C50), and PO2/c Toru Kurita (B12). The fighters were in two groups of six and 16 aircraft. Lt Kanno led the smaller element belonging to Squadron 301 while Oshibuchi led Squadron 701 and Squadron 407. Although Lt Hayashi belonged to the larger group, his division of four aircraft fought independently.

Lt Matsumura had to cancel take-off. He was taxiing with his men on the runway, avoiding the bomb craters which were marked by red flags. Steering slightly to the right, one of his landing wheels caught a hole and the aircraft overturned. CPO Sato and two other wingmen took off safely. Immediately after the accident, B-29s arrived overhead. The exasperated lieutenant watched from the center of the airfield as bombs exploded near the command huts.

At 07.40 hrs, Lt Hayashi was last seen attacking a 19-aircraft formation in collaboration with PO1/c Toshinobu Shimizu from S301. The S407 leader was hit in the engine and crashed along the coast of Akune. Shimizu was also hit in the engine and plunged into the sea west of Warabi Island.

BELOW: Lt Yoshishige Hayashi, photographed whilst with the 343 Kokutai. (H. Hayashi)

LEFT: Lt Yoshishige Hayashi's aircraft was hit by defensive fire from B-29 gunners on 21 April 1945 and he tried to ditch along the beach at Akune in Kagoshima Prefecture. However, as a result of violent impact upon landing, he died of a fractured skull. Two men can be seen examining the wreckage.

RIGHT: The wooden grave marker reads: "Cremation site of Squadron Leader Hayashi." Traditional ceremonial offerings are placed at the base of the marker. Lt Hayashi was the first squadron leader of the 343 Kokutai to die in combat.

FAR RIGHT: Lt Yoshishige Hayashi was so revered by his men that they erected a permanent marker over the wooden one after the war. (H. Hayashi)

Analysis of both the B-29 and 343 *Kokutai's* mission reports reveal why the Japanese interception was ineffective. The *Shiden-Kais* patrolled at a high altitude, between 8,500 and 9,000 m, but the B-29s came in at less than 6,000 m. Furthermore, the three squadrons of the 343rd patrolled separately and seven aircraft from S701 and S301 made emergency landings on other airfields, most likely due to high-altitude patroling.

Lt Kanno did see action. His formation reached 9,000 m at the southern end of Shimabara Hanto and then headed for home. Along the way, they came across a group of B-29s at 4,000 m over Kirishima. They pursued the enemy over the sea off Miyazaki and the squadron leader claimed one victory, although no B-29s were listed as lost that day.

V/Adm Motome Ugaki, commanding the 5th Air Fleet at Kanoya, left words of great dissatisfaction in his diary: *"About 280 B-29s came over our Naval airbases in Kyushu from 06.30 to 08.30 and inflicted considerable damage. In particular, runways at Kasanohara, Usa, and Izumi were bombed and rendered inoperative. Especially their use of time bombs are troublesome. Zero fighters that went up from Kasanohara could not come back to base and landed here and there. Their dismal results are thought to have been caused by their lack of will to fight and inadequate method of interception, as well as the use of* Shidens *(*Kais*)."*

The 343 *Kokutai's* mission summary recorded that *"It is necessary to concentrate fighters on a specific B-29 formation. Accurate and speedy information is indispensable for this purpose."*

Ugaki's displeasure was misguided; the fighter pilots would have taken exception to having their will to fight questioned. Without accurate intelligence from the 5th Air Fleet, Genda was forced to formulate his own countermeasures. He thought that the enemy would come in high, but they had duped him by coming in low. In a countermeasure, Genda changed interception tactics.

On 22 April, the 343rd sortied 14 *Shiden-Kais* from S701 led by Lt Oshibuchi and S301 led by Kanno; two fighters aborted. The mission report stated: *"At 08.10, spotted an eight-aircraft formation heading southeastward east of Miyakonojo at 5,000 meters in a retreat after bombing the Izumi area. Immediately started attack and chased them up to and over the sea southeast of Miyazaki. Damaged one B-29."*

WO Narunobu Ibusuki, 2nd division leader from S701, claimed the damaged aircraft.

Eighty-seven B-29s bombed Izumi, Kushira, Miyazaki, Tomitaka, and Kanoya airfields. One B-29 was lost, but the crash site was not within the Japanese mainland. Because the day's targets were all southern Kyushu airfields, it is possible that the missing aircraft was hit by 343 *Kokutai* fighters.

The 343rd ceased operations from Kokubu Airfield on 25 April and all three squadrons advanced to Omura Airfield, one of the oldest naval bases in Japan. In the short time the unit was based at Kokubu, it was bombed on eight occasions. Even when the group was getting ready to transfer, it was continuously attacked. The Americans kept up the pressure, raiding the airfield daily from the 26th through to the 30th.

Omura, where the 343rd would need time and facilities to rebuild, enjoyed a nostalgic reputation. Many Zero pilots trained there for operations in China, the Indian Ocean, and the Pacific. Most of the 343rd's flight and ground personnel moved to the new destination by ground transport.

V/Adm Ugaki, who had castigated his fighter groups for their ineffective performance on the 21st, was incensed because the B-29 raids continued to delay his *Kamikaze* schedules. Too many airfields and aircraft had been damaged. A group of Jack fighters from the 302, 332, and 352 *Kokutais* were deployed at Kanoya exclusively to intercept B-29s. They had been operating in defense of the airfields in Southern Kyushu when the 343 *Kokutai* moved to Omura. During a two-week period, most of these aircraft were lost or damaged in combat or by bombing.

On 29 April, 121 B-29s were sortied and 111 bombed Miyazaki, Miyakonojo, Kokubu, Kanoya, and Kushira Airfields. CPO Itaru Shikano of S301

BELOW: Omura Airfield, photographed just prior to bombardment by a photo-reconnaissance B-29 'snooper' which always preceded a large bombing raid. These single aircraft were viewed as harbingers of death and terrified Japanese civilians.

recorded that his squadron participated in the interception and that: "*...all aircraft on standby duty took off from Omura.*"

Shikano – a veteran pilot who had once been attached to the carrier *Zuiho* which fought in the Solomons and Marshall Islands – was plagued by a bad engine, but he did not want to abort. He fell far behind the formation when he spotted B-29s over Osumi Hanto. Rather than joining in a formation action, the impatient pilot decided to operate alone. He selected a Superfortress flying at the tail end of its formation and placed it squarely in his gun sight. Then he dived straight down in one of Lt Kanno's classic attacks and opened fire with his 20 mm cannon. Tracer disappeared into one of his victim's left engines. He flashed past the bomber and pulled up to see black smoke spewing from the damaged engine.

Then the nightmare began.

B-29 gunners were tracking Shikano's fighter as he made his run and their aim found its mark; the Japanese took hits in his canopy and rear fuselage. The *Shiden-Kai* vibrated and the pilot lost control and blacked out. When Shikano finally came to, he was floating downwards in his parachute. He was rescued by Army personnel in the mountains of Uchinoura along the northeast coast of Osumi Hanto. His wounds were so serious that he was hospitalized and never returned to flying.

By 30 April, most of the 343 *Kokutai* had completed the transfer to Omura. Engineers and staff officers conferred to establish flight lines. Command huts for S301 and S701 were erected on the east of the airfield near Takamatsu Train Station. S407's tent was set up near the shore of Omura Bay. The mechanics went to work on worn aircraft engines. Replacement officers and new pilots would soon arrive.

During April whilst under 343 *Kokutai*, the 4th Reconnaissance Squadron had lost three Myrts and their crews, and it was subsequently reorganized into the 171 *Kokutai* at Kanoya on 5 May 1945.

The Superfortresses continued their bombing raids into May, flying at about 5,000 m to increase accuracy of bombing. Capt Genda believed that his force of about 30 fighters would be incapable of conducting the effective interception of several large enemy formations. He therefore decided to concentrate the *Shiden-Kais* from all three squadrons against one B-29 formation and destroy it. Genda attached more importance to the adverse psychological effect on the bomber crews by showing them that there was no way out, once a single formation was attacked.

On 4 May, the 314th Bomb Wing struck the airfields at Omura, Matsuyama, Oita, and Saeki. Ten Superfortresses bombed Omura, killing Servicing CPO Yasuo Noda. Ens Teruo Sasaki was badly wounded and died the next day. Fortunately for the 343 *Kokutai*, 36 fighters drawn from all three squadrons had departed the airfield at 06.00 hrs for a mission to Amami Oshima. It was a matter of luck that they were not caught on the ground. Seventeen B-29s came over Matsuyama Airfield just after the last *Shiden-Kai*, bound for Omura, had taken off. The 343 *Kokutai* suffered nine men killed and ten wounded. Barrack buildings were destroyed by more than 70 bombs and many men from a training air group were killed or wounded.

The 314th Bomb Wing again saw action on 5 May and so did the defending Japanese fighters. Fifty-five bombers delivered 276 tons of high explosives on Oita, Tachiarai, Kanoya, Chiran, and Ibusuki Airfields. All airworthy fighters from the 343 *Kokutai* took off from Omura to intercept. With Lt Oshibuchi in overall command, Lt Kanno led S301 while Lt Goro Ichimura took charge of S407 due to the absence of the squadron leader. Oshibuchi was the only one to carry air-to-air rockets, which had been recently delivered to the unit for use against B-29s. He was most eager to test this new weapon.

The Ichimura division spotted a ten-aircraft formation over Northern Kyushu and approached it between 5,000 and 6,000 m. Lt Ichimura dived at a B-29, followed by his wingman, LA Shokichi Kurumoto (post-war name Suzuki). Ichimura recalled: "The flight that had been flying at the tail of the *Shiden-Kai* formation took the first turn to make an attack from straight above the enemy formation. The first sight greeting my eyes was the pure white smoke of the rocket bomb as it exploded in front of the enemy aircraft, and at the same time, the complete disorder of the B-29 formation."

"*The lead aircraft looked to have finished his attack,*" wrote Suzuki. "*The target aircraft filled my gunsight. I aimed at the B-29 to the right side of the tail element. The distance shortened to 400 meters, then 300. I gripped the firing lever. The four 20 mm guns spat fire in unison. Tracers flew, trailing red tails. I aimed at the wing root of the B-29. I saw the shells explode. After finishing the attack, I formed up with the lead aircraft. The B-29 emitted white smoke. I did it! The second and fourth aircraft formed up with us. LA Kinzo Kasuya, our fourth man, was my good friend. We finished our second passes and assembled, but I did not see Kasuya again."*

B-29 42-65305 was one of ten aircraft from the 29th Bomb Group assigned to bomb Tachiarai Airfield. It was most likely jumped by *Shiden-Kais* of the Ichimura Division. The pilot, 1/Lt Marvin S. Watkins, remembered: "On the way back from bombing Tachiarai, I saw a *Kamikaze* appearing from ahead suddenly when we came back about 12-13 miles from the target. I thought it was our last as the method of his attack was a suicide type. Our four gunners returned fire, but he hit us at a dead angle."

PO2/c Kinzo Kasuya of Squadron 407 smashed his fighter into Watkins' B-29. It may have been an accidental collision because 343 *Kokutai* pilots were never ordered nor instructed to ram. Shards of Plexiglas flew as the nose windows shattered and the No.4 engine flamed.

"I ordered my crew to bail out," continued Watkins. "I felt the crisis of our aircraft. I continued to fly with Sgt Teddy J. Ponzca, the flight engineer, but we had to bail out as one wing flew off. I jumped out of the window."

From the stricken B-29, parachutes blossomed. However, one crewman fell to his death, two were shot and killed, one committed suicide and seven were captured. Watkins was sent to Tokyo and survived the war, but six of his men were vivisected at the Medical Department of Kyushu University. PO2/c Kinzo Kasuya crashed to his death near Taketa City in Oita Prefecture. In 1977, a monument was raised to commemorate the American and Japanese victims at the crash site.

The *Shiden-Kais* continued their pursuit of the remaining 29th Bomb Group aircraft. B-29 42-93953 came under very heavy fire and its pilot, 1/Lt Ralph E. Miller, fought his way out to sea with Lt Takashi Oshibuchi and his men in hot pursuit.

PO1/c Toru Kurita penned the following in his memoir: "*Now each division was in a fighting formation. Ens Ibusuki, my formation leader, pointed forward and started his attack. I, his No. 3 man, followed him, catching the enemy bombers to the left ahead of us. We were at 1,500 meters higher than the target. The distance was OK. I adjusted the axis of my aircraft to the target in an upside-down dive, going down vertically and pulling the stick gradually. The distance lessened...1,000 meters, 800, 500. The target bomber rapidly increased in size and was about to stick out of the red ring on the reflector gunsight. Concentrated gunfire came towards me in unison from the bomber formation. Red tracers flew to me, right and left. I ducked in spite of myself. I continued go in with my teeth set. I gripped the firing lever with all my might."*

Kurita opened up with his four 20 mm cannon. Pushing forward on the stick, he started to float upward, but his seat belt held him down. He flashed by the stricken bomber so fast, he just caught a glimpse of the huge tail.

The B-29 gunners were tracking the Japanese fighter, but Kurita skidded his aircraft to avoid the shells coming from behind. He pulled up gradually on the stick while flying at almost 740 km/h. He blacked out and was momentarily out of the fight.

"We crossed the north part of Kyushu and reached Oita, by which time, the B-29s had been decreased to a three-aircraft formation in an escape," recalled Ichimura. "One was lagging about 40 to 50 meters behind with its two of its engines at a standstill. At the time I was short of fuel and ammunition. I closed in beside a crippled B-29 and observed that the crewmen were tossing things out; they would not fire their guns at me. Also at this time, I realized that my wireless phone was out of order. I looked around and saw Lt Oshibuchi and others persistently attacking. During their attacks, I saw a very strange thing happen; when the *Shiden-Kai* was going in a head-on dive at the lone B-29, the other two B-29s, seemingly undamaged, slowed down and made a tight formation, taking positions on either side of the crippled aircraft. Every time we dived on it, the two escorting B-29s opposed us, pulling up their noses just like small aircraft! I first noticed something of the B-29 crews' fighting spirit!"

When Kurita recovered, he pulled the stick and climbed: "I climbed again to take position at a higher altitude. I could see that things were being thrown out of a window of the B-29 which was spewing white smoke from the wing root. Very quickly, the smoke changed to flames. The bomber began to fall behind the formation and nose down. Two crewmen bailed out in succession..."

The two 6th Bomb Squadron B-29s which came to Miller's defense and which had greatly impressed the Japanese were the *City of Springfield* flown by 1/Lt Allie A. Thacker and the *City of Oklahoma City* piloted

RIGHT: 1/Lt Ralph Miller's B-29 going down in flames. (A. Amell)

by 1/Lt Frank Klassen. The two comrades had 'sandwiched' Miller's aircraft with Thacker to the right and Klassen to the left.

2/Lt Alexander 'Sandy' Amell, navigator aboard Thacker's aircraft, described the situation: "Our gunner thought it was the aircraft of 1/Lt Frank 'Red' Klassen and reported this to our AC, Allie Thacker. Klassen's crew and our crew were living in the same Quonset hut; Thacker and Klassen had gone through flight school together, were originally some of the 'Sgt.' pilots and had married girls who were very close friends. So as the fighters jumped on the burning aircraft, Thacker decided to drop back to try and give it some cover. Also, the fire on the wing was such that it was obvious the wing would soon break off, so we were going to attempt to escort them over the ocean, have them bail out and we would bring a sub to them.

"Now, the irony of the story. The crippled aircraft was not Klassen's but Ralph Miller's. When we dropped back, Klassen saw it and thought we were in trouble so he dropped back to provide cover for us. Fighters were still around us, so our gunners were firing at everything and as Klassen's aircraft came closer to us, one of our gunners started firing at it! Looking out of my window, I recognized it as a B-29 and yelled over the intercom to stop firing at it. We tried to raise the crippled aircraft on the radio but got no response. By this time we were out over the ocean and the fighters were backing off. In a while the wing burned off and the aircraft went into the sea. We did not see any chutes."

1/Lt Wade Garton, navigator aboard Klassen's aircraft, saw the aircraft make its horrifying plunge. Klassen and Capt Bobby C. Karther saw eight crewmen parachute. The big machine slammed into the sea at the mouth of the Bungo Channel, between Kyushu and Shikoku Island. None of the crew survived.

Lt Ichimura headed for home and made an emergency landing at Oita Airfield because of fuel shortage. The runway was pockmarked with bomb craters, but he landed safely. He was shocked to learn of the death of PO2/c Kasuya who was No.4 in his division. Just the day before, he had lost PO2/c Magashichi Koma, his No.3 man. Before month's end, he would lose another wingman to become the sole survivor of his original division. Besides Koma, the only other 343 *Kokutai* pilot casualty was Ens Teruo Sasaki from S701. He failed to take shelter when the B-29s came overhead, thinking that they would skirt the airfield. He was killed by shrapnel in the bombing.

BELOW: Lt Goro Ichimura seen whilst serving with 302 Kokutai in 1944. (G. Ichimura)

"Our aircraft was badly shot up and was grounded for a couple of missions for repairs," recalled Amell. "Guess which mission number this was for our crew? Our 13th!"

1/Lt Allie A. Thacker was awarded the Silver Star for gallantry but the 314th Bomb Wing lost three B-29s on the mission to Kyushu.

On 7 May, the 313th Bomb Wing arrived over Kyushu with 41 aircraft to attack the *Kamikaze* bases at Oita, Usa, Ibusuki, and Kanoya. Piloting B-29 44-69887 of the 505th Bomb Group was 1/Lt Richard A. Gray.

Lt Naoshi Kanno, commanding three *chutais* from the three squadrons of *Shiden-Kai* alerted his men by radio. He saw a ten-plane formation heading south at 5,000 m after bombing Oita Airfield. The fighters dropped their belly tanks and climbed rapidly.

PO2/c Tsuneharu Tamura in Kanno's division saw his leader hit a B-29 at the wing root in a diving attack. Tamura was amazed at Kanno's flawless technique. The fighters made repeated passes on the Superfortress, flaming the Nos. 1, 2, and 4 engines.

Down on the ground, people gazed up to the sky to watch the combat between the four-engined bomber and the swarm of fighters. Nao Takaki, the co-author's mother, was a passenger on an ancient charcoal-fired bus lumbering slowly down a dusty road when it came to a sudden stop. She was on her way to visit her children in a neighboring village to take food and clothes. Suddenly, people on the road pointed to the sky and shouted: "*Look! An air battle! Our aircraft are attacking a B-29!*"

The passengers scrambled out of the bus to view the struggle overhead. Mrs. Takaki recalled: "The large B-29 was swarmed by very small fighters looking like horseflies, diving at it and rising from below it."

Up above, WO Mitsuo Hori dived and concentrated his fire into the wing of the doomed bomber to deliver the *coup de grâce*. It suddenly exploded. Jiro Date, a high school student, climbed to the roof of his boarding house in Saeki to watch the action. "One of the B-29's wings broke off," he remembered. "It exploded and fell like a dead leaf. This had not been a ramming attack, but the *Shiden-Kais'* diving attack from straight ahead and above."

2/Lt William Harrell, Jr., a radar operator in a nearby B-29, reported the fate of his comrades: "*Left wing burned off at number two engine. Left elevator came off first, then the whole stabilizer fell off; aircraft lost altitude and left wing burned off as it spiraled down. Aircraft was seen falling in two pieces with four enemy fighters circling. The tail gunner parachuted and another crew member parachuted from the rear; fighters followed the tail gunner's parachute down...*"

CPO Hideo Nakao of Squadron 407 recalled chasing a parachuting crewman from the stricken B-29. One of his comrades was careless and accidentally cut the cords of the parachute, damaging his wing. The bomber crewman fell to his death. The errant pilot was scolded by LCdr Yoshio Shiga upon his return.

Lt Gray's four-engined bomber plummeted to the earth while the people below clapped and shouted at their pilots' victory. The Superfortress crashed at Utogi, Meiji Village near Saiki City. At least one crewman was captured and transported by truck through the main street of Saiki City to the local *Kempei Tai*. None of the other crewmen survived. This bomber was one of three lost on the mission.

The battle was not without cost, however. PO2/c Mutsue Naka's aircraft disintegrated in the air when he climbed again to attack Gray's B-29. Naka bailed out and lost consciousness, but his parachute blossomed automatically. He landed safely and was taken in by personnel from Oita Naval Base. An elderly villager who witnessed the combat was so impressed, he presented his family's treasured *Samurai* sword to the downed pilot. PO2/c Hisao Sugita accidentally collided with a B-29 and was hurled out of his cockpit, but like his comrade Naka, he too survived. Suffering apparently non-fatal damage, the B-29 was able to fly on.

The B-29s continued to assault Kyushu in droves up until 11 May, and they could not be stopped. They bombed Matsuyama Airfield on the 8th and 10th. It is said that *Shiden-Kais* were sortied on these dates, but no B-29s were lost on the last three bombing missions according to the records of the 21st Bomber Command.

The 21st Bomber Command lost a total of 20 B-29s on the bombing missions to Kyushu airfields from 17 April to 11 May. For their sacrifices, they disrupted Adm Ugaki's *Kamikaze* operations by wrecking airfields and aircraft, and causing casualties amongst groundcrew. However, the unpredictable weather over the Kyushu airfields and the use of the B-29 for tactical purposes were not deemed very helpful in the battle to stop the *Kamikaze* attacks.

During the period 17 April to 11 May, Capt Genda stated that a total of 120 *Shiden-Kais* were used against the B-29s. The battles against the four-engined menace were very brief. Genda's pilots claimed 12 B-29s shot down for the loss of three of their own. During this period, the 343 *Kokutai's* casualties were 22 groundcrew and one pilot killed and 24 wounded on the airfields at Kokubu, Matsuyama, and Omura by the bombing. In addition, 15 *Shiden-Kais* were destroyed on the ground.

The 'Iron Fists' of Squadron 407

The systematic destruction of airfields on Kyushu continued without let-up, but the morale in the 343 *Kokutai* never faltered. Indeed the pilots were so full of fighting spirit that LCdr Shiga often had a difficult time keeping them in order! His greatest disciplinary problems were the incorrigible pilots of Squadron 407, whom he described as "...wild beasts in flight suits."

The flyers of S407, known for their bad boy image, lived up to the reputation. Outrageous behavior seemed to be the norm. CPO Eigo Shinoda commandeered a steam locomotive from Omura Station to Takamatsu Station, close to where his squadron was located. He had apparently spent too much time at a *geisha* house and did not want to be late getting back to base. He was caught by the police. LCdr Shiga called him into his office to issue a warning, but the intoxicated Shinoda went berserk and attacked him. Fortunately, the other officers heard the sound of fighting and stepped in to save their executive officer.

There were frequent brawls involving the young NCO aviators from the 343rd. Alcohol, youthful bravado and the pride of belonging to an élite unit were a volatile mixture. LCdr Shiga grew tired of going down to the police station, making apologies, and shepherding his errant men back to base.

"Never lose if you scuffle! But never scuffle!" Shiga would admonish his men on regular occasions.

One night, a group of six flight petty officers from Squadron 407 went into town for a good time. While walking to the Navy Club under a blackout, they encountered three Navy patrolmen led by an older chief petty officer. The CPO rebuked the young pilots for failing to salute him. In the rank-conscious military, the pecking order was strictly observed and often tested, right down through the non-commissioned ranks. If there were two men of equal rank, then the elder usually held seniority. The pilots argued that they had not seen the patrol in the darkness. Tempers flared. The flyers settled the dispute by leaving all three patrolmen bruised and sprawled on the street.

The patrolmen notified the Omura Army *Kempei Tai* rather than calling their own for assistance. Apparently, their motive was revenge, for the *Kempei Tai* was an Army organization serving the Army, Navy, Judicial, and Interior Affairs Ministries. In addition to handling military police matters, it was also involved in intelligence work. Many members of the *Kempei Tai*, especially the non-commissioned officers, became arrogant toward civilians because of the enormous power they held. The *Kempei Tai* was generally regarded with disdain and trepidation and was widely despised for its liberal use of "street justice." There were instances where civilians were beaten by the *Kempei Tai* for offering food or showing kindness to starving American prisoners. Surviving Allied POWs could attest to its *Gestapo*-style interrogation methods. In 1944, some Saeki food merchants were arrested and tortured for bribing naval supply department officers of the Saeki *Kokutai*.

BELOW: CPO Noritsura Kodaka was a veteran of Rabaul and claimed 105 victories during the war. (Maru)

The local *Kempei Tai* dispatched a sergeant and five corporals to bring the six pilots in for questioning. Typically armed, the sergeant carried a sword and pistol, while his subordinates also had pistols and bayonets on their belts. The *Kempei Tai* very rarely resorted to the use of weapons – it was simply unheard of for anyone to challenge their authority.

The pilots received word from concerned patrons of the club that the Army men were waiting for them across the street. Anticipating trouble, they paired up and left the club to walk to Omura Station. Suddenly, a figure blocked their path and a flashlight was shone into CPO Noritsura Kodaka's face. The military police had the misfortune of picking on the wrong man. A hardcore veteran who had survived Rabaul and had faced death on a daily basis, the fearless ace was self-confident, short-tempered, and a champion of the underdog. Kodaka also harbored a personal grievance against the *Kempei Tai*. A few days before, he and a comrade were angered to see them on horseback, galloping down the street with no concern for the safety of civilians. He was critical of their behavior and was determined to punish them if they ever crossed him.

"Can't you salute me?" growled a voice in the dark. In response, Kodaka shone his flashlight right back into the man's face and saw that he was a military police sergeant. Kodaka had just been promoted to Chief Petty Officer on 1 May and thus held seniority. Although his given name was Noritsura, the *kanji* characters in his name could also be pronounced *Tokkan*, which meant "*Charge!*" Everybody called him Tokkan.

LEFT: "Wild beasts in flight suits" were the words used by LCdr Yoshio Shiga to describe the pilots of Squadron 407. Sitting, from left to right: CPOs Hiroshi Takahashi and Shigenori Hirayama. Standing: LA Yoshio Takagi, PO1/c Shigehisa Osawa, and CPO Noritsura "Tokkan" Kodaka, who holds a wooden banner which reads: 'Tenchu-Gumi Pilot's Waiting Hut." (S. Osawa)

"Sergeant," replied Kodaka smugly, "can't you salute a Navy chief petty officer?"

Like a flash of violet lightning, the young pilot floored the leader with a devastating punch. "The Navy's iron fist should teach you a lesson!" he hissed. When the sergeant staggered to his feet to unsheathe his sword, Kodaka and his comrade knocked him into a ditch. The pilots of the *Tenchu-Gumi* (Heavenly Punishment Unit) left the remaining five opponents battered and moaning on the darkened street.

Toward noon the following day, a staff officer from the Sasebo Naval Station arrived at the command center on Omura Airfield and requested Capt Genda to deliver the six ruffians to a court martial. The commander of the Omura *Kempei Tai* was furious that the Navy pilots had assaulted his men. It was unprecedented and it was also extremely embarrassing since the brawlers were unarmed.

An indignant Genda asked his visitor if he was aware who was protecting the sky over Kyushu. He did not wait for an answer: "Our 343 pilots are! If you want me to deliver them, I will do so after they have become bones and ashes!"

The staff officer, who obviously did not enjoy being the messenger for the hated Army organization, did not argue, saluted, and left quickly.

The six pilots were relieved to hear Genda's response, but prepared themselves for a severe reprimand. They were called in and stood at attention. The 'old man' eyed each one as he walked back and forth. The pilots were stood like statues, nervous, and staring straight ahead under their commander's gaze. Then it came.

"I am prepared to allow you to act violently if it raises morale. As air group commander, I will assume the responsibility!" barked Genda.

From that day, the pilots swore an oath to fight even more passionately for their leader. Genda probably realized that some or all of these men would be killed shortly in battle, so he gave them a wide berth and let them be themselves. The word spread quickly of Genda's response and everyone reveled in the *Kempei Tai's* degradation.

Chapter 8

Third combat over the Amami and Kikai Islands

4 May 1945

第三回
奄美大島喜界島制空

The second day of *Kikusui* No.5 started with the 343 *Kokutai* dispatching all three squadrons to Amami Oshima for another joust with American carrier fighters. Lt Oshibuchi, leading 36 *Shiden-Kais*, departed Omura Airfield around 06.00 hrs. The 343 *Kokutai* flew past small groups of slow-moving *Kamikazes* headed toward Okinawa and silhouetted against the sea at much lower altitude to avoid enemy radar.

Squadron 407, led by Lt Goro Ichimura, arrived in the 'killing zone'. He recalled: "The squadron changed course slightly to the left toward the east and came out over the sea east of Amami Oshima. Then the formation made a sweeping right turn and came over Kikaiga Shima when we spotted ten-odd Hellcats orbiting."

In one of the few instances of a successful surprise attack, the 343rd had spotted 12 fighters of VBF-12 from the carrier *Randolph* about 3,300 ft below and about four miles ahead. Armed with two 500 lb bombs each, they had begun to blast military targets on Kikaiga Shima, commencing at 07.45 hrs.

The VBF-12 action report stated: "*After the three divisions on the hop had made their bombing run, two divisions rendezvoused at Angels 12, one N.E. of Kikai, and one west of Kikai. The third division orbited low at Angels 3, S.W. of Kikai about three miles...*"

"Enemy fighters spotted!" Lt Oshibuchi shouted over the radio. At his command, belly tanks were jettisoned from the 1st *Chutai* fighters. Squadrons 701 and 407 prepared to attack while Squadron 301 assumed top cover.

Lt Ichimura: "The *Shiden-Kais* spread out to form a fighting formation of two-aircraft sections. Immediately, I changed fuel supply from the drop tank to the main tank, pulling the drop lever hard. At the same time, the engine throttle was opened all the way, the gunsight was switched on, and the four 20 mm guns were test fired. The combat flaps, of course, had already been changed to automatic. I was ready for combat. My brave wingmen to the right and left were nodding their heads imposingly with oxygen masks. We were all prepared, only to wait for orders to go in."

Lt William R. Jemison, a division leader who had gone back to Wan Airfield to take photographs with his package camera, proved to be an inviting target. As Jemison rejoined his division, Lt Oshibuchi gave the order to attack. At 08.00 hrs, six Georges came screaming down from behind, out of the sun. Jemison was hit in the starboard guns and cowling, but luckily, his wingman, Ens Delmar Johannsen, came to his rescue, shooting down one of the Japanese. The Hellcats began to weave for mutual protection.

RIGHT: Ens Delmar Johannsen shot down his first two enemy aircraft on 4 May 1945 and added three more ten days later to became an ace. Here he sits in Alfred Bolduc's Hellcat for a commemorative photograph taken on 14 May since it already had five Japanese flags painted on it. He ended the war with those five victories. (AFAA)

Battle ensued and Lt Alfred Bolduc brought his men into the fray. The *Shiden-Kais* broke off and formed into two Lufbery circles, from which they would peel off to make an attack and then return. Bolduc and his division made passes from above, with the sun to their backs. In this fashion, the division leader claimed four fighters shot down while the other members downed one apiece. Lt Bolduc achieved acedom in this encounter.

The American radio traffic attracted the attention of two pilots from the *Yorktown* who were in the vicinity. Lt Howard R. Hudson and Lt(jg) William McLaurin of VF-9 rushed to join the mêlée. McLaurin made several runs on two fighters and destroyed one with hits to the engine and cockpit. As he rejoined his wingman, a George fired at him, puncturing his fuselage, wings, and tail. A 20 mm shell exploded in his radio and clanged against the armored seat. Before the George could deliver the finishing touches, Lt Hudson intervened. McLaurin made a hasty retreat back to his carrier.

Lt Hudson went head-to-head with the aggressive Japanese, who had made a frontal run against the pair. The American hauled his Hellcat around and worked his way behind his opponent. A hail of .50 caliber shells hit the *Shiden-Kai's* cockpit and engine. It started to smoke and slammed into the water.

ABOVE LEFT: Lt Alfred Bolduc became an ace on 4 May 1945, ending the war with five victories. (AFAA)

ABOVE: Lt Howard R. Hudson served three tours with Fighting 9 and scored his first victory on 11 November 1943. He shot down a George on 4 May, added two more in June and finished the war with five victories. (AFAA)

"Time had come for the friendly fighters to form up again," recalled Ichimura. "Then, the 16 cover *Shiden-Kais* joined in the fight from above. They dived and drove off the enemy aircraft by using typical formation attack tactics, one-sidedly destroying 12 aircraft. The dreadful attacking power of the *Shiden-Kai* was exhibited to the fullest. I looked down and saw a group of special attack aircraft heading south. I almost felt as if they were roaring toward us to thank us. The commander's aircraft with the white stripes on the fuselage banked as if flapping his wings to illustrate our victory. We assembled one after another, worrying about our decreased fuel supply. After sending our farewell greetings to the strike group who would not be coming back, we set course for Omura."

One group of special attack aircraft seen by Ichimura was the nine Type 94 reconnaissance float biplanes out of Ibusuki. The Japanese Navy had consumed so many aircraft for repeated *Kamikaze* missions during the past month that they were now resorting to biplanes. The biplanes were intercepted about seven miles northeast of Iheya Shima by VF-85 F4Us. Five were shot down, but four fought past the fighter screen to sink the destroyer *Morrison*. On this day, two Hellcats from VF-9 destroyed two Jake floatplanes.

In their encounter with the 343 *Kokutai*, VBF-12 suffered no losses, but did sustain repairable damages to three Hellcats. None of its pilots was wounded. One VF-9 Hellcat was damaged. The 343rd returned to base minus six pilots (two each from each of the three squadrons).

In this combat, the 343 *Kokutai* pilots thought that they had won the day, but in fact they had lost, although they had altitude advantage at the beginning. Aside from the excellent radio communication by the Americans, one of the other critical factors was that of Lt Bolduc's division making "...systematic passes from above, out of the sun." The American anti-blackout suits that "...made maneuvers otherwise impossible such as continued high speed turns" also contributed to their successes.

RIGHT: Seven pilots of Squadron 407 in January 1945. Front row, from left to right: Lt(jg) Tadashi Kawabata (killed 4 May 1945) and CPO Miyuki Shimotsuru. Back row from left to right: LA Toshiharu Hara (KIFA 19 February 1945), LA Hitoshi Kotake, and PO2/c Zenji Asai (both KIA 16 April 1945), CPO Hisaichi Kubo, and LA Akira Hoshino (KIA as PO2/c on 17 May 1945). (Y. Matsumoto)

Chapter 9

Squadron 301 vs. PBMs

11 May 1945

戦闘三〇一対PBM飛行艇

May 1945 was the month of difficult intercept sorties against American PBM flying boats and PB4Y patrol aircraft by the 343 *Kokutai*. As the war moved closer to Japan, these aircraft began sinking tankers, freighters, fishing and ferryboats with increasing regularity. Such was the havoc they raised in the sea lanes of the Tsushima Strait and the Goto Islands, that in early May, the commander of Sasebo Naval Station asked Capt Genda to do something about them.

The first mission by the 343 *Kokutai* to intercept enemy flying boats occurred on 3 May. Four *Shiden-Kais* were scrambled after two PBMs were reported in the vicinity of Onna Jima, but the intruders escaped. A second mission, mounted two days later, was no better; a pair of Mariners got away. This cat-and-mouse game continued until 11 May.

Lt Dick Simms of VPB-21, piloting Martin PBM-5 Mariner BuNo.59031, together with John Hook in No.59046 was ship-hunting near a group of islands off the west coast of Kyushu, known as Goto Retto. The Martin PBM-5 Mariner was a deep-hulled, gull-winged seaplane of incredible versatility. Armed with eight .50-caliber machine guns, it was used extensively for air-sea rescue and anti-shipping patrols near the Japanese mainland. VPB-21 was operating off the seaplane tender *Chandeleur* (AV-10) at Kerama Retto in the Nansei Shoto Islands.

The pair of Mariners spotted seven Japanese destroyers along the Korean coast while flying at 800 ft. As they contemplated their next move, they noticed a bogey flying about seven miles distant, at 2,000 ft altitude. The intruder, later identified as a Rufe floatfighter, closed to within 1,500 yds and flew parallel for a while, but stayed out of gun range. The single-engined patrol aircraft probably belonged to the 951 *Kokutai* based at Sasebo. It was now 11.20 hrs.

Martin PBM-5 Mariner of VPB-21 on patrol near Okinawa. (NMNA)

The VPB-21 mission report stated: "*Suddenly he turned in on a rather half-hearted flat side pass, firing two bursts (as did Hook's PBM), then swerving to his port and speeding right back out again at 4 o'clock. With no apparent damage incurred by either side, the fighter withdrew not to be seen again, apparently having no liking for the Mariners' firepower.*"

At 12.20 hrs, the 343 *Kokutai* received information that enemy flying boats were heading north approximately 65 km west of Uku Shima. One division of *Shiden-Kais* was alerted and placed on standby. At 12.35 hrs, the 343 *Kokutai* HQ received information that two PBMs were heading toward Tsushima. Fifteen minutes later, four *Shiden-Kais* of Squadron 301 were scrambled to intercept. The division consisted of Lt Masaji Matsumura and his wingman CPO Yoshikazu Miyamoto, with 2nd section leader CPO Seiichiro Sato and his wingman PO1/c Mahito Yoshihara.

ABOVE LEFT: Lt Masaji Matsumura's first attempt to use air-to-air rockets was a disappointing failure.

ABOVE: CPO Seiichiro Sato made two rocket attacks on Lt John Hook's Mariner. He failed to score in the first attack and believed that his second attack was no better. Fifty-five years later, he learned that he had, in fact, inflicted severe damage on the American aircraft.

CPO Seiichiro Sato was a hardcore veteran whose presence in the division was reassuring. He was originally posted to the 265 *Kokutai* at Rabaul, and despite his inexperience, became wingman to the squadron leader, Lt Usaburo Suzuki, who was an academy classmate of S701's Lt Takashi Oshibuchi. The young novice became a confident and skilled fighter pilot under the tutorage of Suzuki, a gifted fighter-leader.

After the retreat from the Marianas, Sato was posted to S301 under 201 *Kokutai* following Lt Suzuki's appointment as leader of Squadron 301 at Davao in August 1944. During the Battle of Taiwan, Lt Suzuki ditched his aircraft due to engine trouble on 13 October and was reported missing. When Squadron 301 was placed under the command of Lt Kanno and 343 *Kokutai*, Sato was automatically posted to the new Squadron 301 and became the 2nd man to Lt Masaji Matsumura.

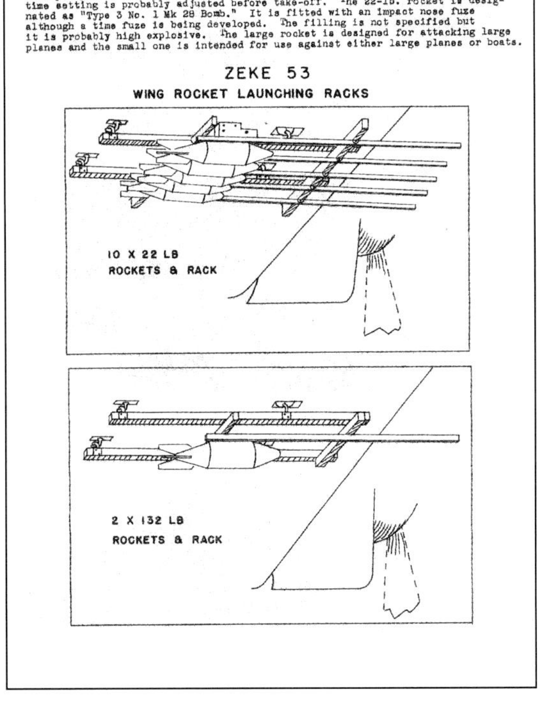

CONFIDENTIAL

A captured document states that ZEKE 53 C is equipped with jettisonable rocket launching racks as shown in the reproduced drawings. The rack appears to employ a channel track to guide the rocket. The propellant is probably ignited electrically. The 132-lb. rocket, listed as "Experimental Mk 27 Bomb," is filled with phosphorus and fitted with a clockwork type aerial-burst nose fuze. The time setting is probably adjusted before take-off. The 22-lb. rocket is designated as "Type 3 No. 1 Mk 28 Bomb." It is fitted with an impact nose fuze although a time fuze is being developed. The filling is not specified but it is probably high explosive. The large rocket is designed for attacking large planes and the small one is intended for use against either large planes or boats.

ZEKE 53

WING ROCKET LAUNCHING RACKS

10 X 22 LB ROCKETS & RACK

2 X 132 LB ROCKETS & RACK

LEFT: A US Intelligence bulletin on the Japanese air-to-air rocket which was simply a modified aerial burst bomb. It achieved little success. (Jim Lansdale)

Lt Matsumura was still relatively 'green' in terms of aerial combat. His first dogfight occurred on the intercept mission over Kikai and Amami Oshima on 12 April. This time, Sato was in a position to tutor the division leader, which was welcomed.

The pilots were also anxious to test a new weapon – the air-to-air rocket – which had been used for the first time by 343 *Kokutai* when Lt Oshibuchi attacked B-29s over Eastern Kyushu on 5 May. Developed by the *Kugisho* (Air Technical Arsenal) at Yokosuka, the rocket was a modified 60 kg No.3 aerial burst bomb fitted with a detonator fuse and timed to explode about three seconds after launch. The rocket casing held 135 small *kodan* (bomblets) filled with yellow phosphorus. A wooden launching rack was fitted under each wing and employed a channel track to guide the rocket which had a range of about 1,000 m. The solid-fuel propellant was ignited electrically. In comparison to the American HVARs, these devices were extremely crude and erratic.

Lt Etsutaro Tanaka, an armament specialist with the Yokosuka *Kokutai*, was placed in charge of testing the rocket against enemy aircraft, but though it was used several times in combat, the results were not very promising. The intended purpose was to attack B-29s by firing it ahead of a formation. A shipment of rockets arrived at Omura Airfield from Yokosuka and was immediately put to use.

"The rocket was a new weapon at the time, and all of us were interested in who would be the first to destroy an enemy with it," recalled Matsumura.

The Mariners of VPB-21 were flying between the northeastern part of Gotto Retto and the northwestern coast of Kyushu when contact was made. The *Shiden-Kais* were detected at 2,500 ft altitude, about 12 miles away and misidentified as Tojos and Jacks. Dick Simms was flying about 1,500 ft off the deck and about 1,000 yards from his wingman. Since it appeared that the situation was about to become 'ugly', the two flying boats closed in for mutual protection. Hook brought his aircraft tight to Simms' port side and both aircraft lowered their altitude to 200 ft. The Japanese fighter pilots orbited out of range to assess the situation. Then Lt Matsumura and CPO Sato suddenly made their move. The division split up into two sections, pushed over, and came in singly about one kilometer apart.

"I pulled the firing lever but the rockets did not launch!" lamented Matsumura. "I tried again. Instead, the lever broke and fell onto the floor of the cockpit! I gave up and was going to fire my guns, but I had no time to use my gunsight. I was already turning to the right, ahead of the target aircraft, without firing."

Matsumura would now have to fight with his rockets still attached to his aircraft. Meanwhile, CPO Yoshikazu Miyamoto was 90 m above Lt Hook and about 900 m out when he launched two rockets. They left a thin trail of smoke and exploded about 76 m in front and below the enemy aircraft. As a PBM crewman later wrote, the initial burst was "*... cherry red in color, then characteristic spidery, white phosphorus fingers reached out over wide area.*"

It was now CPO Sato's turn. He aimed his fighter toward Lt Hook's PBM while PO1/c Yoshihara waited for his leader to fire his rocket. The flight leader was determined to score. Sato pulled the firing lever and the rocket hissed away like a smoking arrow, exploding far behind Hook's aircraft. As for Sato's wingman, if Yoshihara carried rockets, he certainly did not fire them.

In the second round, Sato was determined to achieve success. He fired again and the rocket's trajectory looked good. The resultant explosion was spectacular. Sea1/c Richard C. Kudrna, the tail turret gunner aboard Hook's PBM, was lucky to be alive. Sea1/c Robert W. James, the ammo handler, recalled: "A stick bomb had hit the aircraft just ahead of the tail turret. Shrapnel perforated the tunnel and the intercom was

LEFT: A George circles around for another pass. These photographs were taken by Donald Fruchey from Lt Dick Simms' Mariner on board the seaplane tender Chandeleur (AV-10) (USN)

BELOW: A George makes a run on Lt Hook's PBM. The PBMs' basic defensive maneuvers were to lower their altitude and close in for mutual defense. (USN)

LEFT: An aerial rocket explodes, showing its characteristic spidery tentacles of white phosphorus. (Don Sweet)

BELOW: One of CPO Yoshikazu Miyamoto's rockets explodes close to Lt Hook's PBM. (USN)

out. When Kudrna came out of the turret after the fighters had left, his pants' legs looked like a couple of bulldogs had attacked him."

Sato, however, believed that he had missed, yet his rocket had inflicted more damage than he ever realized. The port and starboard wings of the PBM, as well as the horizontal stabilizer, were peppered with large and small shrapnel holes. The damage report stated: "*Empennage: four inch diameter hole in top of tunnel at leading edge of horizontal stabilizer and almost center of tunnel. The skin is pushed out and away from rivets, caused by concussion of bomb that apparently made the large hole and exploded inside the tunnel. Skin is peppered with tiny fragmentation holes. The aerial bomb exploded inside tunnel and caused severe damage to the fuel jettison tubes and the de-icer lines. The I.C.S.*[1] *lines were also severed. The internal structure of tunnel was damaged and weakened by the bomb explosion. It caused several holes approximately two inches in diameter and about two hundred tiny fragmentations through that section of the tunnel. The parachute at the tail turret has a small hole in it.*"

ABOVE: CPO Yoshikazu Miyamoto, wingman to Lt Masaji Matsumura on 11 May 1945, fired two rockets at American Mariners. They made a "cherry red" burst, but failed to hit anything. (USN)

As the Georges made their passes following the last rocket attack, Lt Hook's gunners opened fire with their .50 caliber machine guns. The tracers from the deck guns could be seen entering the undersides of one fighter, but the aircraft flew by seemingly unscathed.

Lt Matsumura came in for another pass but he turned too early and could not frame his target in the reflector gunsight for a good shot. The four Japanese fighters tried one tactic after another. They flew parallel to the Mariners, then dropped behind to spray their tails with 20 mm cannon fire, but nothing seemed to work. The PBM gunners 'hosed' the fighters. Sea1/c Lavernus Cottrell, the starboard waist gunner aboard Simms' aircraft, was credited with a kill. The mission reported stated: "*He caught the* Tojo *broadside during its recovery from a run at 11 o'clock, pouring hits into and about the cockpit. He saw the* Tojo *pour smoke and Ferganis, J.L., in tail turret saw him blow up astern the PBM.*"

There was utter chaos in the cockpit of Simms' aircraft. A 20 mm shell had exploded in the cockpit, wounding Lt Simms superficially in the lower right leg. The co-pilot, Lt(jg) Bill Graves, was nicked on the bridge of his nose. The cockpit filled with dense smoke, choking and blinding both men for about 15 seconds. The tachometers and manifold pressure gauges on the starboard engine, air speed indicators, and interphone communication system were destroyed.

Lt Matsumura saw white smoke belching from one of his unit's victims. He reported his progress to base. Lt Naoshi Kanno had been listening to the communication between the four pilots and was extremely irritated at the lack of accomplishment. Finally, after 16 minutes of continuous attacks, Lt Matsumura ordered his men to disengage and return to base. The timing was fortuitous for the Americans.

The American mission report stated: *"By this time, Simms was in bad shape. The main fuel lines from hull tanks to wing tanks had been severed by 20MM, fuel (due to pumping at time) pouring out as though from a hydrant. The starboard engine controls were gone, the engine stuck at 2500 RPMs and 44 HG which was rapidly depleting the 400 remaining gallons of fuel which were available (couldn't use hull tanks)."*

Simms communicated with his wingman by blinker light, telling Hook that he was going to ditch. His wingman was in no position to render assistance and had to struggle home; Lt Hook made it safely back to base.

The badly damaged Mariner touched down in rough seas with both engines dead. The entire crew swam away from their sinking aircraft and climbed into rubber rafts. Later, other Mariners from their squadron arrived over the stranded crewmen and flew protection until the next morning. Donald S. Fruchey (chief photographer's mate) had been snapping photographs of the fighter attacks and managed to capture some great 'Kodak moments.' Despite going into the 'drink', he saved the camera. He photographed his comrades bobbing up and down in their life rafts. Old war stories get grander as the years pass, but these photographs would eventually support their amazing stories. A Mariner from VH-3 eventually landed close by and picked them up.

Lt Matsumura and his division returned to base at 15.05 hrs. Although the PBM gunners claimed one kill and some damaged, in fact none of the *Shiden-Kais* had been hit.

Naoshi Kanno was in a foul mood when Lt Matsumura made his report to Capt Genda and LCdr Shiga. Only one PBM was claimed damaged. Kanno, who graduated one year ahead of Matsumura at the Naval Academy, accused his subordinate – in front of their superiors – of failing to bring down even one aircraft. However, Lt Matsumura was not about to accept such verbal abuse and argued back. Fifty-five years later, both Matsumura and Shiga would learn that one of the Americans had not actually escaped. Looking at the aforementioned photographs, Seiichiro Yamamoto (formerly CPO Sato) gave a smile of satisfaction and remarked: "One of the PBMs was tottering and sure enough, it ditched!"

BELOW: Simms' crewmen in their life raft show no hint of their misfortune. (Don Sweet)

RIGHT: Another six Mariner crewmen bob along on the sea following an attack on their aircraft by the 343 Kokutai. Co-pilot Bill Graves can be seen on the far right with a white cloth strip over his face. He received a deep cut across the bridge of his nose from a 20 mm shell which exploded in the cockpit. (Don Sweet)

ABOVE: Rescued at last! Survivors from the 11 May action board a Mariner from VH-3 the following morning. American rescue efforts greatly impressed Capt Minoru Genda. (Don Sweet)

Endnote

1 InterCom System or Internal Communication System.

Chapter 10

Chasing Grummans
13 May 1945

グラマンを追う

Because B-29s were simply not capable of pinpoint accuracy and were more suited for bombing large industrial targets, on 13 May 1945, 12 VF-82 Hellcats from the carrier *Bennington* were launched to provide escort for a group of five SB2C Helldivers and six TBM-3 Avengers which had been briefed to strike Izumi Airfield on Kyushu. *Kamikazes* still posed a threat to American carriers, and the airfields remained a high-priority target.

The small strike force was led by LCdr Edward W. Hessel and when it arrived over the airfield, it proceeded to strafe and bomb installations. There were no enemy aircraft to be seen on the ground. Strafing was the fighter pilot's least-favored job. There was no satisfaction in exposing oneself to a "golden BB" while shooting up hangars and shops.

LCdr Hessel ordered two fighters to remain over Izumi Airfield to snap photographs of the destruction inflicted. He took the rest of his men to regroup, heading down the west coast of Kyushu. The pilots felt a little satisfaction when they strafed two trains; at least these targets were moving. One locomotive blew up while the other was set ablaze. Hessel's formation proceeded to Tanega Shima, a large island off the southern tip of Kyushu. The Americans had been warned about the accuracy of the local Japanese anti-aircraft defence which proved itself by blowing off the wingtip of the commander's Hellcat. Not wanting to tempt fate any further, Hessel decided to call it a day. He assembled his men and headed back to the carrier.

LEFT: Lt Phil Perabo, Jr,. (in leather jacket) gives his narrative of a dogfight to intelligence officer Wally Lyon aboard the Bennington. Perabo was shot down by Lt Goro Ichimura of the 343 Kokutai on 13 May 1945.

ABOVE:
Ens John B. Hoag

However, the mission was not quite over for Lt Phil Perabo, Jr. and Ens John B. Hoag, the two photo pilots. Having completed their task, they observed some small boats nearby, swooped down and sank three of them before flying off to the rendezvous area. Then their troubles began.

Lt Goro Ichimura and his wingman, CPO Tsuneo Ozeki, of Squadron 407 were at that very moment flying back to their base at Omura when something caught their attention. They were 2,000 meters over the coast of Akune when they saw a lone *Shiden-Kai* attacking a weaving two-aircraft section of Hellcats. The pilots of the Hellcats were the two 'Photo Joes' – Lt Perabo, Jr. and Ens Hoag. According to the VF-82 mission report, "*...they were engaged by enemy fighters in numbers. They called for help by the 'Hey Rube' call but on a frequency jammed by some plane's microphone being closed and after all planes had been ordered to shift to another channel.*"

In the skillful way the American pilots weaved their aircraft, Lt Ichimura recognized a very dangerous situation. He immediately ordered Ozeki to top cover position and joined up with the other *Shiden-Kai* to do battle. Ichimura had fought Hellcats before, and even though it was three against two, he knew that they were not going to be easy to shoot down.

VF-82's report continued: "*Three enemy fighters made a run on them, hitting Hoag's a/c in the starboard wing. Almost immediately 10-12 enemy fighters were on them. At this time they were at 4,000 feet, weaving and the Japs were not very aggressive. They would come in individually on a run and pull away out of range. By continued weaving, the two pilots were able to counter the attacks of far superior numbers for a short time and Hoag succeeded in shooting down two Japs and forcing another to crash in the water.*"

"I hit one F6F and saw the pilot bail out," recalled Ichimura. "However, my engine was not in good order and I worried about the fuel. So we didn't chase the other Grumman."

There was a flash-fire in Lt Perabo's cockpit and he immediately opened the canopy and jumped. Lt Ichimura's aim was true. Seven pieces of shrapnel had hit Perabo in the left ankle and his situation only worsened. He failed to clear his aircraft, striking his leg against the tail and breaking his right ankle.

Ens Hoag saw his comrade being hit as they were scissoring. An instant later, he disappeared from sight. Alone in enemy territory, he was followed by four fighters and was worried about his damaged condition. He decided to turn into them, willing to fight, but the pursuers backed off, and he continued his long journey home.

Hoag crash-landed on his carrier's deck. The fuselage split from the engine. The engine stayed on the deck, but the rest of the aircraft slid into the sea. The pilot quickly freed himself from the cockpit and was picked up by a destroyer, suffering from mild shock and nausea. Strangely, Ens Hoag was given credit for three victories, although they could not be confirmed; he had lost his gun camera film and his section leader.

BELOW: Lt Goro Ichimura relaxes between sorties at the Squadron 407 command tent.
(G. Ichimura)

Lt Perabo landed in the sea and had to swim for hours to reach shore, where the Japanese were waiting for him. He was burned and could not walk. His captors carried him to a road where a vehicle took him into town. Surprisingly, he was afforded medical attention. His burned left hand was wrapped in oil-soaked paper and he was given water. "I was later placed in a motorcycle sidecar and driven through town where people took a swing at me!" he recounted.

At the end of the war, Lt Phil Perabo, Jr. was released from Ofuna Prison Camp. In his 80th year and retired from the real estate business, he was asked if he had any words for Lt Goro Ichimura, the pilot who shot him down. He replied: "No, not really!"

Chapter 11

Squadron 701 vs. PBMs
15 May 1945

The pilots and aircrew of Patrol Bombing Squadron 18, based on the *St. George* (AV-16), were working hard to contribute to the demise of the shrinking Japanese Empire. Missions for May 1945 consisted of 86 day searches out of 166 sorties. The unit's confidence was boosted on 2 May when Lts Martinn and Fitzgerald shot down a Kate torpedo aircraft and a fixed-landing gear Type 97 Nate during an enemy raid over the US fleet. Their gunners also damaged two Zero fighters during the mêlée.

At 06.30 hrs on 15 May, two PBM-5s under the command of Lt Marvin Hart and Lt(jg) Irving E. Marr, departed their base at Kerama Retto in the Nansei Shoto Islands. Their mission was an anti-shipping sweep along the coast of Kyushu and Korea. Each aircraft carried four 500 lb and four 100 lb general purpose bombs.

"Lt Hart and his crew had just joined our squadron as a replacement crew," recounted John Martinn. "...this ill-fated flight was his first in a combat zone. Regulations called for an experienced pilot to fly with a new crew on their first flight. Lt Dixon, who was in my crew, was assigned to fly with Lt Hart."

About four hours later, both aircraft found a 200-ton wooden sea truck ten miles off the east coast of Kamino Island and proceeded to bomb and strafe it. They left the sinking ship for fresher game and found a 3,500-ton freighter about ten miles north of their first victim. Lt Marr dropped all seven of his remaining bombs in train, with a 50 ft spacing at 180 knots. Two bombs struck the ship, with smoke and debris rising to a height of 300 ft. When Lt Hart approached to deliver the *coup de grâce*, the ship exploded and disappeared under the waves.

The pair of Mariners went looking for more action. Ten miles from their third victim, a 4,000 ton vessel appeared. Lt Hart took the lead and tried to get on the ship's stern, but the Japanese captain knew what was coming. He maneuvered his ship so tightly, Lt Hart missed with four bombs. The third bomb, a 500 pounder, was a near miss, which probably caused extensive underwater damage. The American gunners sprayed the deck liberally with their .50 caliber guns.

Hart circled the crippled freighter for the finishing blow at 1,500 ft when his aircraft took a bullet in the port engine. Just as he began his attack, his aircraft lost power. Hart quickly salvoed his remaining bombs, climbed for altitude, and ordered his gunners to dump all but 200 rounds from each gun. He called Lt (jg) Marr to let him know of his situation, leveled off at 2,500 ft, and headed for home along the west coast of Kyushu.

Lt(jg) Akio Matsuba (Y. Izawa)

At 11.57 hrs, a watch post at Karatsu reported two American flying boats heading south. Immediately, eight *Shiden-Kais* of Squadron 701 were scrambled for interception. Leading the flight was Lt(jg) Akio Matsuba, a veteran ace. Engine problems forced two fighters to abort. The *Shiden-Kais* groped their way through bad weather for over 30 minutes toward the Goto Islands. Suddenly, Matsuba spotted a pair of PBMs flying from behind the clouds.

According to the VPB-18 combat report: "*As they came abreast of the center of Kamino Shima, about ten miles off, the planes passed through a rain squall which lasted about ten minutes... Coming out of the squall, Lt(jg) Marr called Hart and reported a group of fighters below them. Both planes started for the deck and Marr closed up on Hart's port side.*"

"It was a sudden encounter," recalled CPO Takumi Sugitaki, who was flying 343-C-53. "Two flying boats were heading south at low altitude. At first, we had to chase the two Martins by making a steep turn and attacked from behind."

Lt(jg) Matsuba and PO1/c Nobumitsu Minoura (343-C-66) swept in from 1,000 ft to make a bow run on Lt(jg) Marr's PBM. They set his starboard engine on fire and also hit the cockpit. The bow gunner, AOM3/c

J.V. Carroll, put his sights on Minoura's aircraft and blazed away, hitting his opponent in the engine. Suddenly, Minoura's landing gear came down, a sure sign that he had also taken hits in the hydraulic system. Both aircraft passed and headed down. Minoura's fighter plowed into the sea and Lt(jg) Marr's aircraft went in close behind, sending up a column of smoke and water 300 ft high.

The *Shiden-Kais* now turned their attention to Lt Hart and crew. The VPB-18 report continued: "*The first five runs were all singles – one flat-side pass on each side, and three bow-on runs. The attackers broke off for five minutes or so, and then returned to make a series of five coordinated runs, high and flat-side, with one plane coming in from the port while two others made a pass on the starboard.*"

Lt Hart took his aircraft down to wave-top level. He turned into the attackers as they approached. The gunners had a limited ammunition supply and snapped out short bursts. The fighters kept making bow runs, which kept the bow gunner, AOM3/c Ray Armknecht, very busy.

"They kept coming at us but didn't show much experience in combat or much bravery," recalled Armknecht. "Any one of them could have put us down, like they did to Mr Marr's plane if they had been brave like the first one. After about 30 minutes, a plane that I call the 'mystery plane' was spotted coming at us from way out in front. He was flying about a 15-degree angle to the water, but he was quite a bit higher than us. When he got closer, I gave him a short burst. He had his guns pointed at us, but as far as I could tell, he never fired at us. When he got closer, I fired another short burst and I'm sure I hit him, but no smoke or other indication that he had been hit came from him. When he passed over the top of us at 150 to 200 feet above us, I gave him the third burst to his underside. I know that he was hit that time but he disappeared from my line of vision."

CPO Ryuji Yagi in 343-C-62 left the combat area alone after Armknecht had stitched his fighter. Losing his way in bad weather while trying to return home, he lost contact with his base over Izumi and crash-landed on the coast of Kumamoto.

"This one tried to escape by flying very low, skimming the surface of the sea," recalled Matsuba. "We attacked it from behind and above. I flew in parallel with the flying boat, giving orders and instructions to my men on the radio. Then, a 12.7 mm bullet hit me from the side. Quickly, I stopped the bleeding by using my oxygen mask tube, but I had to return to base because of profuse bleeding from my foot."

This combat was the last for Ens Akio Matsuba, thanks to Ray Armknecht's fine shooting. The 18-victory ace was hospitalized and was not able to return to operational service before the war's end.

The mission report of VPB-18 stated: "*The enemy airmen pressed their attack home aggressively, but displayed little imagination. Instead of staying well above the crippled patrol plane and making strafing runs on it (as it would be natural considering its extremely low altitude), the fighters came right down to the deck with their target and attempted to make standard fighter runs.*"

"It was a fight just over the crest of the waves," described PO1/c Toru Kurita who was piloting 343-C-65. "If you were careless in manipulating the aircraft, you would dive into the water and die. It was very dangerous to make a run at very low level as if skimming the surface of the sea, but it was effective because we would meet with the least enemy gunfire."

With Lt(jg) Matsuba out of action, CPO Yagi would have assumed command, but he had already left the scene of battle, having been hit. CPO Sugitaki was now in charge. Each time the flying boats changed course, he wrote down the time and direction on a chart while holding the control stick between his knees. As a former carrier pilot aboard the *Shokaku* in 1943, he had been trained in navigation. However, he was fearful that he would lose his bearing during combat in bad weather. Frontal assaults from below were most effective and Sugitaki saw one of the propellers of the target aircraft stop.

"*Crew members saw three enemy planes go down with thick black smoke completely enveloping their cowlings and cockpits,*" continued the American report. "*On account of the extremely low altitude (about 25 feet), none of these had time to recover before they hit the drink. Two other planes were also hit, and withdrew from the action, trailing smoke.*"

The Mariner gunners hit their targets but their observations were poor. The Japanese pilots gradually withdrew after a 45-minute fight. The encounter left them with one dead, another wounded, and four aircraft damaged. In addition to hitting Matsuba's aircraft, CPOs Sugitaki, Ryuji Yagi, and Takeo Yamada (343-C-58) also took hits, all courtesy of AOM3/c Armknecht.

Although the Japanese were off his back, Lt Hart still had his hands full. As soon as the last enemy fighter turned away, the PBM's port engine froze from lack of oil, and then caught fire. The prop was feathered, but the elevator controls were shot out, and the aircraft continued to lose altitude.

The mission reported recounted: *"He was heading into the wind and the swell when he first hit, at 110 knots, and the ill-fated engine broke loose from its mount and fell into the sea. As the ICS had also been shot out, Hart had no way of informing the men in the waist of the emergency in time, with the result that the plane shipped a large quantity of water through the open waist hatches. This turned out to be fortunate, for the water kept the nose of the plane high, thus helping to counteract the loss of elevator control."*

"The PBM hit with great force and bounced high in the air, whereupon it immediately started to swing around to the left under the influence of the starboard engine. It was then that Hart discovered that one rudder had also been shot out and was dangling from its cables, so that his directional control was very slight. The plane bounced a second and a third time, this last with such violence that all the instruments popped out of the panel, scattering glass and parts into the pilot's lap. Two more bounces followed, during one of which the port wing float carried away, before the plane came to rest in the trough of the swell, 90 degrees off the wind."

The ditching, whether by skill or luck, was considered a miracle. The waves were between 30-40 ft high. The PBM had hit the sea at approximately 12.30 hrs and Hart gave the crew the order to abandon ship. Ten men scrambled into two of the four-man life rafts. Three other crew members disappeared in the high waves and were lost. After a short while, the surviving crew started vomiting the gas and oil they had swallowed. Monstrous waves battered the rafts and overturned them. If the men thought that they were facing the worst situation imaginable, they were wrong; it was about to get even worse.

A forbidding sight greeted the survivors as they bobbed in the rough sea, clustered in their life rafts. Three Japanese fighters appeared over them flown by CPOs Sugitaki and Yamada, and PO1/c Toru Kurita.

"I made a large turn at low altitude and confirmed that another Martin ditched," recalled Sugitaki. "The crewmen were holding on to the life rafts in the water. They waved to us. I felt pity; they were helpless. They had lost their fighting ability and we did not need to shoot them. I set our course to base."

Sugitaki had been shot down into the sea in the Marshall Islands in December 1943. He was written off as killed in action. However, he had been rescued by natives and returned home after his funeral ceremony had taken place! This experience no doubt influenced Sugitaki to have mercy on the downed Americans.

Having seen the flying boat drifting helplessly in the sea, the three Japanese pilots turned for home. PO1/c Kurita was now anxious about his remaining fuel supply. He glued himself to the leader's aircraft. The worsening weather situation alarmed him; the heavy sheets of rain obscured his visibility.

CPO Yamada moved forward and hand-signaled to CPO Sugitaki that he was going to ditch. Sugitaki saw oil leaking from his comrade's fighter, probably caused by gunfire. Yamada, formerly a member of *Shiden* Squadron 401, was considered a skilled pilot who had claimed four Hellcats on 12 October 1944 during the Battle of Taiwan.

Yamada descended and safely ditched in the sea very near the white, sandy shores of Shijiki on the southern part of Hirado Island. It was the first coastline on the straight course from Uku Shima to Omura heading southeast, leading to the mainland at Kyushu. CPO Sugitaki's navigation was accurate and he banked his wing to his colleague floating in the sea and hurried for Omura together with PO1/c Kurita. CPO Takeo Yamada was able to return to base, but was killed by a P-47 on 28 May 1945.

The two *Shiden-Kais* reached Omura Airfield during heavy rainfall. Kurita could hardly see anything. He watched his instrument panel and the surface of the ground alternately. Manipulating his aircraft carefully, he landed on the sodden airfield.

After landing, CPO Sugitaki went to the squadron's command tent to make a verbal report to his division leader, Lt Ryoichi Yamada. However, neither Lt Yamada or Takashi Oshibuchi was there. A staff officer beckoned him and presented him to Capt Genda in the underground HQ.

The captain spread out a large map on the table and asked the chief petty officer to make a detailed report. This was a most unusual briefing because of the great difference in ranks between the two individuals. The chain of command was sidestepped and the lowly enlisted man reported directly to the commander. Genda expressed his appreciation for the pilot's efforts. CPO Sugitaki was greatly moved. "This was the greatest emotional experience I ever had in the 343rd!" he remembered proudly.

Meanwhile, the ten Americans had been in the chilly 58 degree water for nearly an hour and a half. After capsizing so many times and then scrambling to get back into their life rafts, they allowed them to fill with water to provide stability. Their throats were dry from swallowing so much salt water. Two Jake floatplanes flew close by and the men feared being strafed, but on neither occasion were they spotted.

By 17.00 hrs, the Americans were about 15 miles from land and the currents kept pushing them closer. Two PB4Y Privateers of VPB-21 on anti-shipping patrol, diverted to help in the search, passed about ten miles away. The stranded crew fired two flares, but they were not seen. At 17.30 hrs, a Japanese floatplane was seen circling about ten miles northeast of them. This aircraft was from the 951 *Kokutai* based at Sasebo, which had been dispatched to search for PO1/c Nobumitsu Minoura, who was never found.

The subsequent report stated: "*At about 1830, two PBMs (sent out by the squadron) were sighted eight miles away on heading of south. They fired one Verey shell and numerous thirty-eight caliber tracers. It was getting dark and the shells were seen. The two planes immediately turned in their direction and sighted the survivors. They dropped smoke lights and headed off in a southerly direction. The sea was much too rough for a landing.*"

The two PBMs flown by Lts Stan Kalemaris and Baily began laying down smoke lights around the two rafts at midnight. They also dropped several large flares. Around 03.00 hrs, the waterlogged rafters spotted a searchlight from the submarine *Ray* (SS-271).

Although the rescue submarine pulled alongside them, crossing the final hurdle became another dangerous task. Weakened considerably by their harrowing ordeal, several men became caught between the hull of the submarine and their rafts in the dark, rough sea; others became tangled in the lines. Eventually, strong hands brought all ten survivors aboard. At the time of the rescue, they were only 4,200 yards off the island of Shiro Shima.

The submarine continued its patrol, sinking several ships and getting depth-charged in the process. It was a terrifying ordeal for the PBM crews. Lt Dixon later told his comrade John Martinn who had not been on this mission: "If I ever get shot down again and a sub wants to pick me up, I will say '*No thank you!*'"

The Japanese did not ignore the heroic American rescue efforts. Capt Genda remarked in his memoirs: "*After one or two flying boats were shot down, search planes were observed every day and night looking for downed crews. Such rescue activities were clearly known to us by monitoring enemy radio communications and reports from our watch posts. It was a surprise to us and we were impressed with their making every possible effort to save the lives of pilots and crews. We were short of aircraft in an unfavorable war situation and could not afford to put such rescues into action. I think there was a great difference fundamentally in the way of thinking with respect to human lives.*"

BELOW: This VPB-18 Mariner was hit by Japanese fighters and forced to land. The crew is taking to life rafts. (Bill Ferrall)

A few days after this incident, fishermen found the bodies of Ens Ronald R. Hecht, AMMF3/c Edmund J. Day, Jr. and AOM2/c Richard F. Morey from Marvin Hart's crew. They had been washed ashore on Ukushima and were buried on the beach.

Exactly 56 years after the encounter, Ray Armknecht, who was credited with three victories and a 'damaged', had this to say about his opponents: "They hit our fuselage some, but never did hit us in any vital areas. If we hadn't already been crippled with a bad oil leak, and had had full canisters of ammo, I don't think they would have had any success against us. I don't wish to belittle them, but they got us only by forcing our pilots to use full power on our damaged engine."

Chapter 12

"A tough little fighter…": Squadron 407 vs. Privateers: 16 May 1945

小さいが手強い戦闘機」

On the clear and cloudless morning of 16 May 1945, two Privateers of VPB-118 went in search of the crew of a PBM shot down the day before. The aircraft were piloted by LCdr Arthur F. Farwell and his wingman, Lt Robert M. Finley. They scoured the area northwest of Goto Retto. The Consolidated-Vultee PB4Y-2 Privateer was the US Navy's powerful, low-level patrol bomber, a role in which it was relatively successful despite seeing only limited use toward the end of the Pacific War. Developed from the successful B-24, it was essentially a flying gun platform, sporting twelve .50-caliber machine guns in four power-operated turrets and two waist positions. The most distinguishing feature of the aircraft was its huge tail.

Lt Naoshi Kanno was eager to demonstrate to the members of his unit how he intended to handle enemy intruders. Previous interceptions had left him irritated, especially following his confrontation with Lt Matsumura in front of the CO on the 11th. Anticipating another visit from the Americans, the 343 *Kokutai* had its aircraft on standby. Sure enough, word reached base that two Privateers were 'snooping' in the area where combat took place the day before. At 08.20 hrs, Kanno took off with eight *Shiden-Kais* from Squadron 301 and four fighters from Squadron 407 when the 'Z flag' (scramble alert) was hoisted.

Lt Goro Ichimura's 407th division appears to have operated as the cover flight for the eight fighters of S301 and was composed of CPO Tsuneo Ozeki (No.2), LA Shokichi Kurumoto (No.3), and LA Yoshio Takagi (No.4). Flying at 2,000 m behind the Kanno group, Ichimura saw the Privateers flying north at very low altitude and radioed Kanno. Ichimura then ordered his division to give chase and dived. The four *Shiden-Kais* advanced in formation 1,500 m ahead of the Privateers and prepared for a head-on attack.

At 08.55hrs, the Privateer crews noticed the enemy fighters flying high and approaching from the east. The gunners were ready for their unit's first action against Japanese fighters, which were identified as Jacks. They closed formation, increased their speed to 180 knots, and lowered their altitude to 300 ft. Meanwhile, Ozeki and Kurumoto flew parallel to the PB4Y-2s, dropped their belly tanks, pulled ahead, and climbed.

Lt Ichimura centered one PB4Y-2 perfectly in his gunsight. Praying that enemy bullets would not hit him, he pulled the firing lever of his 20 mm cannon, but they failed to operate. The division leader flashed past his intended victim and left it to his comrades. When flying the Zero, it was customary to test-fire the 7.7 mm guns, but on this occasion Ichimura did not test the *Shiden-Kai's* cannon because it carried less ammunition than the Zero. He flew back toward Omura alone and was amazed to realize that his compass was also inoperative. Due to heavy fog, he could not be sure where he was. He climbed over the clouds and at last saw Kyushu. Following an inspection at base, it was found that his electrical system was malfunctional.

"The enemy aircraft struggled to escape," recounted Shokichi Suzuki (formerly Kurumoto). "We saw Iki Island far ahead. We saw enemy aircraft on the left side and still proceeded to make a frontal attack. There was no other way to attack them. When I commenced my attack pass, I saw CPO Ozeki going in. I waited for a little while. I switched on my gunsight and slipped off the safety catch to the guns. I made an approach so that only one aircraft filled the gunsight. Everything before my eyes was red because of enemy gunfire. I approached, sliding sideways. At 200 meters, I stopped sliding and pulled the firing lever. My four 20 mm guns fired in unison and I kept the lever down. Tracers flew trailing red trails toward the target aircraft as if they were absorbed into it. One pass on one aircraft, and another pass on another aircraft, then it was finished."

"*All fighter runs were identical,*" stated the VPB-118 mission report. "*Started at 12 o'clock, about 1,200 feet higher and two miles distance, the fighters would attack in a long, high-speed shallow dive to a position one hundred feet over the search planes, and then break away in a steep left chandelle.*"

The Japanese pilots came in so close, crewmen could clearly see the two 20 mm cannon firing from each wing. The Georges reportedly made a total of eight passes. In every pass, the gunners reported placing hits. LCdr Farwell's aircraft was damaged between the after-top and starboard waist turret. Lt Finley's aircraft was hit in the port wingtip.

During the last firing pass, LA Kurumoto's aircraft was hit as he looked back to confirm the results of his attack. He received two bullet holes in his rudder and one in the fuselage. A shell had severed a rudder control cable and the wires for his radio equipment. "I felt the nose of my aircraft rise," he reported. "The cockpit became white. I gave up confirming and decided to return to base."

CPO Tsuneo Ozeki was not so fortunate. The American gunners stitched his fighter from nose to tail. He flew straight for about half a kilometer, then stalled, and the fighter slammed into the water. The pilot parachuted, but he did not survive and his body was washed ashore on Iki Island. He was identified from the name on his life jacket.

In the heat of battle, bomber crewmen "saw" things that were not there. Incredibly, the gunners reported a second aircraft also spinning into the sea out of control. Evidently, the two crews saw the same aircraft go in and none of them saw the pilot bail out. The two Privateers turned for home.

Once again, Lt Kanno was denied combat. He and his men continued to patrol over the sea 56 km west of Uku Shima, but did not spot the enemy aircraft because their altitude was too high.

The mission report of VPB-118 concluded: "*The Jack is a tough little fighter and evidently has good fuel tank protection. The most vulnerable spot is believed to be the engine and cockpit area.*"

RIGHT: Six pilots of Squadron 407 photographed in early 1945. Front row, left to right: PO1/c Shigehisa Osawa, PO1/c Kiyoshi Miyamoto (KIA 3 June 1945), and, CPO Sakae Masumoto (KIA 8 August 1945). Standing, left to right: CPO Bin Nakano, LA Shokichi Kurumoto, and PO1/c Kazuhiko Harada (KIFA 8 March 1945). Kurumoto made passes on a Privateer of VPB-118 and was lucky to escape with his life. (Y. Matsumoto)

Chapter 13

"The Japanese pilot certainly did not lack courage..."

Privateers Return: 17 May 1945

「あの日本軍操縦士は確かに勇敢だった」

On 17 May 1945, two PB4Y-2 Privateers of VPB-109 took off from Yontan Airfield on Okinawa for a routine search and patrol mission off the coast of Korea. Lt George D. Fairbanks, piloting BuNo. 59523 *Hogan's Goat*, and Lt William A. Warren in BuNo.59514 *Blind Bomber,* cruised in loose formation around the southern coast of Korea looking for targets of opportunity. AMM3/c Val Higgins, the bow gunner aboard *Blind Bomber*, had just turned 20. He wondered if it was going to be just another routine day. He did not have to wait long to find out.

At 13.05 hrs, while crossing the Korean Strait towards the Japanese coast, the Privateers sighted a large 100-ton tug chugging along at six knots near Tsushima Island. When the aircraft came in to attack, light anti-aircraft fire from the tug and the shore opened up. Gunners from both aircraft fired at the ship, silencing its machine guns. The radio operator on the tug began sending out SOS signals, announcing that the vessel was under attack by two PB4Ys.

Lt Warren made a bombing run from 300 ft and dropped two 500 lb bombs. One bomb missed, but the other scored a bull's-eye, knocking the vessel 90 degrees to starboard. The Americans left the blazing tug to sink and continued with their patrol.

An urgent message was received at Omura Airfield reporting the American attack near Tsushima Island. Staff personnel of the 343 *Kokutai* quickly plotted an interception course and orders were given to arm each aircraft with *Sango Bakudan* (the No.3 aerial phosphorus bomb). The Japanese plan was to race southwest to cut off the intruders before they could reach the East China Sea. At 14.12 hrs (Japan local time), 11 *Shiden-Kais* led by Ens Mitsuo Ishizuka in 343-B-25 roared off from the grass airfield and climbed for altitude. Along the way, CPO Hideo Nakao in 343-B-07 aborted due to engine trouble and returned to base. The ten interceptors proceeded forward without further direction from HQ.

Thirty-five minutes after take-off, the *Shiden-Kais* intersected the flight path of the Privateers flying at 1,200 ft on the homeward leg of their mission about 30 miles east of Onna Jima. The huge tails on the Privateers caught the Japanese pilots' attention whilst Lt Fairbank's port waist turret gunner gave the alarm that enemy fighters had been spotted at 6,000 ft at ten o'clock.

"The wingman closed formation," recalled Fairbanks. "We descended to a very low altitude so that our 24 guns could concentrate fire at aircraft which could not pass underneath without crashing into the sea."

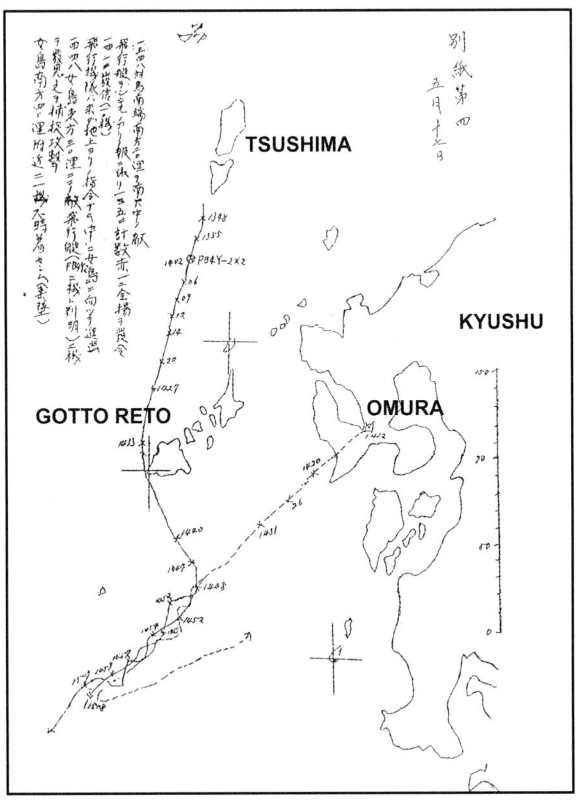

LEFT: The track of the two VPB-109 PB4Y-2 Privateers as they passed Tsushima Island on 17 May 1945, and the intercept course taken by 343 Kokutai from Omura.

RIGHT: 'Hogan's Goat' as seen painted on the side of Fairbanks' PB4Y-2. 'Hogan' was a feisty Irishman and a goat is a cantankerous beast – best left alone! This mascot reflected the crew's attitude toward anyone, including the enemy, who may have been inclined to meddle in their affairs! (G. Fairbanks)

LEFT: Lt George Fairbanks, pilot of 'Hogan's Goat'. (G. Fairbanks)

RIGHT: Val Higgins poses next to his 'girlfriend.' The wording under her name reads: "Can't see a thing, ain't lookin' for nothin.'" (Val Higgins)

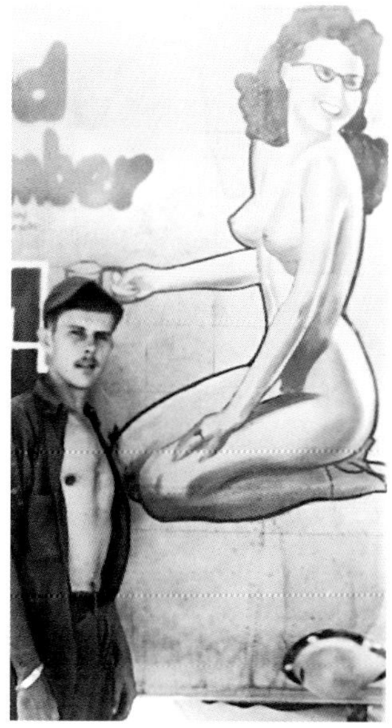

LEFT: The crew of 'Hogan's Goat.' Front row, left to right: AOM3/c Donald Hamilton, AOM3/c Kasper E. Weigant, ARM3/c Marion 'Pete' Kinser, and ARM2/c Robert L. Berry. Standing, left to right: AMM3/c Alfred V. Sandquist, ARM3/c Dean R. Johnson, Ens Robert J. Groce (co-pilot/navigator), Lt George D. Fairbanks (aircraft commander), Ens Abraham Shore (co-pilot/navigator), AMM2/c Robert E. Lally and, Richard Jenkins. Missing from the photograph is AMM3/c Franklin Shlyk. (G. Fairbanks)

RIGHT: The crew of 'Blind Bomber' posing by another aircraft. Front row, left to right: AOM2/c Floyd D. King, ARM2/c Earl W. Newton, AMM3/c Val M. Higgins, AMM3/c Lowell E. Tiller, and AMM2/c Harold J. Carter. Standing, left to right: ARM2/c Richard T. Coleman, AOM3/c William R. Smith, Ens Leo F. Haas (co-pilot), Lt William A. Warren (pilot), Ens Paul E. Geyer (navigator), AMM2/c Chester E. Rosell, and ARM3/c Richard P. Edson. (Val Higgins)

Val Higgins remembered the first moments of the contact: "Upon being jumped, both Privateers went down to very low altitude to protect our underbelly, probably around 500 or 600 ft."

Ens Ishizuka took two sections and jockeyed into position. He released his aerial burst bomb and the others followed suit. The VPB-109 mission report stated the following: "*Almost immediately after the third section of fighters was sighted, one plane at 10 o'clock high dropped what appeared to be a belly tank, which hit the water 300 feet ahead and to port of the planes, bursting into flames. However, five seconds later, three phosphorus bombs were seen to explode in the 9 o'clock position, 150 feet below the search planes' altitude of 700 feet. Ten phosphorus bombs in all were seen to drop, three being duds, but the other seven fairly accurate, streamers striking the wing surfaces of the search planes with each burst.*"

ABOVE: Three pilots in this photo participated in the attack against 'Blind Bomber' and 'Hogan's Goat.' Two are seated: Ens Mitsuo Ishizuka (left) and CPO Toshio Tanaka (right). Standing, left to right: PO1/c Kazuo Yotsueda (KIA 16 April 1945), LA Kinzo Kasuya (KIA 5 May 1945), CPO Tomio Yamamoto (participant on the 17 May sortie), LA Ikuji Matsumura, and CPO Tadayoshi Aizawa (KIFA 16 March 1945). (Y. Matsumoto)

AOM3/c William R. Smith aboard *Blind Bomber* recalled: "Three of the planes bracketed our plane with one each outboard of our port and starboard sides and one above... all just out of firing range. When it was reported that the planes over us had dropped something, evasive action was taken when the object exploded in a shower of phosphorus. After that, the Japanese started making their runs at us."

The aerial bomb attack had failed. The fighters started into individual runs, pulling up in front at eleven and two o'clock at 90 m range, streaking between the pair of Privateers and breaking off to the rear.

Higgins continues: "All of the fighter attacks, to the best of my knowledge, were either head-on or bow runs or from above. The two Privateers had closed ranks in order to maximize our firepower and were almost wing-tip to wing-tip. However, one of the fighters flew almost between us and just a matter of probably 25 to 50 feet higher than our planes. I believe he paid for this daring, for it is my belief that he was one of the pilots who failed to return from the mission. The plane that flew practically between us was somewhat typical of the gunfight in that the Japanese pilot certainly did not lack courage. One or more of them may have lacked a little common sense but not guts!"

Minutes went by and the running 'gunfight' intensified. With PO2/c Akira Hoshino down in flames, the Japanese pilots did not intend to go home 'empty-handed'. Aggressive attacks were reported by the American gunners, but intense return fire from the Privateers forced the *Shiden-Kais* to break off to starboard at a range of 120-150 m.

The Japanese became desperate and resorted to head-on runs. Five bow runs were counted with pairs of aircraft participating on each run. Aboard *Hogan's Goat*, Lt Fairbanks saw the dreaded sight of two fighters coming straight at him, guns twinkling. In a blur, they flashed underneath his aircraft and were gone. His aircraft had taken hits along the leading edge of the wing between the fuselage and the No.2 engine and his bow turret was knocked out. A 20 mm explosive shell detonated inside the fuselage, wounding radioman ARM3/c Dean R. Johnson. He received over 100 pieces of shrapnel in his back. Ens Abraham Shore, the co-pilot and navigator, suffered lacerations on his right arm and back. ARM3/c Marion E. Kinser sustained minor wounds to his foot. According to the mission report: "*Radio communication was destroyed, and throttle control of the #2 engine was destroyed, necessitating feathering the engine.*"

The bow runs by the Japanese fighters also proved devastating on Lt Warren's aircraft. A 20 mm shell hit the vertical stabilizer and destroyed the main spar, causing considerable structural damage. *Blind Bomber* was in danger of losing its tail at any moment.

LEFT: PO2/c Akira Hoshino flew head-on into the two Privateers. Val Higgins in 'Blind Bomber' and the bow turret gunner in 'Hogan's Goat' set the George on fire. The tail and top turret gunners added their parting shots and Hoshino crashed into the sea. (Y. Matsumoto)

ABOVE: The PB4Y-2 could be easily identified by Japanese pilots by its enormous tail. It was developed from the B-24 Liberator. Known as the Privateer, it proved to be a deadly flying gun platform with a dozen .50 caliber machine guns, four of which were in power-operated turrets. It could carry a maximum bomb load of 12,800 lb. It saw limited action towards the end of the war. (G. Fairbanks)

As the two Privateers fought their way home, the Japanese attacks began to lose steam. Gunners aboard both Privateers reported hitting and smoking several enemy fighters. Val Higgins received credit for one and a half kills (shared between himself and top turret) while the other half credit went to bow and top turret gunners on Fairbanks' aircraft.

After about 35 minutes, Ens Ishizuka gave the signal to abandon the fight. Individually, and in pairs, the *Shiden-Kais* peeled off for home, minus two of their comrades. PO2/c Hoshino died when his aircraft exploded in mid-air, while CPO Tomeshiro Hiro went missing. Although Ens Ishizuka claimed one aircraft destroyed, the final report of the 343 *Kokutai* to Sasebo Naval Station stated: "*...one aircraft damaged, smoked both engines on one wing and made it ditch in the sea about 19 km south of Onna Jima.*"

"*Five minutes after the last enemy plane had cleared the area, two PBM search planes of Fleet Air Wing One arrived on the scene,*" recounted the VPB-109 mission report. The PBMs had intercepted the contact report and diverted from their adjoining sector to render aid if needed.

"Soon after the fighters left us," commented Higgins, "two US flying boats intercepted us and offered to escort us back to Okinawa if we so wished; that sounded awfully good to us in case we went down, especially the other Privateer."

The two PB4Y-2s returned to their base at Yontan Field on Okinawa at 17.15 hrs.

A post-mission debriefing of the two Privateer pilots indicated that the Japanese pilots showed lack of coordination and small variation in the majority of their runs. They believed that the average attack was thrown off by heading into the threat and turning slightly after the fighters were committed to their runs and were unable to change.

The 343 *Kokutai* made 16 intercept sorties against PBMs and Privateers from 3 May to 1 June 1945. During these missions, there occurred four engagements on 11, 15, 16 and 17 May, in which a total of two PBMs and one PB4Y were claimed destroyed, two PB4Y probables, and one PBM damaged. However, following their research, it is the authors' assessment that three PBMs were downed (one destroyed and two ditched), one PBM damaged, and four PB4Ys damaged, at the cost of four pilots from the 343 *Kokutai*, and six of its aircraft.

On 25 May 1945, the 343 *Kokutai* was attached to the 72nd Air Flotilla at Oita under the command of the 5th Air Fleet.

Formation composition for attack on PB4Y-2s on 17 May 1945

1st *Shotai* led by Ens Mitsuo Ishizuka, Leader

1st Division

B25 Ens Mitsuo Ishizuka
B36 PO1/c Shigeru Aoyagi
B34 CPO Tomio Yamamoto
B01 PO1/c Yoshinori Matsumoto

2nd Division

B10 CPO Toshio Tanaka
B07 CPO Hideo Nakao, aborted
B32 CPO Hisamitsu Watanabe
B06 PO2/c Akira Hoshino. KIA

2nd *Shotai* led by CPO Sanpei Shiono

3rd Division

C64 CPO Sanpei Shiono
C65 CPO Tomeshiro Hiro, MIA
C21 CPO Junichi Miyake
A19 PO2/c Masayuki Tashiro

Note: The 1st *Shotai* was from Squadron 407 and the 2nd *Shotai* (3rd Division) was from Squadron 701. *Shiden-Kai* 343-A-19, flown by PO2/c Masayuki Tashiro, is now on display at the National Museum of Naval Aviation at NAS Pensacola, Florida, USA.

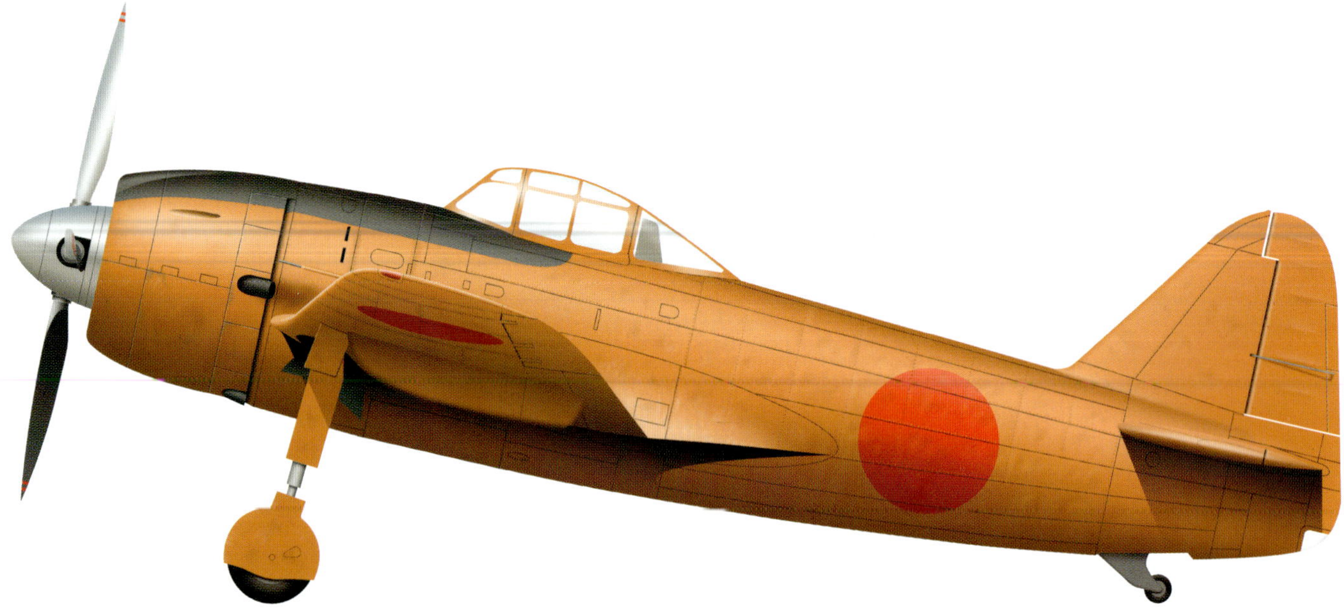

Prototype Shiden: Kasho Ichi-go Kyokusen (Provisional Type Designation – Interceptor No. 1), Itami Airfield, December 1942

The profile shows the experimental Shiden, known by its designation Kasho Ichi-go Kyokusen (Provisional Type Designation – Interceptor No. 1). The prototype was fitted with a 1,825 hp Homare engine and the subsequent mass-production type – the Kawanishi N1K1-J Shiden 11 (Allied codename 'George 11') – with a 1,990 hp Homare 21 engine. The Shiden prototype was first flown by Lt Takumi Hoashi at Itami Airfield on 31 December 1942 and afterwards by Lt Yoshio Shiga. Later, the experimental Shiden was often flown by Lt Shiga at Itami and Naruo Airfields.

(The prototype Shiden-Kai was first flown one year after the prototype Shiden, on 1 January 1944. The type designation was Kawanishi N1K2-J Shiden 21 or 'Shiden-Kai' (Allied codename 'George 21').

Kawanishi N1K2-J Shiden-Kai, 343-B-20, 343 Kokutai, Hikotai 407 'Tenchu-Gumi', Matsuyama Airfield, March 1945

Based on photographic evidence, this profile depicts 343-B-20 of Squadron 407 with a large number 20 painted in its Hinomaru as seen at Matsuyama in March 1945. Though a photograph shows CPO Hideo Nakao standing by this aircraft (see Chapter 22), it is not known if he flew it. The two names in the maintenance box on the tail as seen in the photograph were those of the maintenance crew. The fate of this aircraft is unknown.

Kawanishi N1K2-J Shiden-Kai, 343-B-17, LA Ikuji Matsumura, 343 Kokutai, Hikotai 407 'Tenchu-Gumi', March 1945

This aircraft was flown by LA Ikuji Matsumura in the battle over Matsuyama on 19 March 1945. LA Matsumura lost one eye in an accident on a long-distance test flight on 28 March 1945. CPO Hisamitsu Watanabe flew this aircraft on the B-29 intercept mission on 21 April 1945. The fate of the machine is unknown.

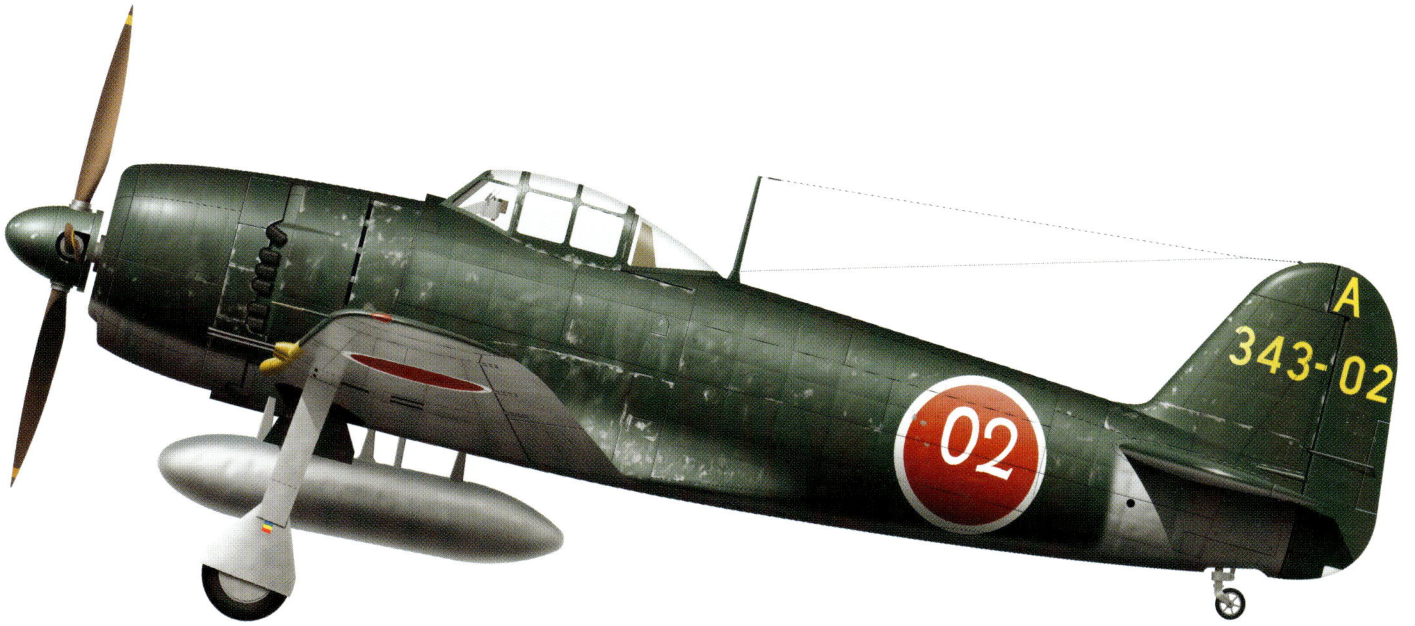

Kawanishi N1K2-J Shiden-Kai, 343-A-02, 343 Kokutai
Matsuyama Airfield, March 1945

This profile is based on photographic evidence taken at Matsuyama Airfield circa March 1945 prior to the 343 Kokutai's baptism of fire on the 19th of that month. It was flown by Lt Masaji Matsumura on the mission to the Amami and Kikai Islands, 12 April 1945. He claimed one victory, but on his return he was forced to make an emergency landing at Tanega Shima. An American fighter followed him down and damaged the aircraft in a strafing attack. Subsequently repaired, this aircraft was later flown by CPO Katsue Kato on the next mission to Kikai on 16 April 1945, but the pilot and aircraft went missing in action.

Nakajima C6N1 Saiun, 343-04, 343 Kokutai, 4th Reconnaissance Hikotai 'Kihei-Tai', Matsuyama Airfield, March 1945

On 19 March 1945, this aircraft was crewed by Ens Mitsuru Takada as aircraft commander, with CPO Minoru Endo as pilot, and CPO Hiroshi Kageura as radio operator. They spotted a large armada of American aircraft making its way toward Kure that day and alerted their base. Subsequently, Takada's aircraft, suffering engine trouble, was attacked by an enemy carrier aircraft and crashed. Since there is no American report of a victory over a Myrt, a possible explanation is that the victor was shot down on the raid and no report was filed.

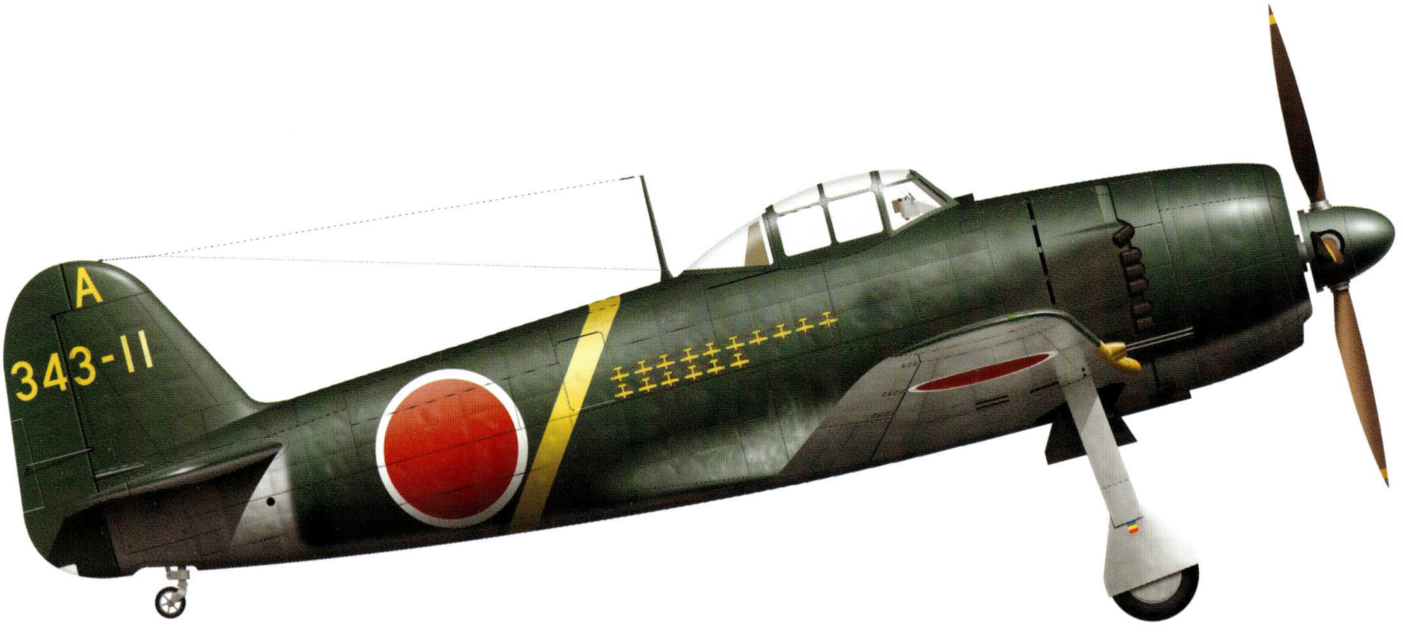

Kawanishi N1K2-J Shiden-Kai, 343-A-11, CPO Shoichi Sugita and others, 343 Kokutai, Hikotai 301 'Shinsen-Gumi', April 1945

Following CPO Shoichi Sugita's death, this aircraft was flown by Lt Masaji Matsumura on the mission to the Amami and Kikai Islands on 16 April 1945. Lt Kanno would also fly this aircraft when his regular machine was unavailable, for example on 21 May 1945. The large number of victory markings were applied as a morale-boosting exercise.

Kawanishi N1K2-J Shiden-Kai, 343-A-15, Lt Naoshi Kanno, 343 Kokutai, Hikotai 301 'Shinsen-Gumi', Omura Airfield, May 1945

Lt Naoshi Kanno, the commander of Squadron 301, chose this aircraft as his personal mount. He ordered his crew-chief to paint two diagonal fuselage bands with the idea that they would attract enemy fighters to him, so that he could then destroy them. The bands were crudely applied and were the same color as the tail number. Originally, there was a large number '15' painted inside the red Hinomaru, but in this profile it has been depicted as having been painted over by a mechanic. There is debate as to whether the number was applied in a temporary whitewash or in paint. The aircraft was photographed on Matsuyama Airfield on 10 April 1945 with the large '15' clearly visible in the Hinomaru. However, there were no victory markings painted on the aircraft at this time.

The profile depicts Kanno's aircraft in May 1945 whilst at Omura Airfield. The silhouettes of three B-29s and two fighters represent Kanno's personal victories. The squadron commander flew this aircraft on 12 and 16 April 1945 on the long-range mission to Kikai and Amami Oshima, and also on 16 May whilst hunting for enemy flying boats. The ultimate fate of aircraft 343-A-15 is unknown. Kanno was posted missing with 343-A-28 on the mission against B-24s on 1 August 1945.

Kawanishi N1K2-J Shiden-Kai, 343-B-03, WO Hiroshi Ohara, 343 Kokutai, Hikotai 407 'Tenchu-Gumi'

This aircraft was to be flown by WO Hiroshi Ohara of Squadron 407, on the 16 April 1945 mission to the Amami and Kikai Islands, but Ohara was forced to abort his take-off due to engine trouble. The profile of this aircraft is based on photographic evidence. The significance of the single white fuselage band is unclear, though there is speculation that aircraft marked in this way were flown by Kutai (division) leaders in Squadron 407. Ohara was injured in a crash following engine trouble on 12 May and never flew again. The fate of this aircraft is unknown.

Kawanishi N1K2-J Shiden-Kai, 343-C-45, Lt Takashi Oshibuchi, 343 Kokutai, Hikotai 701 'Ishin-Tai', Kyushu, May 1945

Lt Takashi Oshibuchi was the Squadron 701 Leader and overall commander in the air. For illustrative purposes, this aircraft has been depicted as carrying underwing 60 kg No. 27 aerial burst rockets. Lt Oshibuchi used this weapon for the first time against American bombers over Kyushu on 5 May 1945. The profile is based on verbal descriptions by surviving veterans and mission records denoting the tail number. The fate of this aircraft is unknown. Lt Oshibuchi was posted as missing in action with another aircraft, C-13, on 24 July 1945.

Kawanishi N1K2-J Shiden-Kai (US Navy designation TAI 001), (formerly 343 Kokutai), possibly flown by LCdr Yoshio Shiga, Omura Airfield, 16 October 1945

One of the Shiden-Kai 21s photographed on the flight line at Omura Airfield (see Chapter 23). The tailcode 'TAI' (Technical Air Intelligence) was painted prior to the ferry flight to Yokosuka on 16 October 1945. This aircraft is one of the three Shiden-Kais in existence in the United States, but its exact location is unknown.

In these three depictions by acclaimed model-builder, Shori Tanaka, Lt Yoshishige Hayashi's aircraft is clearly identifiable by the two white fuselage bands denoting a squadron leader. Shiden-Kais of the 343 Kokutai deployed in two-aircraft sections and four-aircraft divisions. These deployments were quickly identified by the unit's American opponents.

In this faithful depiction by model-maker Shori Tanaka, CPO Shoichi Sugita struggles to gain altitude over Kanoya Airfield on 15 April 1945, following an encounter with a Hellcat flown by LCdr Robert A. Weatherup of VF-46, but in vain. Weatherup pursues and scores hits on Sugita's aircraft at 400 ft altitude.

In another depiction by Shori Tanaka, Lt Naoshi Kanno, in 343-A-15 with double diagonal bands, leads a formation of 343 Kokutai aircraft toward Kikaiga Shima. Even with belly tanks, the aircraft could dogfight for no more than 15 minutes before having to break for home.

An officer of the 343 Kokutai in typical tropical service dress.
Commissioned officers wore a soft cap with two stripes; enlisted men wore one stripe. This officer carries a flight navigation tool bag which could be used to store charts, logbook and personal items. He carries a kai gunto (Navy military sword), which would have been purchased through the Navy Officers Club. The sword was carried as his symbol of authority. An officer paid for his own uniform and sword. The color of the collar tabs indicated the officer's specialty, in this case, red piping, indicating that he is a flight medical officer. (Model: Junichi Gondai Photo: Gary Nila)

A JNAF pilot in typical post-1943 flight gear.
This pilot wears the Model Type 97 parachute harness and a kapok-filled float vest. He also wears a late winter flight helmet with standard goggles and winter flight gauntlets to withstand the cold weather. All pilots wore a scarf, usually obtained from discarded parachute silk. The black boots were standard footwear for Naval pilots during this period. The national flag is sewn onto the float vest in accordance with a directive issued in February 1945 due to an incident in which a pilot, badly burned, took to his parachute. Unable to communicate because of his wounds, he was mistaken for an American pilot and beaten to death. Over enemy territory, pilots carried pistols. They were called 'suicide weapons' and were not meant for personal protection. However, in the 343 Kokutai, at least one pilot carried his pistol as protection against mistaken identity in case he was shot down. JNAF officers did not carry swords with them on missions. Their Army counterparts sometimes did as a symbol of good luck. However, in many formal photographic portraits for their families, Naval pilots often posed with swords as befitting a Samurai warrior. (Model: Eren Ganta Sakaida Photo: Gary Nila)

Chapter 14

Thunderbolts strike twice

28 May 1945

雷は二度落ちる

The P-47 Thunderbolt was the USAAF's 'flying locomotive,' powered by a monstrous 2,800 hp Pratt and Whitney engine, capable of more than 467 mph at 32,500 ft. Armament was provided through four .50-caliber machine guns in each wing. The *Shiden-Kai* was a medium-weight in comparison to the P-47, which was more than five tons heavier when fully loaded.

Against the P-47, the pilots of the 343 *Kokutai* were at a great disadvantage. They had never encountered the Thunderbolt before and when they did, several misidentified it as the P-51 Mustang. The key to victory was to force the Americans to fight on the 343rd's terms by goading the P-47s into a close-quarter dogfight and making them fight defensively. The long-awaited first encounter between the Japanese *Shiden-Kai* and the American Thunderbolt occurred on the morning of 28 May.

A dozen P-47Ns of the 19th Fighter Squadron took off at 08.16 hrs from their base at Ie Shima on a 'heckler' or nuisance mission led by Maj De Jack Williams. It was the unit's first foray over the Japanese mainland. Three days earlier, the unit had engaged in a 'turkey shoot,' downing 28 enemy aircraft in the vicinity of northern Okinawa. The mixed bag of Japanese Navy and Army aircraft had comprised *Kamikaze* aircraft and their escorts of inferior skill.

On Omura Airfield, several pilots were hung over from a night of drinking. May 27 was Japan's Navy Day, commemorating Admiral Togo's destruction of the Russian Baltic Fleet at Tsushima in the Sea of Japan in 1905. Pilots who were not scheduled for duty the following day drank heavily. Lt Kanno of S301 knew he should be on

LEFT: Squadron 407 pilots receive a briefing from division leader Lt Goro Ichimura (squatting with back toward camera) in May 1945. Standing on the far left is CPO Motoo Isogawa who was killed in action on 28 May 1945. (H. Harada)

RIGHT: Lt Takashi Oshibuchi thought that the Shiden-Kai could take on the P-47, but the Thunderbolt was simply too much for the Japanese fighter.

duty together with his men, but nevertheless joined in the celebrations. It was (and still is) a Japanese custom that a senior in the party receives every cup of *sake* (rice wine) offered to him by his junior, as a matter of courtesy. With so many men in the squadron offering him a toast, the young commander fell drunk. By morning, most of the squadron was 'wasted.' May 28 would have to proceed without Lt Kanno and his men. Thus, Lt Takashi Oshibuchi, the senior squadron leader, took off leading 20 fighters from Squadrons 701 and 407 on the day's first intercept mission.

The 19th Fighter Squadron arrived over Southern Kyushu around 09.30 hrs. The dozen Thunderbolts overflew Kanoya and Kushira Airfields at 13,000 ft. Nine Zero fighters were observed taking off from Kasanohara (East Kanoya Airfield). The Japanese took off toward the east, turned north, and headed up a valley, gaining altitude.

Maj Williams could not believe his luck. Five enemy aircraft consisting of a three-ship and two-ship formation were flying at an altitude of about 100 ft, totally unaware of the Americans' presence! Williams' radio had malfunctioned, so he could not communicate with his wingmen. Assuming that they would follow his lead, he jettisoned his wing tanks and nosed down from 11,000 ft, carefully aiming for the leader of the three-ship formation. He opened fire with all eight of his guns. His victim winged over to the right as Williams flashed past and was lost from view.

The 19th Fighter Squadron mission report stated: "*Maj Williams and Lts. Jackson and Sitton climbed to 15,000 feet where all the flights were ordered to rendezvous. At about this time Maj Williams' radio came back on and he ordered Capt Loflin to go across the airfields again. As Capt Loflin started towards the fields, Lt Mathis called in bogeys to the northeast.*"

The 'bogeys' were *Shiden-Kais* of the 343 *Kokutai.* They had just taken off from Omura Airfield and were flying in the vicinity of Mt. Kirishima at around 6,100 m. Looking down, Oshibuchi could see the P-47s attacking the Zeros at around 200 m altitude. It would be a favorable encounter, or so the squadron leader thought. His opponents were at very low altitude and he had the element of surprise. The enemy would have to fight defensively and he lowered his formation to make an approach from behind. However, he had guessed wrong. The Americans had spotted him and climbed steeply to intercept, using water injection.

When Oshibuchi saw the P-47s climbing, he ordered his men to climb higher to maintain altitude advantage, which later proved to be a tactical error. The 'Jugs' were unlike the Grumman Hellcats he had fought before. As altitude increased, the difference in climbing performance became evident – the Thunderbolts' capabilities were formidable. Looking back on this combat, Ryoichi Yamada, a division leader in Squadron 701, recalled: "At this moment, the situation reversed."

The American fighters climbed up to 26,000 ft. Both groups dropped their belly tanks simultaneously. The P-47s gained altitude advantage with their superior high-altitude performance but the *Shiden-Kais* turned into the Americans from their inferior position and launched an aggressive attack.

RIGHT: 1/Lt Stanley J. Lustic achieved his first victory in November 1944 in the Philippines. On 28 May 1945, he downed five more enemy aircraft and became an 'instant ace.' He ended the war with six victories.

Four Japanese pilots pursued Capt William Loflin, but his element leader, 1/Lt Stanley J. Lustic went to his aid and scattered the Japanese who then dived for the deck. Lustic followed right behind, firing at one of the Georges until it disintegrated in mid-air and fell in flames. The other Japanese then made a counterattack, split-S'ed, and fled. Lustic clung to him and shot him out of the sky at 3,000 ft altitude.

2/Lt Vernon Wischer, the No.4 man in Loflin's element, attacked two Georges. One of them was Lt Ryoichi Yamada and the other was his wingman, CPO Takeo Yamada. Wischer's head-on attack proved fatal to the wingman, who spun down in flames. Now Lt Yamada's situation worsened. He had not fired a shot when his engine started to vibrate. At 7,900 m, the engine seized and his propeller stopped turning. He had no recourse but to put his aircraft into a dive and try to escape.

Wischer spotted Yamada trailing a plume of white smoke. Believing that he had inflicted mortal damage on the Japanese fighter, Wischer decided to go in for the kill. Yamada could see two enemy fighters rushing toward him. He misidentified them as P-51 Mustangs and with no time to lose, he quickly unbuckled his shoulder harness, slid back the canopy, and prepared to bail out. But suddenly, the two P-47s made a U-turn and headed away! Yamada had no time to ponder his luck. He quickly settled back into his seat and steered his crippled fighter downwards: "Looking down, there was the Sendai River below me, which often floods, and a bridge. I wondered if I should go over the bridge or underneath it. I ended up going over at about 30 meters. I landed on a sandy area on the river bank."

ABOVE LEFT: Lt Ryoichi Yamada in January 1945. He endured a harrowing experience in the air and on the ground. After the war, he married the sister of his squadron leader, Lt Oshibuchi.

ABOVE: 2/Lt Vernon Wischer was the 'mystery pilot' in Lt Ryoichi Yamada's mind for over 42 years. (V. Wischer)

Lt Yamada endured a harrowing ordeal and was lucky to be alive. He saw some 50 villagers running toward him carrying spears, swords, and tools. Hardly a welcoming committee! The young lieutenant prepared himself.

Japanese home defense pilots did not, initially, carry pistols since they fought over their own homeland. However, it all changed because of an ugly incident on 17 February 1945. During the great raid over the Tokyo region by US carrier aircraft, a pilot from the Yokosuka *Kokutai*, CPO Takashi Yamasaki, parachuted from his stricken aircraft with severe burns. He was not able to speak. Surrounded by angry civilians who mistook him for an enemy pilot, they beat him to death. Pilots from both branches of the military were shocked and disgusted. Yamada was also aware of Lt Kanno's encounter with a farmer when he had parachuted into a field on 19 March. He saved himself when he cursed at the old man who advanced on him with a pitchfork.

Lt Yamada hunkered down in his cockpit with the canopy tightly closed. It would protect him from flying rocks. He took out his pistol and was prepared to use it if they came too close. The villagers surrounded his aircraft from a distance and the stand-off lasted about 30 minutes before an Army captain arrived on the scene. He understood the situation perfectly and approached the fighter without concern. Yamada slid back the canopy and both men introduced themselves. Turning to the locals, the captain yelled that the pilot was friendly. Many of the civilians, some embarrassed and all relieved, turned and headed away. A few approached and offered apologies. They explained that they thought a Grumman Hellcat had crash-landed with fake Japanese markings. The lieutenant was incredulous.

BELOW: CPO Takeo Yamada fell victim to 2/Lt Vernon Wischer on 28 May 1945.

Lt Yamada remained in the cockpit and proceeded to call his base. The reception was crystal clear, thanks to the improved Type 3 wireless which had been adopted by the Navy in 1943. Yamada explained that he had made an emergency landing and asked headquarters to recover his aircraft.

The young lieutenant was mentally and physically exhausted. He was also feverish from the complications of tonsillitis. The civilians escorted him to the village master's house and toasted him with cups of *shochu* – a type of distilled spirit made from Satsuma sweet potatoes. Satsuma is an old name for the Kagoshima area where Yamada had landed, and where this sweet potato is grown. Shochu is also a famous Kagoshima product. There was no escaping native hospitality. They simply could not do enough to show their gratitude.

The following day, Lt Yamada thanked his hosts and left the village of Kusamichi to board a train to Sendai Station on the Kagoshima railway line. He would arrive back at Omura with an incredible story to tell.

Two divisions, led by Lt Masaji Matsumura and WO Mitsuo Hori, took off on an interception mission from Omura shortly before noon. Lt Matsumura aborted over Mt. Unzen due to an oil leak and later, WO Hori was forced to return because of engine trouble. The next senior pilot, CPO Seiichiro Sato, took charge and headed south for Kagoshima at 6,000 m altitude.

CPO Sato soon spotted a group of small aircraft heading north over Kanoya. They appeared to be P-51s but he was mistaken; there were no Mustangs over Kyushu that day. Sato gave orders to drop belly tanks, but the third and fourth aircraft failed to release. They continued to climb on their approach. The No.3 man departed from the formation and dived away. PO1/c Mahito Yoshihara, No.4, was unable to jettison his tank, so Sato ordered him to leave, but his order was not acknowledged.

CPO Sato was about to engage when four enemy aircraft turned to attack his No.4. PO1/c Yoshihara was not an experienced flyer. Sato felt responsible, so he reversed his direction to come to his aid. He approached firing, but he was suddenly hit from below.

Sato's mount was struck by .50 caliber shells which instantly ignited his fuel tank. When the George exploded, he was thrown out of the cockpit unconscious. Luckily, the static line connected to the interior pulled his parachute out of the pack. He plunged into the sea on the north coast of Kagoshima Bay and drifted. His life jacket kept his head above water and prevented him from drowning. He was fished out by a coast guard launch and accommodated in a tunnel shelter where he regained consciousness two days later.

The near-death experience was simply too much for the young pilot. Before he lapsed into unconsciousness again, he had a vision of his mother. He had experienced seeing such a vision once before when he had crashed at Mabalacat in the Philippines following an encounter with Hellcats on 19 October 1944. He would eventually recover but never flew again. To this day, he still believes that a Mustang got him. His sacrifice saved the life of PO1/c Mahito Yoshihara.

When the dogfight ended, the Americans had claimed seven Japanese aircraft destroyed with two probables. Miraculously, not one Thunderbolt was hit. It was said that the 343 *Kokutai* claimed four destroyed, but Lt Yamada – who believed he knew the truth – thought that it was a complete defeat that no enemy aircraft had been shot down.

After the *Shiden-Kais* landed, Lt Oshibuchi reported that their opponents had been P-47s, known for their superior performance at high altitudes. LCdr Yoshio Shiga told the exasperated squadron leader: "You should not have been tricked by such tactics. You should have pulled them down to a lower altitude and forced them into a dogfight."

Lt Oshibuchi was not pleased with the unit's performance. However, the American pilots were very impressed with his men. "*Jap pilots very aggressive and showed excellent ability...*" stated Mission Report 5-20 of the 318th Fighter Group.

BELOW: CPO Seiichiro Sato was shot down by Thunderbolts, but believes that his attacker was a P-51 Mustang.

The 343 *Kokutai's* casualties for the day's combat included CPO Motoo Isogawa (S407), CPO Takeo Yamada and PO2/c Masaru Morikawa (both S701), and CPO Seiichiro Sato (wounded). Isogawa was one of two pilots in the air group who 'returned from the dead' (the other was WO Minoru Honda). He was a *Kamikaze* who returned to base on foot one month after his sortie. Since he had already been cited for his *Kamikaze* service (posthumous promotion of two ranks), the 'dead' pilot could not return to Japan alive. He did finally return to the mainland, though reduced in rank, and was posted to S407 toward April.

In 1987, almost 42 years after the dogfight, retired Gen Ryoichi Yamada of the Japan Self Defense Air Force made an inquiry to determine why his life was spared that day. In a letter to his old adversary dated 15 July 1988, Vernon Wischer of West Bend, Wisconsin, penned the following explanation: "*From memory, we engaged one another over Kanoya. I believe your group, possibly four aircraft were above us and to our right at about three o'clock. I remember advancing to full throttle and climbing at a steep angle. I had an aircraft in my gunsight and fired. The Japanese aircraft started to smoke as it if was struck in the engine... At this time I was on the verge of a stall and I broke off engagement with the Japanese aircraft. I sighted Capt. Loflin diving away from the engagement. I dove to intercept him and to offer air cover for him. When I caught up with him, I observed his windshield covered with oil. Also I could not contact him by radio. We proceeded to Ie Shima.*"

Chapter 15

"We learned some lessons today"
2 June 1945

「紫電改」コルセアを圧倒

Kamikaze operations in the Okinawa area continued to be a threat to advancing American warships as they drew ever closer to the mainland of Japan. According to LCdr Yoshio Shiga, the 5th Fleet Command suggested to Capt Genda that his air group also contribute suicide aircraft to operations over Okinawa. The captain asked Shiga for his views. "I will go first!" responded the executive officer in a suspiciously exuberant tone. "And next, Naval Academy graduates should follow! And you, Captain, will go last! But I have one condition and that is that a 5th Fleet staff officer accompany me in the rear seat of my aircraft!"

Needless to say, the 5th Fleet never brought this subject up again.

The top priority for American fighter pilots was the destruction of Japanese airfields in Kyushu. Because of the generally inferior quality of Japanese pilots encountered in the air, American pilots became overconfident – with fatal consequences. Every American pilot wanted to return home as an ace and was more than willing to push the envelope.

On 2 June 1945, the Corsair pilots of VF-85 and VBF-85 aboard the carrier *Shangri-La,* sat down to a breakfast of baked beans, well-done toast, grapefruit juice, and scrambled eggs. The two units had only 33 victories to show for their deployment since January and most of these were over *Kamikaze* aircraft flown by pilots of marginal skill. Lt Joe Draper Robbins had become their first ace when he downed two Zeros on 11 May.

At 06.00 hrs, the first of 33 Corsairs roared off from the *Shangri-La's* deck. They were a mixture of F4U-1Cs and Ds, and FG-1Ds. Their mission was to sweep the *Kamikaze* staging airfields at Chiran, Kagoshima and Izumi on Kyushu. The strike was under the leadership of CAG-85, Cdr W.A. Sherrill, while LCdr Warren W. Ford commanded the squadron aircraft.

The sweep started off badly. Lt(jg) William H. Marr, in one of the first aircraft to take off, crashed into the sea. He extricated himself from his aircraft and inflated his life raft, but just as he was about to be picked up by a destroyer, he was sucked under and never resurfaced. Lt(jg) William R. Clarke, who was the 'stand-by pilot' as well as a close friend of Marr, was immediately launched in his Goodyear-built FG-1D. He would not know of his friend's fate until he returned from the mission.

The Corsairs assembled with fighters from the *Yorktown* and the *Independence* and made their 325 mile journey to the mainland of Japan. The weather was poor with a ceiling between 300 and 400 ft and visibility from 50 ft to two miles. The pilots flew on instruments until they made Yaku Shima, a large island about 40 nautical miles south of Kyushu where the weather cleared. It is most likely that their presence was detected by spotters on the island, as well as radar along the coast of Kyushu.

Back at Omura Airfield, Capt Genda received word that a large group of enemy carrier aircraft had been spotted en route, heading for the southern tip of Kyushu where the *Kamikaze* airfields were located.

Chiran Airfield was only eight miles inland from the coast. Strangely, there was little – if any – activity on the airfield. The mission report of Fighting 85 recounted: *"The group hit Chiran Airfield at about 0800, strafing planes in the revetments, many of which were believed to be dummies. Damage was not assessable. No planes were seen to burn. Moderate heavy and medium antiaircraft fire was encountered. Evasive action was taken and one of our planes was damaged. The heavy anti-aircraft fire was accurate in altitude but not in deflection."*

RIGHT: VF-85 F4Us flying combat air patrol. The unit started its tour on 17 January 1945 aboard the Shangri-La and fought until the end of the war. The carrier's tail emblem has not yet been applied. (Wm. Clarke)

Fighting 85 wrapped up its attack on Chiran, re-formed, and proceeded further north towards Kagoshima Airfield, but the second target proved no more alluring. A few aircraft were scattered around the strip and were deemed to be not worthy of attention. A decision was made to proceed to the third target, Izumi. One group orbited the airfield and was searching for suitable targets when one of its aircraft was hit by antiaircraft fire, sustaining moderate damage. The second group pounced on the only aircraft seen on the field, a twin-engined type, and destroyed it.

Capt Genda was fully aware of the attacks on Chiran and Kagoshima, but he was not in a hurry to counterattack. "*The enemy had overwhelming strength,*" he wrote in his memoir. "*Formations of about 30 aircraft would come in, one after another, within a short interval. It would not have been effective to have brought our squadron of only 20 or 30 aircraft in front of such large enemy waves, even though the Shiden-Kai group was the cream of the Naval fighter force. If we hit the first enemy group, we would be surrounded and slaughtered by other groups that would follow in succession. Therefore, I thought it would be advantageous for us to attack the enemy's tail end group. Upon departure following the raid, its formation would be disorderly and its pilots would be in a mood to hurry home. It would take much more time for the preceding groups to come back to aid the tail end group being attacked by our fighters.*"

At 08.45 hrs, Capt Genda decided to put his plan into action. Lt Keijiro Hayashi, newly appointed squadron leader of Squadron 407, was ordered to take charge of the interception. His unit contributed eight fighters. Flying under his leadership were Lt Ryoichi Yamada of Squadron 701 with eight aircraft, and Lt Masaji Matsumura from Squadron 301 with five aircraft.

Lt Hayashi was a very capable leader and a veteran of the air battles at Balikpapan, Borneo. A graduate of the Naval Academy, he was a classmate of S301's Lt Naoshi Kanno. Capt Genda was favorably impressed with the new squadron leader. Hayashi had been an ensign assigned as a navigator to the carrier *Akagi* when he crossed paths with the CO who was a staff officer of the 1st Fleet. Hayashi later commanded S602 at Balikpapan, Borneo, before being transferred along with ten of his men to the 343 *Kokutai* in mid-May. He was brought in to replace Lt Yoshishige Hayashi who was killed in action against B-29s on 21 April.

BELOW: Fighting 85 Corsairs on the Shangri-La late in the war. Folding wings greatly facilitated aircraft storage and allowed the carrier to carry more aircraft. (NMNA)

Twenty-one *Shiden-Kais* took off at 08.45 hrs. Eight aircraft from Hayashi's S407 and another two-division group were led by Lt Ryoichi Yamada from Squadron 701. A top cover flight of five aircraft was commanded by Lt Matsumura. It was airborne within minutes, with a mix of aircraft drawn from all three squadrons.

"*It took about 30 to 45 minutes for our Shiden-Kais to reach the southern tip of Kyushu,*" wrote Genda. "*During that time they reached an altitude of 6,000 to 7,000 meters. I wanted our pilots to be in a good position when encountering the enemy aircraft. This was ideal though we could not always do it like that every time, as I wanted.*"

The Corsairs flew aimlessly for some time at 8,000 ft. Having been 'cheated' at three airfields, they simply had too much time and fuel on their hands. At 09.00 hrs, they were about to leave the area for home when they received a report of

Hellcat pilots in distress. Two pilots from VF-9 – Lt Frank V. Scott Jr. and Ens Roy G. Kueller – and Lt Merle M. Hershey from VBF-87 were treading water in Kagoshima Bay. A PBM Mariner rescue aircraft had been called out. The covering Hellcats left for home as soon as the Corsairs arrived to take over.

"*Lieutenant Roy Fuller and his division were over Ibusuki Auxiliary Seaplane Base when he observed some Emilys on the water,*" stated the mission report. "*He was at about 4,500 feet going about 180 knots. He asked for permission to strafe but was told to join in orbiting the downed pilots. He pushed over and after a very short time in his dive, very accurate medium anti-aircraft fire came up and hit him. He crashed into the bay. The rest of his division pulled out of the dive and acted as cover for the orbiting planes.*"

"The first plane to arrive was a PB2Y of VPB-13 on patrol, and when he landed, he tore off a wing float and capsized the plane!" remembered George M. Blair. "Now his whole crew plus the two fighter pilots were in the water. This group was picked up by a PBM, but he had only one JATO (jet-assisted take-off) unit and was unable to rescue the *Ticonderoga* pilot, who was about 20 miles away. However, he was picked up in the afternoon by another Mariner. After we had been lazily circling for about 30 minutes, Fuller must have gotten bored and decided to strafe a couple of beaten up Emilys on the water. He was hit by AA and crashed at the edge of the breakwater."

Twenty-one *Shiden-Kais* finally arrived over Kanoya at an altitude of almost 6,100 m at 09.55 hrs local time. The weather was fair and visibility was good. The sky was virtually cloudless and there were no enemy aircraft to be seen as they approached Kagoshima Bay. With no fear of being jumped from above, Lt Hayashi wondered where the Americans had gone. Suddenly, he saw them far below; 16 gull-winged aircraft heading south to the right of their direction of flight.

Lt Hayashi, leading Squadrons 407 and 701, dived on the orbiting Corsairs of Cdr Sherrill's group. He worried that their excessive diving speed would cause them to overshoot their prey and negate the element of surprise. But the Corsairs scattered as Hayashi's formation came screaming down from twelve o'clock high, with a 2,000 ft altitude advantage.

"For a few minutes, we maintained discipline and organization, but soon there were sections and singles all over the sky," recalled Blair. "My transmitter was out, so I could not give any directions. My wingman, John Moore, stayed right with me, but Chernoff, a substitute section leader, went off on his own to be a hero and was shot down. Moore and I started for altitude and were immediately looking ahead on a Frank(?) section. We both fired a short burst, and at the last second, I broke under the leader, sure he was going to hit me. I heard his engine as he passed overhead! The Japs really knew the game. They'd make section runs about four at a time, leaving eight or 12 above us as cover, and all of them would re-form after each pass."

"It was a complete surprise attack," recalled Genda. "The enemy aircraft had no means to cope with it. Most of them were unable to enter into an ordinary dogfight. Our *Shiden-Kais* glued themselves to the enemy's tails or dived at them and destroyed one after another. Enemy fighters flamed, wings flew off, and spiraled down..."

Four Georges hit Lt Allie W. Callan Jr.'s division. Three of his aircraft were hit and one was seriously damaged. Lt(jg) Harold R. Kennedy was forced to make a water landing on his way home. His fellow aviator, Lt(jg) Charles N. Kirkham, refused to leave him stranded and continued to orbit in foul weather. Kirkham ran out of fuel and was forced to ditch. Neither was ever recovered; Kirkham received the Navy Cross for his valor.

Three Georges went after the CAG, but Sherrill's wingman, Lt(jg) John Dunn, turned into the trio and fired at them head-on. One enemy fighter smoked and some flames were

BELOW: Enjoying free time on Mog Mog (Ulithi Atoll) just prior to their departure towards Japan, five VF-85 pilots standing, left to right: Lt George M. Blair, Ens Robert Bloomfield, unknown, William R. Clarke, and Bill Shinn. Sitting: Lt Allie Callan Jr.. (Wm. Clarke)

ABOVE: Lt(jg) William R. Clarke (Wm. Clarke)

observed coming from the engine wall. The Japanese pilot dived into clouds and exited the scene. Dunn received a credit for a probable, his only claim for the war.

Four Georges jumped Lt George M. Blair's division in an overhead pass. One of them came at Blair in a head-on pass and was determined to score a kill even if it meant ramming, but the Corsair pilot was not about to accommodate his opponent's death wish. He inflicted minor damage to the Japanese, then broke off to avoid a collision. Lt(jg) Saul Chernoff, flying with Blair, broke off his dive after this George, but he was jumped from above. Chernoff's Corsair was hit and it was last seen burning and spinning down through the clouds.

Lt Lawrence Sovanski, then credited with 2.33 victories over twin-float trainers, was hoping to add some fighters to his notch. However, the *Shiden-Kais* were much more than he had bargained for. Several fighters jumped his division and his mount was battered. Sovanski returned to the carrier with only ten gallons of fuel remaining.

Lt(jg) William R. Clarke became separated from the CAG's division after it was jumped. However, a George crossed his shooting path below and he took advantage of the situation. His opponent made a shallow wingover and began to climb. Clarke glued himself to the enemy and continued firing. The George's cowling flew off and it started down, burning and disappeared into clouds, out of control.

Clarke scored his first victory and now he was going for his second. Lt(jg) R.L. Meltebeke and Lt(jg) H.W. Fuog observed a George 500 ft above them and the two pilots made a head-on attack but inflicted no damage. Clarke weaved onto this fighter's tail and scored hits in the engine, cockpit, and wing roots.

"My attack was from slightly above in a starboard turn with apparent speed advantage," recalled Clarke. "I opened fire with all six .50 caliber at about 350 yards and 30 degrees deflection. I was closing distance rapidly and while still firing bursts at about 100 yards from a 5:30 position, what I thought were parts of the enemy plane came off. I immediately broke sharply to starboard to avoid collision with the parts and the aircraft. I noted considerable smoke from the George but did not observe fire. Upon breaking away, I lost sight of the aircraft. Fellow pilots noted the parachute and my gun camera film analysis determined that it was the pilot and not the aircraft parts that I avoided collision with."

Clarke's victim was CPO Jiro Funakoshi of Squadron 701.

The *Shiden-Kais* of Squadron 301, on cover duty, descended upon a group of eight Corsairs led by LCdrs Ford and Hubert. They were orbiting a downed pilot in the southeast position. The Japanese believed that these Corsairs were not aware of the other group's situation. Genda wrote later: "*If they had been alerted and joined in the combat, the situation might have changed. The 301st division rushed into the new group of eight Corsairs. The fight was as one-sided as the first one. Five of the eight were shot down.*"

LCdr Ford and his division flying cover over the rescue aircraft down below watched as the PBM Mariner was able to take off around 11.00 hrs using JATO. It was just in time because fuel supply was dwindling: the entire Corsair force had stayed more than an hour and a half longer than planned. Ford's division flew for about ten minutes before it received a call for help from one of the PB2Y Coronados still in the area. It turned around and went back to help even though time was not on its side. Luckily the four-engined seaplane found refuge in the clouds, so assistance was not needed.

Ford's men turned back into the bad weather front for the trip home. Ceiling and visibility were near zero as they groped their way through thick cloud on instruments for over an hour. Lt(jg) Wallace Atkinson was forced to ditch due to loss of oil pressure. Lt(jg) R.F.Toenges landed his Corsair near a picket destroyer with only 12 gallons of fuel left. He had been forced to use an excessive amount of fuel to stay in formation due to the buffeting of his aircraft which carried two wing tanks.

When the group landed, inventory was taken, and it was appalling. Nine Corsairs were lost: two had definitely been shot down by enemy fighters; two ditched on their return flight; one had run out of fuel while orbiting a downed pilot; two were hit by antiaircraft fire; one crashed on take-off; and another was stripped on return and stricken from the roll. The pilot toll was even worse: five men were missing in action and one died in the take-off accident.

On the credit side, Fighting Squadron 85 put in claims for two destroyed, one probable, and two damaged. The day's high scorer was Lt(jg) William R. Clarke with two victories, followed by Lt(jg) John C. Dunn with a probable, with a damaged each to Lt George M. Blair and Lt(jg) Donald P. Grau.

On the Japanese side, two pilots failed to return – CPOs Eiji Mikami from Squadron 301 and Jiro Funakoshi of Squadron 701. So wild was this encounter, the 343 *Kokutai* claimed 18 victories, with Lt Hayashi's men claiming 13!

According to the aircraft action report, the Corsair pilots were more than impressed with the caliber of their opposition and its aircraft: "*The Japs were most aggressive and were experienced pilots. They made section passes and attacked only when they had the tactical advantage.*" The George compared favorably with the Corsair in speed, had good diving characteristics, and could outmaneuver it.

The report focused on the lessons learned that morning: "*1) Don't orbit airfields when anti-aircraft is known to be present; 2) Don't leave your section leader to become a hero; 3) Don't be taken in by decoys so as to be easy meat for a/a or enemy fighters; 4) Don't try to cover without obtaining sufficient altitude to prevent being jumped from above; 5) If outnumbered or at an altitude disadvantage, immediately start defensive weave. Japs apparently will not attack.*"

Lt George Blair's last entry in his diary for the day summed it up best: "*We learned some lessons today, but the tuition was too high.*"

During the year 2000, locals in Kagoshima Prefecture were cultivating land in the vicinity of Kanoya and found an aircraft machine gun. They dug further and discovered a name chop (family seal) inscribed with the name "*Mikami.*" Name chops are used by the Japanese to imprint the family seal on documents and letters. This further confirms that Lt(jg) William R. Clarke hit CPO Jiro Funakoshi, who parachuted into the sea. His bleached bones were found along the shores of Takeshima and his identity was confirmed by the name written on his life jacket. He had no relatives and died alone.

LEFT: George Blair was an honored guest of the 343 Kokutai Association at its annual gathering in Tokyo in 1987. Blair is flanked by retired generals Ryoichi Yamada on the left and Masaji Matsumura on the right. On 2 June 1945, both Yamada and Matsumura fought against Blair and his comrades. (Jiro Yoshida)

RIGHT: Mr and Mrs George Blair with Gen and Mrs. Genda at the 343 Kokutai Association meeting. (Jiro Yoshida)

Chapter 16

Clash over Kanoya
3 June 1945

鹿屋上空の遭遇戦

The 343 *Kokutai* was feeling very satisfied about its previous day's operations against the American Corsairs over Kagoshima. Capt Genda was extremely proud of his men. VF-85 from the *Shangri-La* had taken a beating, but he knew that the Americans would be back for a rematch and he was to be proved correct.

At 06.00 hrs on 3 June 1945, it was VBF-85's turn to strike at the airfields on Southern Kyushu. A sweep comprising a mixed group of 20 F4U-1Cs, Ds and FG-1Ds led by LCdr S.B.Strong, VBF-85's CO, was launched to attack the airfields at Chiran, Kagoshima, and Tojimbara. At the same time, 36 Hellcats of VF-9 from the *Yorktown* and 12 from VF-46 off the *Independence* took to the sky. Their targets were the airfields in the Kanoya area. Of the *Yorktown* 'Cats, 32 carried fragmentation bombs. The American formation made its way toward Kyushu in extremely bad weather. Heavy rain buffeted the fighters and the ceiling was between 300 and 500 ft, though conditions cleared when landfall was reached.

At Omura, 28 *Shiden-Kais* of the 343 *Kokutai*, led by Lt Naoshi Kanno, had taken off at 08.30 hrs to intercept enemy aircraft. They arrived over Kagoshima and soon spotted an American formation, probably the Corsairs of VBF-85. The Hellcats of VF-9 had just finished their attacks on Kanoya Airfield and were heading home while Lt Howard R. Hudson's division remained behind to make photo runs to assess damage. After taking photos, they proceeded to the next target. About 12 miles southeast of Kanoya, while the Hellcats were at 7,000 ft altitude, they were jumped by a pair of fighters diving down from 9,000 ft. The Japanese approached the second section of Hudson's division head-on.

Lt(jg) William E. Brewer fired on a Zeke and his opponent's engine immediately flamed. The Japanese pilot tried to run, but Brewer followed him down and hit him again. The Zeke's propeller was slowly windmilling as it plunged toward the earth. Brewer received credit for the destruction of this aircraft. Although the identity of the Japanese pilot is unknown, the other pilot was PO1/c Kiyoshi Miyamoto of the Squadron 407.

BELOW: PO1/c Kiyoshi Miyamoto of Squadron 407 was shot down by Lt Howard Hudson of VF-9 on 3 June 1945. (Photo credit: Y. Matsumoto)

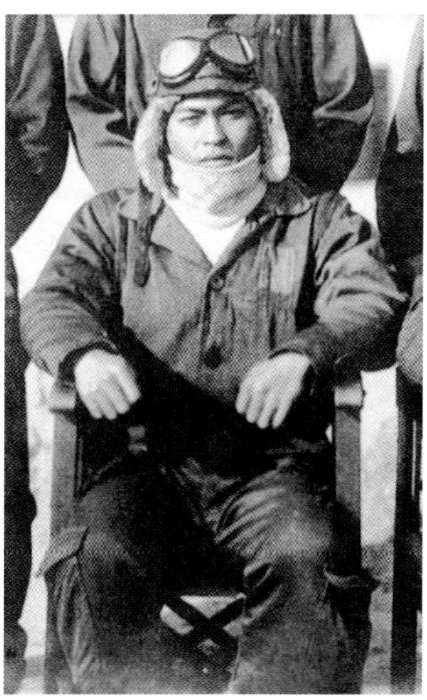

Miyamoto failed to hit his target and was chased by Lt Hudson and his wingman, Lt(jg) Bill McLaurin, for over 15 miles. When Hudson closed the distance and fired, Miyamoto executed a wingover, but the Americans were tenacious and stayed on his tail. At 1,500 ft, the Japanese executed another steep wingover, but it was not enough to save him. Lt Hudson directed his fire into the George's cockpit and the left wing tank exploded. Kiyoshi Miyamoto crashed into Kagoshima Bay about five miles southwest of Kanoya Airfield. The photo division completed its assignment and headed out to sea.

The Corsairs of VBF-85 were over Tojimbara when they spotted a pair of fighters approaching at four o'clock below from the southeast, which were identified as Tojos. They were in fact *Shiden-Kai* from Squadron 301. The aggressive Japanese commenced a low side run on Lt J.H. Shroff's section. Shroff and his wingman made a hard right turn to counter and fired a few rounds as they crossed over the Georges. One of the Japanese fighters started to smoke from the wing root. While the Americans were preoccupied with their targets, two more Georges commenced a low side run on them.

According to the VBF-85 mission report: "*Lieutenant (jg) Douglas J. Dok and his wingman in the second section turned right from their port position in the formation to counter... In doing so, Dok got a snap shot at one of the first two Tojos. A second pair of Tojos from the northwest pursued Shroff and his wingman, Lieutenant (jg) Charles W. Fox, and was pursued by Dok to the south for some distance until two more enemy planes got on Dok's tail. Dok had a 6 o'clock shot at one, which trailed some smoke from wing root. Fox went into a 60 degree dive to the right, losing Shroff on his left as well as the enemy planes. Shroff was heard to call Mayday, and was not heard nor seen thereafter."*

The battles were a combination of hit-and-run skirmishes and pursuits. The fighting was spread over a wide area. According to Lt Goro Ichimura of Squadron 407, the squadron was attacked by F6Fs and F4Us and forced to flee at full speed. After regrouping, it went on the offensive and fought against the Corsairs.

When the shooting was over, VBF-85 pilots claimed three Tonys destroyed and one damaged, a Zero shot down and one damaged, and a Tojo probable and three damaged. Evidently, the *Shiden-Kais* of the 343 *Kokutai* were not the only opposing aircraft to take part in the combat. On the American side, VBF-85 lost three aircraft. Lt J.H. Shroff and Lt(jg) Edward Dixon were shot down by fighters and were declared missing. Lt Sigurd Lovdal crashed after shooting down a Tony when his tail snapped off, probably due to an aileron overbalance.

The records of the 343 *Kokutai* for this action do not survive. However, the unit's only casualty on this date was PO1/c Kiyoshi Miyamoto. Undoubtedly, the 343rd pilots claimed this encounter as another 'turkey shoot.'

Chapter 17

"Splash one, I'm shot to hell..."
22 June 1945

「一機海中に撃墜、われ被弾…」

The goal of almost every American fighter pilot was to shoot down five enemy aircraft and return home as an 'ace', for an aerial victory was an ego-boosting validation of his skill. Unfortunately, with so many American fighters in the air, it seemed as if there were not always enough Japanese aircraft to go around! Some pilots were so desperate to score a victory, they adopted extreme measures. 1/Lt John W. Leaper of VMF-314 was a classic example of this.

According to Leaper's hometown newspaper, on 22 June 1945 "*...in the heat of an all-day air battle over Okinawa and surrounding waters, Marine Lt. John W. Leaper of Hopkins, Minn., deliberately crashed into an enemy plane Friday and then parachuted to safety... Leaper was dueling with a Japanese Zeke fighter in a sky dotted with enemy aircraft when he suddenly found he was out of ammunition. He broadcast a radio message that he intended to chew the enemy's tail off with his plane's propeller and then bail out. He did just that, and was picked up a few minutes later without a scratch.*"

When Leaper returned to his unit a few days later, he was arrested for "*...willful destruction of government property.*" His comrades complained that he hogged all the action and left nothing for them. Later, the press got hold of the story. To avoid a public relations disaster, the Marine Corps ordered the commanding general to drop the charges and award him a medal! 1/Lt John Warren Leaper was credited with the destruction of two Zeros and a Betty, and was duly awarded the Navy Cross.

The 22nd of June – the day the Americans captured Okinawa – also marked the last day of the *Kikusui* Operations, which totaled ten in number. It was a pitiful last effort with only three small groups of suicide aircraft launched – six Betty bombers carrying *Ohkas* from Kanoya; eight Zeros, also from Kanoya; and 11 Army Franks from Miyakonojo East Airfield.

Kikusui No.10 marked the final long-range mission of the 343 *Kokutai* to Amami Oshima and it would pit the unit against American land-based aircraft for the first time. Take-off began at 07.45 hrs from Omura Airfield. Lt Oshibuchi commanded a 50-aircraft formation representing the entire airworthy force of fighters on hand. However, deteriorating servicing and supply problems forced 19 *Shiden-Kais* to abort or cancel take-off.

The Marine Corps fighter pilots of Fighting 113 landed their F4Us on Ie Shima on 24 May, led by Maj Hensley Williams. They spent their time strafing and bombing targets on Okinawa and conducting air patrols. The few Japanese aircraft they shot down were *Kamikazes* or escorts that did not put up much of a fight, but intelligence reports indicated an unusually heavy build-up of enemy aircraft in Southern Kyushu, which meant more *Kamikaze* attacks could be expected.

BELOW: 2/Lt Alson C. Frazer (A.C. Frazer)

Twenty Corsairs from VMF-113, led by Maj Orvin H. Ramlo, took off at 08.46 hrs for a routine barrier CAP to Amami Oshima. Two aircraft pulled out of the mission when one pilot suffered engine problems and his wingman peeled off to provide escort back to base. The remaining 18 aircraft flew on. Two divisions and the section remaining from the broken division (ten aircraft) took up stations at 15,000 ft while the remaining two divisions took their position at 6,000 ft.

2/Lt Alson C. Frazer, flying an FG-1D in one of the top cover divisions, was bored: "I began manipulating my mixture control and rpm, playing games with my machine to maximize fuel economy. I managed to drift out of position from the leader, so I began a very slow close up. Hunchar, my wingman, became impatient with the rate I was closing, so he broke off and zoomed off to join up with Johnson, our division leader. About that moment, I detected a sun glint. I watched it for a while and realized that it wasn't my imagination. So I tally-hoed the object and Johnson passed the command to me and joined up as my second section. At that time, Hunchar rejoined me as my wingman. I closed on the object, and as I got closer to it, Johnson also picked it up. It was a group of four aircraft."

ABOVE: 1/Lt Russell Hunchar (Olga Hunchar)

The unidentified aircraft at first appeared to resemble natural aluminum P-47 Thunderbolts. 1/Lt Russell Hunchar watched the bogeys to the starboard side, flying on a parallel course, but became shocked when he saw the glaring red meatball on their fuselages. Looking to port, he was even more electrified to see six four-aircraft enemy divisions. Then another comrade broke in on the radio to report still more enemy aircraft to starboard.

Frazer: "We proceeded to drop our belly tanks, charge guns, and turn on the gunsight. But as I learned later, my tank failed to release. We tried to close to a firing range, but they just kept out of range. My recollection was that they were flying a circular path. At that point, they seemed to gain altitude, and at that instant, we looked up and realized that we were coming head-on into a formation of Japanese aircraft with an estimated 4,000 feet altitude advantage. There were five sections of four planes flying in a V pattern, each four planes forming a V, and there were five layers."

The *Shiden-Kais* were flying at high speed and passed directly over the Marines. The Japanese jettisoned their belly tanks in unison, almost striking several Corsairs below. The Americans saw the enemy fighters coming down in two groups – one from twelve o'clock and the other from six o'clock. 1/Lt James D. Johnson ordered his men to climb for attack.

1/Lt Hunchar also had a vexing problem much like 2/Lt Frazer, his section leader. "Unfortunately, my tank would not release!" he recalled. "This concerned me, since I still had some fuel in it and I knew I was vulnerable. While we were dodging the belly tanks dropped by the Japanese flying toward us, I was on the extreme right side of our division and saw a Japanese plane on our left, making a run on us."

The four Corsairs became enticing targets for the seven aircraft from Squadron 301, acting as top cover. Lt Kanno should have maintained his position, but he jumped the gun and proceeded to dive into the Americans before the attack order was given. The Americans thought that he was a decoy.

"Lt Kanno was really brave, but he was a bit reckless," explained Ryoichi Yamada who flew on this mission. Lt Oshibuchi, the overall commander, had intended to go in from behind the Corsairs.

"I immediately turned up and towards him and fired head-on," recalled Hunchar. "I thought we were going to collide, but a large amount of black smoke came from his engine. He turned sharply to the right and down. I doubt if anyone in my flight saw this. At the time I saw tracers going by me from behind. I turned hard right and up; the Jap plane turned right and leveled out directly in front of me. I was a short distance behind him and fired a short burst. I saw flames coming from the port side of his engine. I fired again but was unable to follow through because I was hit from behind. My plane seemed to just flip over and I was losing altitude. I saw that ammunition belts were out of the cans and resting on the wings. Large holes were blown in both wings in the area of the ammo cans. I checked my guns – no fire. I recharged – no fire. At this point, I called Lt Johnson, reported my plight and asked for help. Lt Johnson called back stating that they were also in trouble and that I was on my own."

Hunchar dived down in a split-S towards the cloud and headed for home. His departure and the sheer numbers of Japanese aircraft broke up the Corsair division.

Frazer: "We turned and Johnson took his squadron and I took a position to his right and below, in a position to weave and protect as we had trained to do. We turned to form the attack to their rear and they, in turn, came back on us and proceeded to dive on us in an orderly way whereby four planes would peel out of their pattern and pass through us, and four more behind that. About the time the second four peeled off, the orderliness of the day collapsed. Johnson opened fire and was gaining significant hits on the incoming aircraft. I was

LEFT: Lt James D. Johnson's wing shows the damage inflicted by 20 mm hits from a George. (Olga Hunchar)

BELOW: A 20 mm shell struck Lt Johnson's aircraft in the horizontal stabilizer. Small dents in the rudder show the impact of shrapnel from the explosive round. (Olga Hunchar)

BELOW: 2/Lt Alson C. Frazer's Corsair as it looked immediately after its return. (Olga Hunchar)

watching to his rear, trying to weave with him and protect the rear from attack. I was then aware that Hunchar was not on my wing. I wasn't sure where he was and couldn't find him in my mirrors, so I joined up with Johnson in a closer pattern and tried to stay with him as a wingman, with Allcroft still flying his other wing. We were a three-plane group for a short period of time."

1/Lt Johnson suddenly made a very sharp climbing turn to the left, which Frazer could not follow. He had spotted a George on Frazer's tail and had swung around to attack it. The Japanese aircraft pulled off Frazer, executed a split-S and headed down, with Johnson on its tail. The George pulled into clouds. Johnson had only one gun firing and he desperately tried to recharge another. By the time his opponent exited and entered the clouds again, Johnson was shooting with two guns. The George disappeared, trailing a heavy plume of smoke.

ABOVE: Lt Keijiro Hayashi (Kikuyo Hayashi)

As Johnson went after his victim, Frazer flew wide in a high climbing chandelle and was coming back in an attempt to rejoin Johnson. Suddenly, a George came into view on his right, flying to the left, below him at almost a 90 degree angle. Frazer led his prey with about 80 mils lead in his gunsight and opened fire. The George's canopy shattered and a large chunk of metal flew off the right wing root. It is possible that Frazer hit Lt(jg) Keijiro Hayashi who was leading the 1st division of the 2nd *Chutai*.

Hayashi had attacked four F4Us heading north over the area between Takara Jima and Kikaiga Shima. He reportedly shot down one Corsair, but, having been hit in the right wing root, he was trailing white smoke. While falling behind his formation, he was attacked by F4Us with altitude advantage and proceeded to 'hide' in clouds. After this encounter, Lt Hayashi was posted missing.

While Frazer was gunning down his opponent, a George was shooting at him: "I began to notice that my left wing was disintegrating. I heard no shots, I heard no gunfire. But I became visually aware that my left wing was coming apart. So I realized that I had been hit, so I broke off my left turn and rolled out and pitched over and commenced to dive. I remembered being quite frightened. My adrenaline was undoubtedly very high but my mental capacities were good. I just felt that I've been hit, OK, he won, I lost, now let's go home!"

The fight was not over for 2/Lt Frazer. A George had followed him down and was firing at him. The Corsair entered clouds and the pilot turned sharp left. The George was right behind him. "I could see the attacking aircraft over my left shoulder," Frazer recalled. "Either he could not come to bear, turn as tight as I could, or had too much speed. We kind of looked at each other for what seemed like eternity. We probably made at least one complete circle. He was unable to bring his guns on me, and he broke off and disappeared on me." Frazer left the dogfight and landed safely in his damaged fighter.

The Marines got more than they had bargained for; it was four against 31. There were simply too many enemy aircraft in the air. 1/Lt Johnson was jumped by three fighters, but he decided boldly to meet them head-on. The trio pulled away, allowing another trio to pull up from astern and pepper his Corsair. Johnson was hit in the propeller, elevators, fuselage, and left wing. He popped his flaps and one of his attackers overshot, but by then there were three more on his tail. By making judicious use of cloud cover, Johnson lost his pursuers and returned home safely. He was credited with one aircraft destroyed and one probable.

BELOW: CPO Sueo Ishizue was one of four 343 Kokutai pilots missing in action on 22 June 1945.

2/Lt Roger D. Allcroft had two Georges on his tail, but they overshot and Allcroft made them pay for this mistake. After gunning down two opponents, he was last seen riding on the tail of a Japanese with all guns firing. His last transmission was: "*Come on up here, I've got a Tojo and two Zekes cornered!*"

While the mêlée was in progress, another division of VMF-113 under the leadership of Lt Hollingsworth was about 40 miles east of Amami at 18,000 ft. 2/Lt Clinton H. Jones tally-hoed a lone George in the distance going north at 13,000 ft. Spotting the Corsairs, the Japanese turned west but when the American division was about a half mile away, the George started to turn towards the Americans. Jones opened fire from 400 ft and his opponent exploded. The enemy pilot either jumped or was blown clear of his aircraft; his parachute opened, but he was seen to fall out of it at 9,000 ft. He may have been Lt Keijiro Hayashi. In his damaged fighter, escape was impossible, so his only option would have been to fight.

2/Lt Francis B. Smith, Jr. and 1/Lt W.C. Green, forming the remaining section of the broken division, did not know their location when the dogfight erupted. Smith informed his wingman that he could see a hole in the clouds and was going down to check their position. Suddenly, Smith called out: "*For God's sake, help me! There's a million of them! There's four on my tail!*" Twenty seconds later, he radioed: "*Splash one, I'm shot to hell and heading back to base!*" He was never heard from again.

The 343 *Kokutai* returned to base with four of their own missing: Lt Keijiro Hayashi, and CPOs Sueo Ishizue, and Takashi Yanagisawa of Squadron 407, and PO1/c Susumu Tsuchiya from Squadron 701. The unit claimed seven F4Us downed in combat.

VMF-113 lost two pilots and suffered three aircraft damaged. The unit's aircraft action report gave this assessment of its opposition: "*The enemy planes and pilots engaged were by no means of the type usually encountered. Their planes were good and the pilots skilled and aggressive. A very good formation was flown and maintained. Deception was used by sending in one division as a decoy. The planes waiting above were grouped in four-plane divisions, sections roughly abreast, and divisions stacked. They attacked by divisions in the following manner: The whole division executed a chandelle to the left, thus starting a diving approach on opposite course with altitude advantage. The division then split into two sections, one going to each side of the target planes and executed high side runs simultaneously, completing the run by tailing in and joining.*"

The costly *Kikusui* operations drew to a close. The next morning – 23 June – the defenders of Okinawa, LtGens Mitsuru Ushijima and Isamu Cho, committed ritualistic suicide by *hara-kiri,* disemboweling themselves with their short swords. Seven other staff members shot themselves. In all, over 110,000 Japanese soldiers perished on Okinawa as well as 75,000 civilians.

RIGHT: Squadron 407 pilots in May or June 1945. Sitting: Lt Keishichiro Hattori (left) and Ens Mankichi Sawada (right). Standing, left to right: CPO Hisachi Kubo, PO1/c Ichimaru Naruse (KIA 2 July 1945), LA Yoshimi Daito, CPO Eigo Shinoda, CPO Takashi Yanagisawa, KIA 22 June 1945), and CPO Hiroshi Takahashi. (Y. Matsumoto)

Chapter 18

"I had never seen so many Japanese fighters"
2 July 1945

初めて見る日本機の大群

By 2 July 1945, the Marine pilots of VMF-224 and VMF-311 were restless and frustrated with lack of action. The men of Fighting 224 had not scored a victory since 22 June, while the situation for the VMF-311 was drier still; no 'kill' since the 11th. When they heard that there was going to be a fighter sweep of Kyushu, they felt elated. If they did not account for any *Kamikazes* along the way, then the enemy mainland would certainly be a fertile hunting ground. Little did they know that their fortunes would be beyond their wildest expectations.

Capt Joseph P. Lynch was a seasoned veteran who had scored 3.5 victories in January 1943 at Guadalcanal with VMF-112. His first tour of duty had ended in August 1943 when his unit was sent back to Miramar, California, for reorganization as a carrier squadron. The Bostonian was shy of 1.5 victories from acedom. On 2 July 1945, however, Lynch's squadron had just been moved up from Yontan, in central Okinawa, to its most northern tip, in order to act as a vanguard unit against the *Kamikazes*.

At 07.30 hrs, Lynch took off from Chimu Airfield with 16 F4U-1Ds to patrol over Kyushu. The Corsairs were 'hot' – equipped with water-injection engines. They cruised leisurely toward Japan, passing dozens of picturesque islands leading straight to Kyushu. It was a beautiful morning.

Maj Mike Yunck from VMF-311 had taken off at 08.00 hrs with 13 F4U-1Cs, only to have one crash on take-off and eight turn back (six due to operational problems together with two escorts). The popular 26-year-old fighter-leader from Detroit had had a spectacular combat debut when he had shot down three floatfighters on 3 December 1942 at Ramada Bay, New Georgia, but he had been scoreless since.

Maj Yunck and his men were going into enemy territory heavily armed. Their F4U-1Cs (only 200 of which were built) carried four 20 mm M2 cannon instead of the usual six machine guns. Their mission was to sweep Ronchi Airfield on Kyushu.

The Corsairs made landfall, but there was thick fog and a solid overcast to 1,300 ft. They flew for another 50 miles up the east coast of Kyushu, then turned back and flew west across the southern tip of the island. Yunck's group was flying at 23,000 ft while Lynch and his men were at a lower altitude. Just north of Pt. Kaimon, Yunck spotted an estimated 20 Japanese fighters below on a reciprocal course. The VMF-311 mission report recorded: "*Two divisions of enemy aircraft totalling seven planes were in a tight formation with the balance of the flight some 2,000 yards to side.*"

Mike Yunck in a post-war photo (R.F. Toliver)

At almost the same time, pilots of VMF-224 stumbled upon the scene. Joseph Lynch remembers the shock of seeing the opposition en masse. "We were up at 20 to 21,000 feet altitude. On one of our southern legs, shortly after we had arrived on station and with Mike Yunck's flight off to my right, I was searching the sky in all directions when suddenly, I saw a tremendous group of Japanese fighters about 3,000 feet below our flight. They were streaming from dead ahead and back toward my seven or eight o'clock position. I had never seen so many Japanese fighters in one group since my days on Guadalcanal and it was the first time that I ever had altitude advantage in the entire war!"

The enemy aircraft were 24 *Shiden-Kais* from all three squadrons of the 343 *Kokutai*, led by Lt Naoshi Kanno. As replenishment of aircraft had ceased since Kawanishi's *Shiden-Kai* factory was bombed on 9 June, this was the largest force the 343rd could muster. The seven fighters seen by the Marines belonged to S701, which was flying top cover under the leadership of Lt

ABOVE: Capt Joseph Lynch at Guadalcanal in 1943. (USMC)

Ryoichi Yamada. The group was returning to base from patrol duties when they crossed paths with the Corsairs.

CPO Takumi Sugitaki, Lt Yamada's 2nd man, looked up and noticed a formation of what appeared to be taper-nosed fighters flying about 1,000 m above. He thought initially that they were Army Type 3 Tony fighters (strangely, pilots of VMF-224 at first identified their opponents as Tonys). From their flight maneuvers, Sugitaki then reconsidered them to be P-51 Mustangs. Presumably, since he had seen the Corsairs from below, he had not noticed the peculiar form of the gull wings.

Sugitaki shouted a warning over the wireless. At such a warning, the leader would normally order: "*Form up for combat!*" and start maneuvering for position. But on this occasion, Lt Yamada showed no response – possibly because of bad radio communication.

Sugitaki saw American aircraft making quick rolls and diving. He could not do anything to avoid the attack because he was in the midst of a tight formation. He knew he was a sitting duck and that the opposition would hit their formation hard.

Simultaneously, Marines from the two units charged down. Maj Yunck's men went after the top cover Georges while VMF-224 went after the rest further below. Yunck struck first, setting his gunsight on the tightly formed lead flight. He was followed by his wingman, 1/Lt Fred Edison.

"About ten rounds flashed into my cockpit and six to seven hit the dashboard." recalled Sugitaki. "All the shells looked like tracers which dazzled me. I felt the whole body of my aircraft absorb hits."

The fuel tank took an incendiary round and flames blew into the cockpit like a blowtorch. Sugitaki had no time to think since the aircraft exploded and he was ejected from the cockpit. His parachute automatically popped open and the lifeless form slowly drifted down to the sea.

According to his unit's mission report: "*In his first run, Major Yunck scored hits in the fuselage of a Zeke from 8 o'clock above firing from 800 to 300 feet, and the enemy aircraft rolled over out of control, started to burn, and disappeared in the overcast below.*"

"We had a word – harmonization – which meant convergence of bullet trajectory at a certain distance, for example, 180 or 210 meters," Yamada recalled. "My engine, fuselage, and tail wings were not hit, only the main wings got holed. My plane must have been at a position behind the harmonization or convergence point. I saw the face of the enemy pilot. I made a quick roll and dived."

The VMF-311 report continued: "*Lieutenant Edison, Major Yunck's wingman knocked down another with many hits in the fuselage and wings which sent the enemy down burning. Credit for the third Zeke went to Lieutenant Driscoll, who hit engine and left wing root in a run from 7 o'clock above. The Jap started to burn, rolled over, and disappeared below in the overcast.*"

Maj Yunck was not through yet; he became separated from his division, but that did not stop him from scoring again. He chased another enemy fighter, closed to 1,000 ft, and knocked its right wing off. This was his final victory of the war and elevated him to ace status.

The surprise attack by the Americans scattered the S701 pilots and they fled without putting up a fight. This was confirmed by the American action report, which stated that: "*...none of the enemy aircraft encountered displayed any enthusiasm for fighting. They scattered and for the most part dove for the deck.*"

Lt Yamada headed home at an altitude of 300 or 400 m above the sea. After landing, he started counting the bullet holes and stopped at 26 or 27, then asked his groundcrewman to finish the tally. Fifty-six years after this combat, Ryoichi Yamada would learn that his aircraft was hit by Maj Yunck.

While VMF-311 was tearing up the enemy's top cover, Capt Lynch and his comrades prepared to attack Lt Kanno's formation comprised of S301 and S407. Lynch recalled: "As soon as I spotted them, I 'tally-hoe'd'

LEFT: CPO Takumi Sugitaki (standing left) with fellow pilots of the 302 Kokutai at Atsugi. (T. Takaishi)

and told everyone to drop their belly tanks. Just as I made my release, someone radioed: '*You dropped your tanks, Joe!*' I rocked my wings for the attack signal and rolled left into the formation. The three other pilots in my division were Truex, Smiddy, and Schleicher. As we dove on them I couldn't perceive that they had spotted us yet since most of the formation had passed beneath us before I attacked with my division. They had not made any move which would indicate that they had seen us. I tried to get my sights on the leader but to do so would have caused me to pull flat in the midst of their whole group, so I settled on another which may or may not have been an element leader and I managed to hit that plane in the engine from above as we dove through the formation."

The Japanese dispersed in all directions. Lynch picked another victim and hit him in the right wing and belly area: "I flew almost straight up and past him and saw another enemy fighter slightly off to my right and in a slight right turn. Finally after scoring a great number of hits, that fighter caught fire. It showed cherry red flames under the belly. I think that the pilot might have parachuted because I have a distinct recollection of the pilot leaving the right side of his cockpit and getting out over the right wing. That plane had taken many hits in the fuselage. As so often happens in a dogfight, at one instant, the sky was full of planes and almost as if by magic, the sky clears. From my viewpoint, that seemed to happen and suddenly the whole mêlée was over."

As Capt Lynch started searching for a comrade with whom he could join up, a George pilot fired an accurate burst of 20 mm shells at him. The Corsair's right wing exploded. Three .50 caliber machine guns were blown out of their mounts and the ammunition caught fire and exploded. The right aileron appeared to be shattered. To make matters seemingly hopeless, a 20 mm shell had also exploded behind his armored seat. His predicament continued to worsen: "The wing became totally engulfed with flames and I remember choking from the cordite fumes which filled the cockpit and seeped right through my oxygen mask. The plane went into a 45 degree dive and I slid the canopy back in order to bail out if I had to... As the plane fell, the flames subsided and the fire went out."

The Japanese pilot, probably thinking that this F4U was finished, did not deliver the *coup de grâce*.

"The trip back was slow and uncomfortable, and uneventful until a Corsair joined with me as I neared Okinawa," Lynch recounted. "He looked me over and said that the plane was riddled with shell holes and he didn't know what was keeping it in the air. I didn't know either because as I looked through the hole in my right wing, I could see the ocean below me and I figured that the hole was so large, a six-footer could have dropped through it length-wise without touching head or toes on the jagged metal."

Then came the really frightening news. The other pilot told him that his drop tank was still attached! "Evidently, it had not released... Talk about sitting on a time bomb! That tank was probably full of fumes. The way my Corsair had been blazing should have exploded that tank. My guardian angel was working overtime. By the time I reached Okinawa, I had managed to climb to only 4,500 feet and I had the plane head directly at a small mountain just beyond Chimu. I had been nicked by small particles of shrapnel on my right side and my right leg was trembling from the sustained effort of pushing full right rudder for several hours, so I happily baled out. Attempting a landing was impossible. I dove out toward the trailing edge of the right wing and as I fell clear of the plane, I saw how badly it had been shot up. Chance-Vought really came up with a good one when they designed the Corsair – it brought me home. I landed hard in my 'chute a few hundred yards from the fighter strip at Chimu and by the time the flight surgeon got through treating my injuries, the story of our dogfight had already been told."

Back at his fighter base, Capt Lynch penned his remarks concerning the enemy pilots in the official mission report: "*...flew good formations and made skillful runs. Enemy pilots were aggressive and used very good tactics.*"

Long after the war, he added: "I don't know who they were or what unit they belonged to but I'll say this – they were among the very best fighter pilots that I ran into during the war. The Zero pilots at Guadalcanal were skilled, tough, and aggressive, at least the ones I faced were. The pilots at Kyushu on 2 July 1945 were every bit as good. They were tough customers."

With his two victories, Capt Joseph P. Lynch finally became an ace. His harrowing 325 mile struggle to make it home was an inspiration to his comrades. 2/Lts Denver Smiddy and Lowell Truex were credited with one victory apiece and the trio had pushed the squadron's final wartime score to 110 victories.

VMF-311 also claimed four victories: two for Maj Mike Yunck, with one each for 2/Lt Joe Driscoll and 1/lt Frederick W. Edison. These were to be the unit's last victories of the war. In almost three months of combat, it had claimed 71 victories.

Back at Omura Airfield, four Japanese pilots were missing: CPO Shojiro Ishii and PO1/c Ichimaru Naruse from S407, shot down by VMF-224, and CPOs Takumi Sugitaki and Takashi Sakuma from S701, downed by VMF-311. Of the four missing Japanese pilots, only Sugitaki was found alive. An Army watch post reported: "*One aircraft exploded in the air and a parachute descended into the sea eight kilometers off Kagoshima Hanto.*"

It was around 03.30 hrs the following morning when CPO Sugitaki regained consciousness as he drifted in the sea. He had sustained burns to his face, both hands, and bruises on the back of his head and back. His watch had stopped at 10.30 hrs. Though the month was early July, it was very cold in the sea. The current eventually washed him toward a steep cliff. He rode the powerful crest of the waves and was able to grasp at the branch of a tree. He climbed up the cliff and came upon a peasant who rendered him aid. When told what time it was, he determined that he had been in the water for more than 17 hours. This was Sugitaki's fourth ditching in the sea.

CPO Sugitaki received treatment at Kagoshima Naval Hospital for three weeks. When the doctor would not release him, his strong sense of duty compelled him to escape back to his unit without full recovery.

Lt Keishichiro Hattori recalled with fondness, his faithful wingman Ichimaru Naruse whom he lost on this mission: "He was always my third man for more than a year at Balikpapan... We returned to the homeland and were assigned to Squadron 407 together. He again flew in the No.3 position to me. Coincidentally, after he was killed in action on 2 July 1945 south of Kyushu, on the afternoon of that day his parents came up to Omura to see him. I met them with a broken heart and told them: '*CPO Naruse is absent now as he has gone to Matsuyama and it will take some days for him to come back. It may be better for you to go home, as air raids have become intensive.*' They stayed at a nearby inn. I visited them there and told them about their son's activities and gave them some rations that we airmen had. After I arranged to obtain train tickets for them, they went back home a few days later."

Lt Hattori found himself unable to tell Naruse's parents that their son was missing from the morning combat.

Chapter 19

Shiden-Kai vs. P-51 Mustang
5 July 1945

「紫電改」対 P-51ムスタング

The 343 *Kokutai's* situation was not good. Its pilot contingent was rapidly diminishing as the seventh month of 1945 approached. Cdr Takahide Aioi, attached to the 343 *Kokutai* HQ Staff, was very concerned. He recalled: "I remember I sent a request to GHQ in Tokyo in June 1945. We needed replenishment of pilots who could be used immediately in combat."

At that time, 'A' and 'B' class pilots were on the decrease and 'C' class pilots occupied a greater part of overall strength.

B-29s had 'visited' Omura Airfield between March and May. The neighboring 21st Air Arsenal had been bombed by Superfortresses from China several times since the previous year and almost all of its accommodation had been destroyed. Major aircraft repairs were conducted in buildings covered with tarpaulins on iron frames. There was no effort to rebuild or repair such buildings since such an undertaking would invite more bombing. Barracks were not used by groundcrews for fear of air raids, and men were quartered in civilian residences around the airfields.

To save aircraft from destruction, they were dispersed and hidden after fuel was removed. When standby duty was over, the aircraft were hauled by men along the bumpy gravel-covered roads to the dispersal areas about four kilometers away. In the early morning, they were hauled back to the airfield again. Consequently, the servicing situation deteriorated.

The *Shiden-Kai* remained the latest frontline fighter in the JNAF's arsenal. In their duels against F6F Hellcats, F4U Corsairs, and P-47 Thunderbolts, the pilots of the 343 *Kokutai* still held confidence in their

LEFT: P-51 Mustangs of the 35th Fighter Group in 1945. This photograph was probably taken at Clark Field in the Philippines. Three squadrons from the group converted from P-47s to P-51s in March 1945. The Mustang proved to be too strong an adversary for the Shiden-Kai on 5 July 1945. (J. Stanaway)

ABOVE: A P-51D of the 41st Fighter Squadron in mid-1945. The Mustang was considered one of the the best single-engined fighters of the Second World War. (J. Stanaway)

aircraft. However, reports and rumors had been circulating since late 1943 about a serious USAAF challenger known as the North American P-51 Mustang which had made a name for itself in the China-Burma-India Theater.

The P-51 had been known to Japanese aircraft manufacturers since September 1941. It had originally been designed and built for Britain's Royal Air Force in 1940. A great improvement was made when North American installed British-designed but Packard-built Merlin engines in two experimental XP-51Bs. The aircraft achieved a maximum speed of 441 mph.

When the USAAF landed at Ie Shima, it was equipped with the P-51D, a major production variant of the Mustang, with a new bubble canopy and six .50 caliber guns. Its 1,490 hp Packard engine delivered a maximum speed of 437 mph. The sleek Mustang was heavier and much faster than the *Shiden-Kai* and had a service ceiling of 41,900 ft. The two opposing aircraft were serious contenders in the dogfight arena. On 5 July 1945, the pilots of 343 *Kokutai* would test their skills against the Mustang for the first time.

The Mustang jocks of the 40th Fighter Squadron, 35th Fighter Group felt like third stringers on a football team. From the New Year to 3 April 1945, the squadron had claimed only seven victories. Sweeping airfields on Kyushu in preparation for the invasion of Japan was not the glorious job the fighter pilots looked forward to. It was now 5 July 1945 and no kills had been registered since 3 April. Then, at 10.42 hrs, four divisions of P-51s took off from Ie Shima for a sweep in the Kyushu area. Along the way, three aircraft aborted.

On Omura Airfield, members of the 343 *Kokutai* were engaged in a lively baseball game. The game was introduced to Japan from America in 1869 by Henry W. Dennison, when he arrived as a clerk at the American Consulate at Yokohama. Between 1867 and 1868, Dennison had played for a professional baseball team, the Olympics, in Washington, DC. In 1871, Horace Wilson arrived as a teacher at Tokyo Middle School No.1 and taught his students baseball. The Japanese immediately fell in love with the American sport and it gained tremendous popularity. Up until the massive air raids on the mainland, the Japanese continued to play baseball.

Lt Keishichiro Hattori of S407 recalled the events of that morning: "On July 5th, we had a baseball game between the 407th and the 701st at Omura Airfield. Such was a very rare event. The game ended with so many scores that we would have needed an abacus to add them up, including Lt Oshibuchi's three base hits and Lt Kinoshita's two base hits."

Eight pilots from Squadron 701 were on standby duty that morning, playing and sweating in the hot sun with full flight gear. Just as the game ended, the Z flag ('scramble') was raised. Word had reached base that enemy reconnaissance aircraft had been detected over Sasebo. The eight pilots ran the 250 meters to their aircraft. Led by Lt Kazuchika Kinoshita, within minutes, they were airborne.

Lt Kinoshita was an experienced pilot who had fought with Squadron 602 at Balikpapan, Borneo. A student of economics from Tokyo University, this young officer gained a reputation as a tough instructor. Despite his harsh approach to training, he was generally sympathetic towards his men.

At approximately 12.45 hrs while flying at 9,000 ft, Capt Henry P. Rettinger and his fellow pilots of the 40th Fighter Squadron spotted four Georges at 8,500 ft, flying east, just south of Sasebo, over Nishisonogi Peninsula – they were the *Shiden-Kais* of the 1st Division led by Lt Kinoshita. The Japanese pilots flew in alert fashion, searching for the enemy. PO1/c Sato noticed P-51s over the clouds and also below at much lower altitude, but before the Japanese could take any offensive action, 13 Mustangs fell into battle formation, with Rettinger leading his men down to 8,500 ft, right behind the division of Georges.

Capt Rettinger selected his opponent, closed to about 150 yards, and opened fire. His two-second burst hacked into his opponent's tail; the George turned to the left and went down. Another burst knocked off the right wing tip. The stricken fighter rolled over and plunged into the water spewing black, greasy smoke.

The 40th Fighter Squadron combat report recorded: "*During this action, Capt Rettinger had not dropped his wing tanks. He then dropped his wing tanks and climbed to the right. The three Japanese planes which had pulled up to the right, crossed about 500 feet above and headed down towards Capt Rettinger at about 90 degrees. Capt Rettinger made a right turn and started firing at 55 degrees. He fired a five-second burst and noticed direct hits on the canopy that was blown off. The engine was smoking badly. The plane went straight down at about 45 degrees and crashed into the water.*"

The two remaining Georges split up with one going to the left. 1/Lt John W. Ebert latched onto this aircraft and gave it a two-second burst which hit along the cockpit. His opponent immediately executed a split-S down to the water, and Ebert was given credit for a probable.

1/Lt Irving C. Pine, the No.4 man in Rettinger's Red Flight, fired his guns at one George, then became separated from his flight and also lost contact with the enemy. He climbed for altitude and looked back to see a George about 2,000 yards behind him. When the Japanese entered cloud to lose him, Pine turned into him and fired a ten-second burst at 40 degrees. The shower of .50 caliber shells connected and the George started smoking. It disappeared into cloud, but Pine was not about to let it get away.

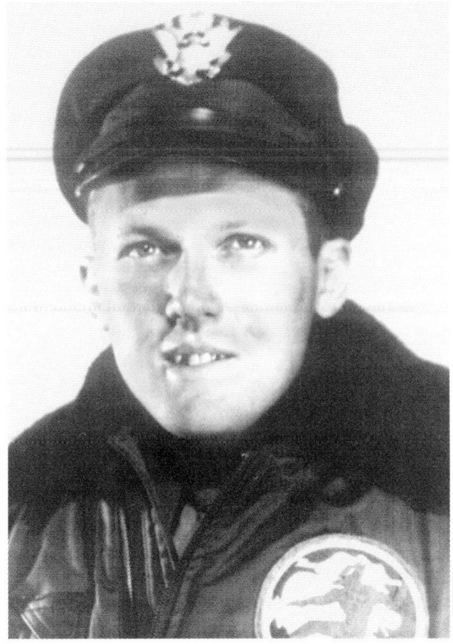

ABOVE: 1/Lt Ellis Baker, Jr. (AFAA)

The two aircraft were lost in the cloud. Pine flew on instruments for a few minutes, and when the cloud broke, he found himself behind his target, not more than 50 yards away. A five-second burst took off a large part of the left wing and the George entered a cloud belching black smoke. Since no one saw this aircraft crash, 1/Lt Pine was given a probable.

1/Lt Ellis Baker, Jr., leading Green Flight, flying about 500 yards above Red Flight and three-quarters of a mile behind, spotted a flight of two Georges and they were aiming for him. He immediately executed a Split-S, came around on their tails, and fired a two-second burst from 25 degrees. His opponent was hit along the wing root and fuselage, and tried desperately to escape. Another well placed burst sent Baker's opponent down into the sea. He then climbed to 7,000 ft and saw another George directly ahead.

Twenty year-old PO1/c Teruichi Sato tried to lose his pursuer by taking his *Shiden-Kai* into the clouds, but Baker was close behind. At a distance of about 300 ft, a stream of shells struck the George along the fuselage. Sato entered the clouds once again to escape, but he knew that his aircraft had received serious damage. With so many American fighters in the air and fearful of being shot while baling out, Sato delayed parachuting until his engine caught fire and spewed flaming oil in the cockpit. The Japanese jumped from his burning aircraft with burns to his face. During his free fall, Sato fainted but was jerked back into consciousness when his parachute blossomed. While floating downwards, Sato could see his airfield being bombed and strafed. Now the Mustang pilots turned their attention to the helpless figure dangling in his parachute. At first, four P-51s, then seven or eight more, came around and shot at him. Luckily, their aim was poor and the frightened pilot was unhurt. As soon as he touched water, he engaged the quick-release mechanism on his chest, and the harness and parachute came away. He tried to swim away from the floating white silk which would make him a target again but probably due to extreme fright and burns, Sato fainted.

BELOW: PO1/c Teruichi Sato (T. Sato)

When PO1/c Teruichi Sato regained consciousness, it was in the home of a fisherman. He did not know what type of boat had rescued him nor how many men fished him out of Omura Bay but the grateful pilot stayed one night. The following day, a launch was sent out to take him back to base. His silk parachute had three bullet holes in it. He would end the war as a patient at nearby Omura Naval Hospital.

Sato believed that the *Shiden-Kais* had been 'stalked' and 'ambushed', but in reality, Capt Rettinger had spotted them inadvertently as the Mustangs had approached Omura to make a strafing run. Although it was not an intended surprise attack, the 343rd pilots were indeed surprised to come under fire as they delayed in climbing.

The final tally for the Mustangs was two each for Capt Henry P. Rettinger and 1/Lt Ellis Baker, Jr; 1/Lts John W. Ebert and Irving C. Pine received one probable each. 1/Lt Baker became an ace in this engagement. These were the last victories of the 40th Fighter Squadron. The parachute shooting incident was never mentioned in the squadron's official mission report.

The Japanese lost three pilots: Lt Kazuchika Kinoshita and CPOs Akira Takahashi and Kiyoshi Toyohara. PO2/c Teruichi Sato was wounded. One Mustang was claimed destroyed. Although the P-51 pilots believed that they had fought two divisions of *Shiden-Kais*, in reality, they had only fought four Georges of the 1st Division, all of which were shot down. According to Sato, the other division of four aircraft was ordered to land and did not see combat. It appears that the 2nd Division landed at Oita Airfield because its base was being raided.

Lt Ryoichi Yamada, a senior division leader in Squadron 701, believed that the Japanese were beaten and outmaneuvered because the Mustangs had the element of surprise.

On the airfield, 14 groundcrew were killed and 11 wounded during a surprise air raid by B-24 Liberators and their Mustang escort. Lt Ryoichi Koga, servicing officer, recalled with disdain: "The enemy did not use large bombs to attack Omura Airfield. Rather, they used small bombs with a spontaneous fuse and they blew out many small steel shrapnel fragments. They mowed down everything within a ten meter radius as soon as they came into contact with the ground. I clearly remember an incident where a crewman was killed by such a bomb. He was turned into mutilated flesh moments after we had talked and parted."

Capt Genda lamented "...after the heavy and continuous B-24 raids beginning in July, our serviceability rate went down continuously."

Chapter 20

Duel over the Bungo Straits

24 July 1945

豊後水道上空の決闘

With the fall of Okinawa and Iwo Jima, the Allied Forces peeled away the last defensive layer to the Japanese mainland. These strategic outposts of the once-mighty Japanese Empire crumbled under the Allies' massive and sustained assaults. B-29s from Saipan, Guam, and Tinian now had a clear path to Japan. P-51 Mustangs, P-47 Thunderbolts, and B-24 Liberators moved onto airfields in Okinawa. Iwo Jima provided an emergency landing strip for the B-29 Superfortresses and their escorts. The Pacific War was now in its final stretch.

Task Force 38, part of R/Adm William 'Bull' Halsey's Third Fleet, was the Allied 'divine wind' building up its tremendous energy to pulverize the enemy mainland. Organized into three task groups, it comprised nine large and six light carriers, nine battleships, three heavy and 16 light cruisers, and 62 destroyers. The British Pacific Fleet joined the Third Fleet on 4 July, contributing four large carriers, one battleship, six cruisers, and 18 destroyers. Together, they struck Tokyo on 10 July, then steamed up to Hokkaido. Between 14 and 15 July, they sank 140 ships. They also destroyed five train ferries running between Hokkaido and Honshu. On the 17th and 18th, they bombarded Tokyo again and headed out to sea for the staging area. They were now preparing for the 'big one.'

To fuel and sustain this mighty force was a supply group consisting of 26 fleet oilers, one store ship, four tugs, six escort carriers, 13 destroyers, and 19 destroyer escorts. The greatest sea replenishing effort in history took place between 21 and 22 July, about 650 nautical miles from Japan. The armada took 379,157 barrels of fuel oil to tank off, plus 555,000 gallons of aviation fuel for its aircraft and 6,369 tons of munitions. Fresh from American farms came butter, sides of beef, slabs of bacon, apples, and oranges. Beans, potatoes, milk and eggs, wheat flour, sugar, and ground coffee were delivered to each ship on palleted crates. There was no escaping the George A. Hormel Company's infamous 'Spam' luncheon meat in the industrial 5 lb cans. It followed the troops everywhere.

Morale aboard the ships was at an all-time high, thanks in part to the efficiency of the military postal system. The fleet post office delivered 5,822 sacks of mail. Letters from home were a godsend. Family photographs and the latest gossip were shared amongst comrades. Homemade cookies, baked by mothers, sisters, and girlfriends arrived in smashed boxes. Copies of *Stars and Stripes* and *Life* magazines were eagerly sought and read cover to cover. Hometown newspapers publicized the contributions of their native sons.

Six Essex-class carriers at anchor at Ulithi Atoll in March 1945 prepare for attacks on Japan. (USN)

The greatest problem was keeping the pilots busy and in shape. Many men did calisthenics topside while others worked out with medicine balls, punching bags, or basketball played on one of the elevators. On some carriers, diehard football players organized into teams and played tag ball. Serious card players plied their trade in the ready room to relieve their comrades of cash and cigarettes.

The ships' stores were always busy. A carton of cigarettes was only 50 cents and a bottle of Coca-Cola was a nickel. An almost infinite variety of stale candies were available. Toiletries, seldom seen by their compatriots in the mudholes, were shamefully plentiful and cheap.

ABOVE: R/Adm William 'Bull' Halsey of the 5th Fleet (left in cap) with V/Adm Raymond Spruance (right, wearing pith helmet) of the 3rd Fleet, photographed in April 1945. Both men were dynamic carrier task force commanders who finally brought the war to Japan. (USN)

The Hollywood movie industry did its part to entertain the troops at sea. Ens Roy 'Eric' Erickson of VBF-10 recalled: "The biggest off-hours activity was watching movies in the ready room. Movies were traded between other carriers, tankers, and ships of the fleet so that we seemed to have an endless supply. The most appreciated films were always war films and especially anything that had aircraft in them."

On Omura Airfield, Capt Genda was stressed at the worsening war situation. "A B-24 group from the European Theater was stationed at an Okinawa airfield, and they began coming to attack us almost every day. They would raid Omura after coming northward along the west coast of Kyushu and then making a 90-degree turn over Nishisonogi Peninsula. The frequent bombings by the B-24s coming from a relatively short distance had a great effect on the deterioration of our serviceability rate."

The morale of the 343 *Kokutai* was lifted with the arrival of well-known ace Ens Kaneyoshi Muto. On 26 June, Ens Muto was transferred from the Yokosuka *Kokutai* to Squadron 301. In exchange for Muto, training officer Ens Saburo Sakai of Squadron 701 went back to his former unit. Genda told Muto to protect Lt Kanno. "Now that I am here, I will *guarantee* his life!" proclaimed the newcomer.

Ens 'Kinsuke' Muto was a popular figure whose great sense of humor was enjoyed by all. He gained his first victory on 4 December 1937 over Nanking. He ended his tour of duty in China with five victories and a rare commendation for distinguished service. On the first day of the Pacific War, he flew with the 3rd *Kokutai* and attacked Clark and Iba Airfields in the Philippines. During 1942-43, the diminutive 1.6 meter ace fought in the Solomons. Against incredible odds over Iwo Jima in June-July 1944, Muto downed several Hellcats and survived to see his homeland again.

Ens Muto became famous when TF58 carrier aircraft raided the Tokyo region on 16-17 February 1945. On the first day of the formidable aerial attack, Muto went up with a *Shiden-Kai* from the evaluation department of the Yokosuka *Kokutai*. According to press accounts of the day, he single-handedly took on 12 Hellcats over Atsugi and shot down four. In fact, Muto was one of ten pilots fighting VF-82 and four of its Hellcats were shot out of the sky. However, the nation needed a hero in those dark days and the military propaganda machine created one. Even to this day, the '12 against one' myth has become 'fact' in Japanese combat aviation history. His personal victory tally was close to 30.

During the morning hours of 24 July, TF38 carriers began launching their aircraft for strikes against shipping in Kure Harbor. It was almost a repeat of the 19 March attack on Kure.

Amongst those participating in the morning strike were VF-49 Hellcat pilots Lts(jg) Jack A. Gibson and George M. Williams from the carrier *San Jacinto.* The duo had arrived together on the carrier in May and became wingmen for the life of the squadron. Gibson was a country boy who came from a long line of farmers in Boydell, Arkansas. Anyone with the surname Gibson was slapped with the nickname 'Hoot' after the pre-war Hollywood western cowboy hero. 'Hoot' Gibson studied at Louisiana State University and had dreams of becoming a professional baseball player. While playing the sport in Kansas in 1940, he broke his ankle and had to return to the farm. At the start of the war, he joined the Naval Air Service with five of his fraternity brothers from Mississippi State.

RIGHT: In a typical scene, fighter pilots relax in the ready room aboard the carrier San Jacinto. (J. Gibson)

'Willy' Williams was a city boy from Madison, New Jersey, who fell in love with naval aviation when his mother took him to Navy Day in 1937. He enlisted in the Navy out of Rutgers University and passed through NAS Pensacola to become a fighter pilot.

Eight F6F-5 Hellcats and nine TBM Avengers were launched from the *San Jacinto* starting at

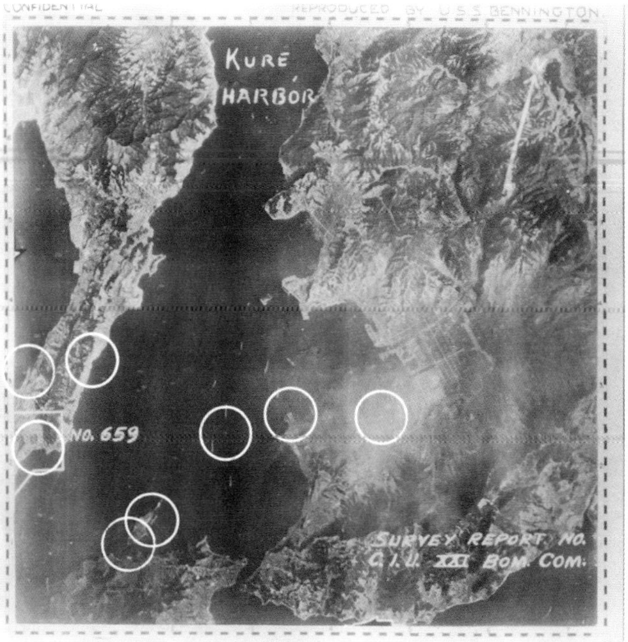

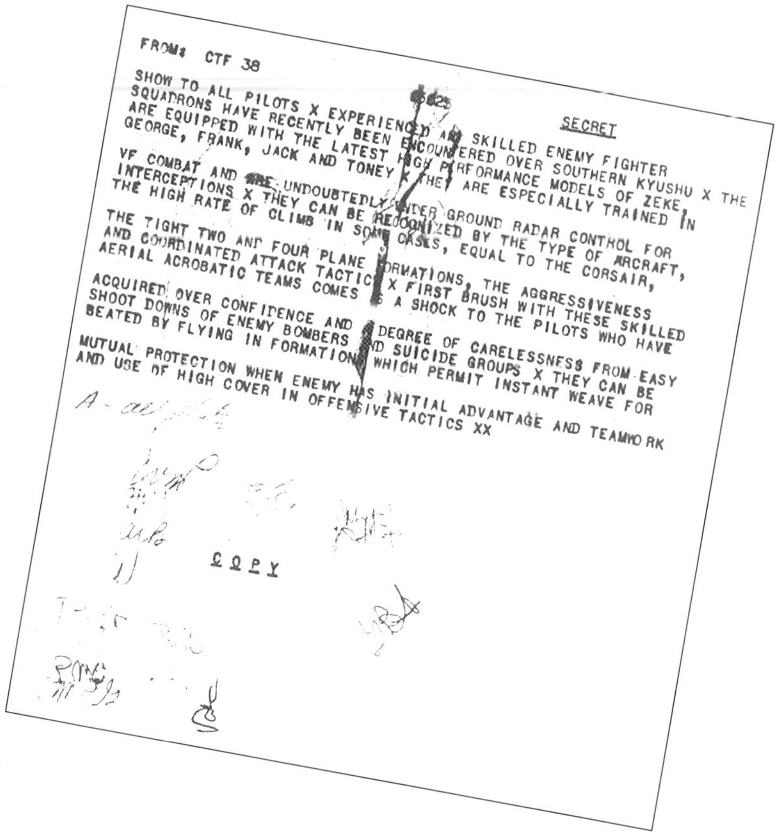

FROM: CTF 38

SECRET

SHOW TO ALL PILOTS X EXPERIENCED AND SKILLED ENEMY FIGHTER SQUADRONS HAVE RECENTLY BEEN ENCOUNTERED OVER SOUTHERN KYUSHU X THE ARE EQUIPPED WITH THE LATEST HIGH PERFORMANCE MODELS OF ZEKE, GEORGE, FRANK, JACK AND TONEY X THEY ARE ESPECIALLY TRAINED IN VF COMBAT AND ARE UNDOUBTEDLY UNDER GROUND RADAR CONTROL FOR INTERCEPTIONS X THEY CAN BE RECOGNIZED BY THE TYPE OF AIRCRAFT, THE HIGH RATE OF CLIMB IN SOME CASES, EQUAL TO THE CORSAIR, THE TIGHT TWO AND FOUR PLANE FORMATIONS, THE AGGRESSIVENESS AND COORDINATED ATTACK TACTICS X FIRST BRUSH WITH THESE SKILLED AERIAL ACROBATIC TEAMS COMES AS A SHOCK TO THE PILOTS WHO HAVE ACQUIRED OVER CONFIDENCE AND A DEGREE OF CARELESSNESS FROM EASY SHOOT DOWNS OF ENEMY BOMBERS AND SUICIDE GROUPS X THEY CAN BE BEATED BY FLYING IN FORMATIONS WHICH PERMIT INSTANT WEAVE FOR MUTUAL PROTECTION WHEN ENEMY HAS INITIAL ADVANTAGE AND TEAMWORK AND USE OF HIGH COVER IN OFFENSIVE TACTICS XX

COPY

ABOVE: This aerial photo of Kure Harbor was taken by a B-29 and supplied to the US Navy. Priority targets are circled.

RIGHT: This secret memo was distributed to the TF38 carrier air group commanders. In a reference to the 343 Kokutai, it described the appearance of "... skilled enemy fighter squadrons."

BELOW: The survivors of Squadron 701 in a photo taken sometime between late July and 8 August 1945 at Omura Airfield. Front row, left to right: Ens Misao Kishimoto, Lt Ryoichi Yamada, Capt Minoru Genda, LCdr Yoshio Shiga, Ens Haruo Segi, and Ens Kazuo Muranaka. Second row, left to right: CPO Masamori Ono, CPO Kotaro Koyae, CPO Ryuji Yagi, CPO Mitsuo Anraku, WO Masao Sasakibara, CPO Yasuo Matsumoto, WO Hiroshi Okano, CPO Junichi Miyake, PO1/c Tsutomu Yamada, and CPO Masao Ono. Third row, left to right: Unknown, PO2/c Masayuki Tashiro, PO2/c Sueji Imamura (KIA 8 August 1945), CPO Sanpei Shiono, PO1/c Tsutomu Murayama, PO1/c Morita, CPO Yasuo Furuzumi, CPO Hisao Nishimoto, CPO Kaemon Yokobori (KIA 8 August 1945), PO1/c Tanezo Inoue, PO2/c Ryoji Niwa, and PO2/c Masao Hasegawa.

ABOVE: PO1/c Kaneyoshi Muto married his wife Kiyoko in October 1942. However, he was in combat at the time and the two families performed the ceremony without him. This wedding photograph was created by pasting two photographs together.

07.58 hrs. Seven of the Hellcats carried 1,000 lb bombs, for in addition to their duties as top cover, they were also to be employed in dive-bombing.

Against hundreds of American aircraft, Capt Genda could only muster a token force. His bitter experience at Midway had taught him always to take the offensive. The incoming escort fighters would be fully fueled and armed. It would be extremely difficult for his *Shiden-Kais* to get at the bombers. Still believing in the offensive strategy, he gave orders to attack them while they were heading back to their carriers. Their dwindling fuel supply and possible battle damage would make them ripe for easy pickings.

The superiority of the Americans in numbers alone meant that Genda would be throwing his men into a potential bloodbath. He had feelings for his flyers and did not believe in wasting precious resources. He walked a fine line between duty and affinity to his subordinates. Many able officers on both sides suffered under this vexing situation. Genda knew the end was near and he now concentrated on leading his unit to an honorable conclusion.

Twenty-four *Shiden-Kais* were warmed up and readied for action. At 09.04 hrs, Lt Takashi Oshibuchi took off from Omura Airfield, leading Squadron 701. Lt Takao Mitsumoto followed directly behind them with Squadron 407. He was a classmate of Lt Naoshi Kanno at the Naval Academy and had also flown the *Shiden* in fierce combats over the Philippines in 1944. As division leader of Squadron 402 under the command of the 601 *Kokutai*, Mitsumoto had fought over Kikaiga Shima during the *Kikusui* Operations before joining the 343 *Kokutai*.

The last to take off was S301, led by Lt Masaji Matsumura. Squadron leader Lt Kanno was off duty this day and did not participate. Engine problems forced two aircraft from Squadron 301 to abort, including Lt Matsumura's wingman.

Enemy aircraft were getting to Kure and going back unhindered. Oshibuchi could have picked a fight anywhere.

"At the time, enemy groups of about 30 aircraft each were coming north for attacks on Kure and going back in a long column," Capt Genda wrote. *"Their movements were clearly known to us from the information sent from the air defense network. The only problem was to judge which formation we should attack with our* Shiden-Kais. *We needed to make a swift retreat after dealing a hard blow before we could be attacked by the next enemy formation."*

Lt Oshibuchi and his men circled around Cape Sata at 6,000 m while receiving information and instructions from Capt Genda. The composition of his division included his wingman CPO Jiro Hatsushima, with Ens Kazuo Muranaka leading the other flight with his wingman, CPO Yoshio Takahashi, who lagged behind because of a violent headache. Muranaka was a veteran from the China War and a survivor of the Pearl Harbor and Midway battles, with close to ten victories.

Oshibuchi finally decided on a course of action and took his men south into the Bungo Strait at 10.20 hrs. Ryoichi Yamada, Oshibuchi's 2nd flight leader, recalled: "Groups of enemy aircraft came into my sight like scattered sesame seeds! Gradually, they became large enough to be identified as enemy aircraft – about 250 of them. They continued to fly south in an orderly formation as if they did not notice us."

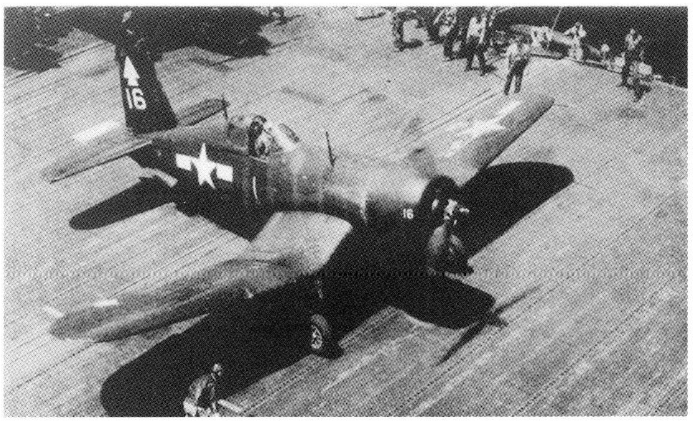

RIGHT: A Bennington Corsair commences its long journey toward Kure on 24 July 1945. (R. Applegate)

Oshibuchi focused his attention on a group preceding the group from the *Bennington*. Hellcats and Corsairs were guarding the Helldivers and Avengers at an altitude of 9,800 ft. The Squadron 301 went up as top cover while Oshibuchi took his group to hit the escort fighters from behind.

"Enemy spotted! This is Matsumura No.1!" yelled Lt Masaji Matsumura, the 3rd *Chutai* leader from S301. Matsumura was flying at the right end of the *Shiden-Kai*

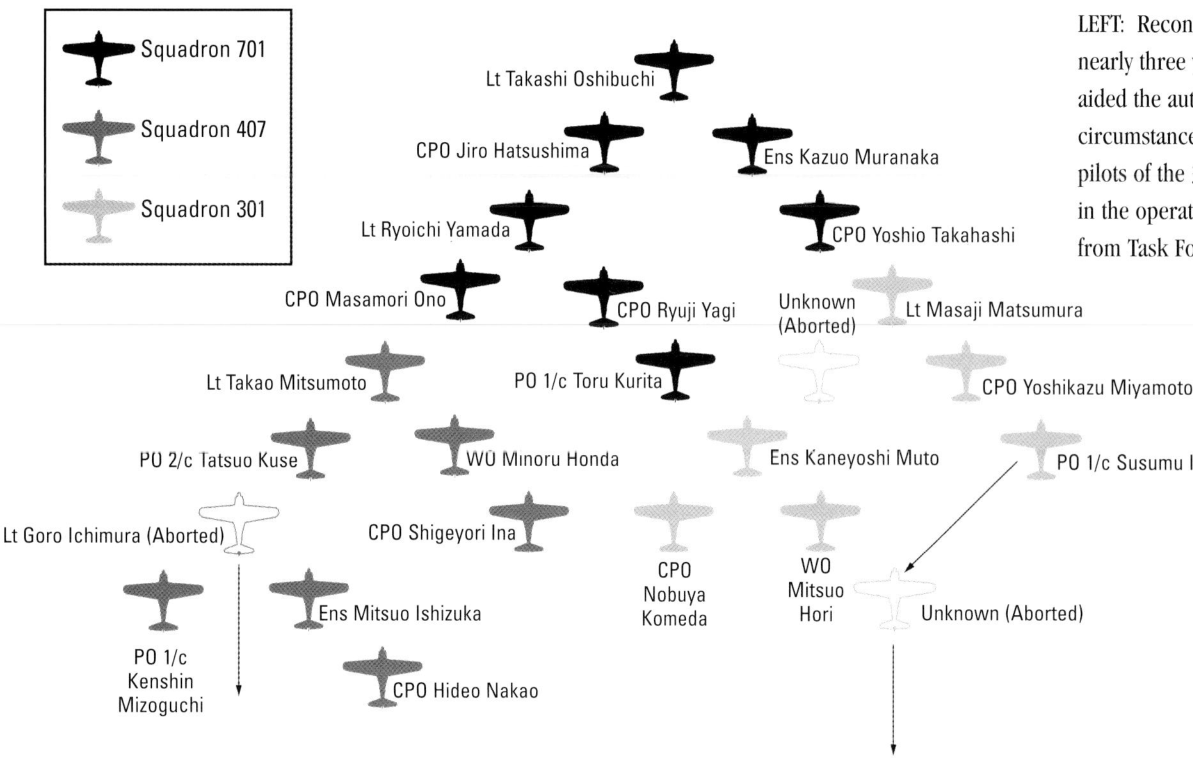

LEFT: Reconstructing this diagram took nearly three years of intense research. It aided the authors in determining the circumstances of how each of the six pilots of the 343 Kokutai met their deaths in the operation against carrier aircraft from Task Force 38 on 24 July 1945.

formation, and saw another enemy group coming into view from behind. Back came the response from Oshibuchi: "*This is Oshibuchi No.1! Go in immediately!*"

Lt Oshibuchi commenced his dive and his men followed. Lt Ryoichi Yamada and his wingman, CPO Masamori Ono, had too much altitude, which resulted in an ineffective attack on their first pass. Yamada recalled: "Flying side by side with Lt Oshibuchi, I fired a first burst at an enemy aircraft during my first pass. I took my position again and started my second firing pass, seeing Lt Oshibuchi taking his position for his second pass. At this time the enemy formations had lost order. The fight became a mêlée as soon as the 301st directed its attack on an enemy group that approached in succession."

Lt Yamada had an ominous feeling that he would not be seeing Oshibuchi again.

CPO Ono did not like the situation as they commenced attack: "I felt that it was a bit early but I followed the lead aircraft as he went in. I did not fire my guns in the first pass. A wingman would not shoot because he was responsible for covering the leader. When I pulled up, I lost sight of Oshibuchi's section. I fired at an enemy aircraft in my second pass. As the enemy strength was overwhelming, our favorable position turned into unfavorable and we fell into a hard fight."

BELOW: CPO Yoshio Takahashi flew as wingman to Ens Kazuo Muranaka and was in Lt Oshibuchi's division on 24 July 1945. He suffered from a violent headache as he climbed to engage aircraft from the Bennington, and fell back, unable to cover his squadron leader. (Y. Takahashi)

Bad luck would plague Lt Oshibuchi as he headed into combat. He was without his 2nd flight when Ens Muranaka fell behind due to a loose engine cover. Muranaka's wingman, CPO Takahashi, was unable to climb higher than 4,500 m and was left far behind. The squadron leader had no time to dwell on his problems and commenced his attack. He completed three passes before his engine took a hit. He took his 343-C-13 into a dive to escape. His faithful wingman, CPO Jiro Hatsushima, flying in 343-C-89, clung to him. With a damaged aircraft, the young lieutenant had no choice but to pull out.

At the same moment, two divisions of Air Group 49 Hellcats from the *San Jacinto* were headed south. They bombed the last big Japanese carriers *Amagi* and *Kastsuragi*, which were heavily camouflaged and moored at Mitsugo Jima, and then strafed some oilers. Many of the pilots still had ammunition left and were eager to give the Japanese some parting shots. There was a tall lighthouse on Mizunoko Shima which stuck out like a sore thumb in the middle of the channel leading to the open sea and which was often strafed by American carrier aircraft. It was pockmarked with .50-caliber shells on its outside wall for it was standard procedure to hit the lighthouse with a burst on the way out.

RIGHT: Cdr Hubert Harden in his Corsair.

"We heard a call from some fighters ahead that they were jumped by enemy planes," reported Lt(jg) Jack A. 'Hoot' Gibson. *"I immediately joined up and started to their aid. As we approached the planes ahead, I noticed two planes coming toward us, and I went with my wingman, Lt(jg) George M. Williams, to investigate."*

Oshibuchi and Hatsushima saw the Hellcats and immediately dived for the water. In his damaged fighter, Oshibuchi was at a great disadvantage and decided to make a run for it. *"I'm fixin' to attack!"* radioed Gibson and he made a sudden 180-degree turn, followed by his wingman.

"Our section split and I followed one, while Lt(jg) Williams closed on the other," Gibson recorded. *"The F6F-5 had no trouble at all in closing with the addition of water injection."*

CPO Hatsushima saw Williams closing in rapidly and yanked his fighter around to get on the American's tail.

"He probably knew the Hellcat could not turn with him," explained Williams. "His full-bore turn to the right just as I had him in range was timely. I can clearly recall .50 caliber hits in the cockpit and wing roots, but I could not stay with him."

In the tight turn, Williams blacked out momentarily. When he came to, he saw his opponent in a tight spiral trailing a great cloud of smoke. Just before the George hit the water, two FG-1Ds of VBF-1 arrived on the scene. Cdr Hubert Harden and his wingman, Lt(jg) Charles R. Moxley followed the smoking George down and fired into the enemy aircraft.

Lt(jg) Gibson narrowed the distance to Lt Oshibuchi's fighter. He could see a white band around the fuselage, forward of the horizontal stabilizer, indicating a flight leader. *"The Jap tried to turn inside of me, but every time he started a turn, I would fire a short burst in front of him which would always bring him back straight and level,"* stated Gibson in his 1945 mission report. Fifty-five years later, he added: "I did a lot of hunting before the war and was involved in a lot of deflective shooting which helped me with aerial gunnery. The estimated range of my opponent was right in the convergence zone of my guns and where they converged was the lethal zone of machine guns on fixed-wing airplanes. When I got in effective range, my first burst knocked his wheels down which decreased his speed tremendously. I overran him and had to pull up in a wing over to drop back on his tail. I had to drop my flaps and my landing gear to be able to stay a distance behind him so I could hit him with my machine guns. He then tried skidding turns without success. By this time I began to get my bursts in the cockpit and he commenced to smoke. He fell over on one wing and went straight into the water, without burning."

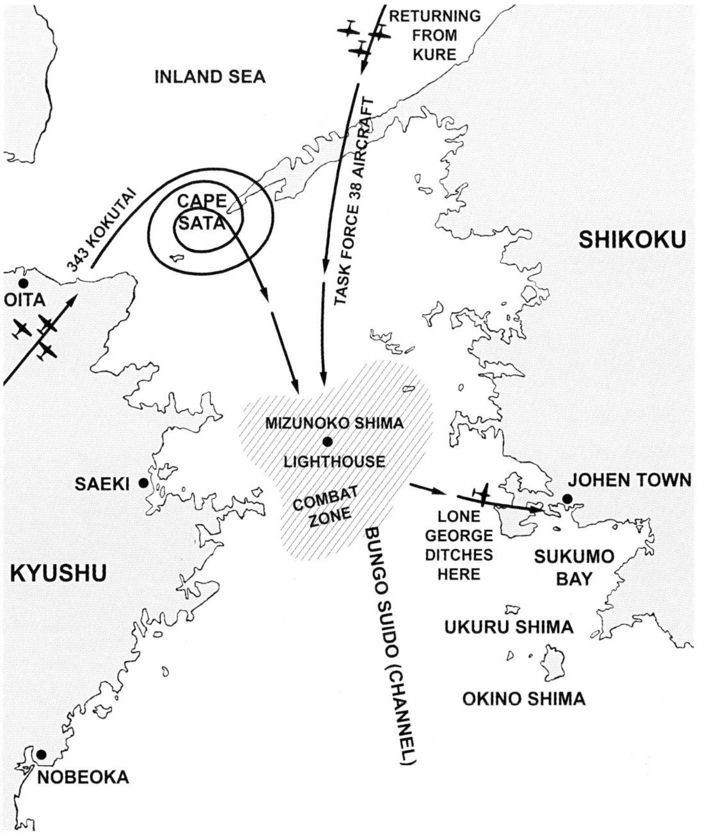

Lt Takashi Oshibuchi, senior squadron commander of the 343 *Kokutai*, crashed to his death in the vicinity of Mizunoko Shima. The circumstances of his last fight are revealed here for the first time.

Ens Walter A. Yancy of VF-49 looked to the rear of his group and saw seven Japanese fighters on the Kyushu side of the Bungo Straits near Mizunoko Shima. At this point in time, the two divisions from S407, led by Lt Takao Mitsumoto, had become seven aircraft. Lt Goro Ichimura, the 2nd division leader, had left the formation due to engine trouble, leaving command to Ens Mitsuo Ishizuka who was leading the other section. Just before contact, Mitsumoto had to flee when his belly tank would not release. Ens Minoru Honda, a hardcore veteran, had to take over a three-aircraft division.

RIGHT: Lt Takashi Oshibuchi was reported missing in action. Fifty-five years later, the details of his last combat surfaced.

FAR RIGHT: CPO Jiro Hatsushima, seen here in the cockpit of a Zero, was Lt Oshibuchi's faithful wingman. He died while trying to protect his leader.

LEFT: On 24 July 1945, Lt(jg) Jack A. Gibson, seen here being decorated for distinguished service, was forced to drop his flaps and landing gear to maneuver behind Lt Oshibuchi, but finally shot him down. (J. Gibson)

RIGHT: Lt(jg) George Williams being debriefed by an intelligence officer. His wingman, Lt(jg) Jack A. Gibson, is visible on the extreme right. (National Archives)

ABOVE:
PO1/c Kenshin Mizoguchi.

"I Tally-ho'ed and Ensign E.A. Case, leading the division, made a 180 degree turn and started climbing," reported Yancy. *"We started closing on a single bogey but he turned and dove away."*

Yancy dropped his belly tank, closed, and fired a long burst at a George below him. His opponent started smoking and executed a very sharp climbing turn to starboard. Yancy made a slight climbing turn to port and found PO1/c Kenshin Mizoguchi in his sights. The George pilot had also failed to release his drop tank and lagged behind his formation. The Hellcat closed and fired a short burst. The stricken *Shiden-Kai* dived down from 8,000 ft. Yancy followed him down and gave him a longer burst. The George entered a steep spiral and slammed into the sea.

"Another enemy plane was splashed close by, and Ensign Yancy believed it was the first plane brought under his fire as no other planes were observed in the neighborhood," stated the VF-49 mission report. He received a probable for this aircraft in addition to one confirmed victory.

Lt Allen W. Lindsay, also of Fighting 49, latched onto a George and gave it a long burst. He saw his shells go into the aircraft, but lost it when it made a tight turn. Since it was not seen to crash, Lindsay was credited only with a damaged.

Yancy encountered Lt(jg) George Williams flying back alone. "I never did see 'Hoot' again until we got back to the ship," recalled Williams. "On the way out, I joined with Walter Yancy, and we flew back to the fleet as a twosome. I remember that well because my homing device was out, and I was really pleased to find him somewhere at the mouth of the Inland Sea."

As Lt Oshibuchi and his men entered combat, Lt Matsumura ordered his men to release their belly tanks, which fell down toward Muto's division.

"Ens Muto must have been surprised to see our tanks fall. I should have begged his pardon!" recalled Matsumura. He then dived at the TBM formation, but lost his division in the process. Luckily, Matsumura did not attract the attention of the escorting Hellcats, but his 2nd flight leader and his wingman did. Chased by F6Fs from VF-49, CPO Yoshikazu Miyamoto and his wingman, PO1/c Susumu Imai, made a run for it, with Miyamoto losing his canopy in the process. Imai then joined up with WO Mitsuo Hori.

RIGHT: Five VF-49 pilots of Division 3. From left to right: Lt(jg) George Williams and George Pompeo. Standing, left to right: Jim Shealy, Lt Allen "Pop" Lindsay and C. Smolsky. (G. Williams)

"My first pass was unsuccessful," explained Matsumura. "In addition, my method of attack resulted in disorder to my formation. Fortunately, the enemy fighters were faithful to their assignment to cover the bombers and did not oppose us; they only interfered with our actions. I plowed through the tracers crossing intensively from their rear gunners. I looked back and after pulling up, saw formations collapsed and fighting had started here and there. Most of the friendly aircraft were above the enemy aircraft. I struck at the Avengers only. The fighter counterattack was not so bad because I thought they had expended so much ammunition while striking Kure."

For Ens Kaneyoshi Muto, this was his first mission with the 343 *Kokutai* and it would be his last. The last man to see Muto was Lt Matsumura, who stated that their flights separated as they prepared to attack. Analysis of the American mission reports, and elimination of the George pilots not returning from the mission, suggests that Ens Muto took his division to attack a pair of Corsairs lagging behind a larger formation up front. The laggards were Lt(jg) Robert M. 'Apple' Applegate and Ens Robert J. Speckmann of VBF-1.

ABOVE: Lt(jg) Robert M. Applegate (R. Applegate)

ABOVE: Ens Robert J. Speckmann (R. Applegate)

ABOVE: Lt Rodney C. Tabler (R. Applegate)

ABOVE: Fighting 88 Hellcat pilots on the Yorktown. From left to right: Lt(jg) Ken Neyer, Lt Malcolm Cagle, Lt(jg) Henry Cleland, and Lt(jg) Raymond Gonzalez. (M. Cagle)

Lt(jg) Applegate was a leader of the second section of Lt J.W. Rowley's division. Their target at Kure was the cruiser *Aoba.* "We didn't have dive brakes on our planes," recalled Applegate. "We were supposed to lower our landing gears as we dive-bombed! I wasn't about to do that. I just wanted to drop my bomb and get out of there fast."

Lt Rowley's thousand pounder hit the target about 15 ft off the starboard bow but Speckmann scored a hit on the bow and Applegate was certain he hit the vessel too. After completing the attack, Applegate strafed several buildings, a sampan, a lighthouse, and several small boats. Though he reveled in the awesome and destructive power of his six .50 caliber machine guns, the 23-year-old Oregonian would later seriously regret this indiscriminate skylarking. At the rendezvous point, Lt(jg) Applegate joined up with his wingman. Lt Rowley had been hit by anti-aircraft fire over Kure; he and his men remained with the TBMs at a lower altitude to cover the bombers. Applegate and Speckmann were flying as the odd section at the tail end of the FG-1D formation.

"We hadn't seen any aerial opposition and I think we let our guard down," explained Applegate. "I put my plane on maximum lean so that I could conserve fuel. We weren't in a hurry to get back to the carrier anyway because they never landed you immediately. They'd make you circle around and around."

As the Air Group One formation approached Mizunoko Shima lighthouse, a voice was heard on Applegate's radio calling out bogeys. It was a transmission from another flight flying behind them. Since he was leading a three-aircraft flight, Applegate sent the one odd man forward, while he and Speckmann made a 180-degree turn to investigate. Applegate saw four aircraft high overhead at 12,000 ft, crossing the group up ahead. Initially, he thought they were Grumman Hellcats flying high cover, but then felt that they did not "look right". He kept his eyes on them. In fact, the unidentified aircraft were Georges from Squadron 301.

BELOW: PO1/c Susumu Imai.

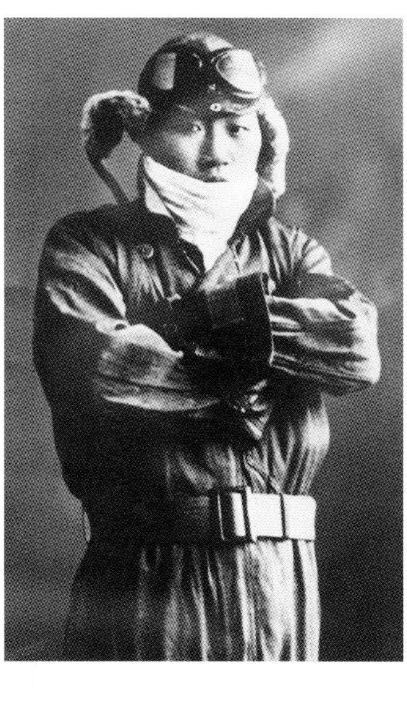

Ens Kaneyoshi Muto's wingman was CPO Nobuya Komeda, a veteran who, in 1944, had served in the 265 *Kokutai* in the Marianas, and later with Squadron 301 at Luzon. He had also taken part in all of the 343 *Kokutai's* major combats. Leading Muto's 2nd flight was veteran ace WO Mitsuo Hori with his wingman, PO1/c Susumu Imai. It was a formidable team and Muto maintained altitude advantage at 12,000 ft.

Up ahead, Lt Rodney C. Tabler and Ens Ed Barzyk, in Cdr Hubert Harden's second FG-1D section, also heard the call of bogeys but stayed with the formation to protect it. However, the pair dived into clouds when two Georges made an overhead run on them. When 'Barz' Barzyk pulled out of the clouds, he watched in horror as his wingman descended under control but flaming at the wing roots. A flash of passing tracers slapped him back to reality and he dived down through the clouds and escaped. Lt 'Tab' Tabler, a 23-year-old Californian from the small farming town of Reedley, never returned.

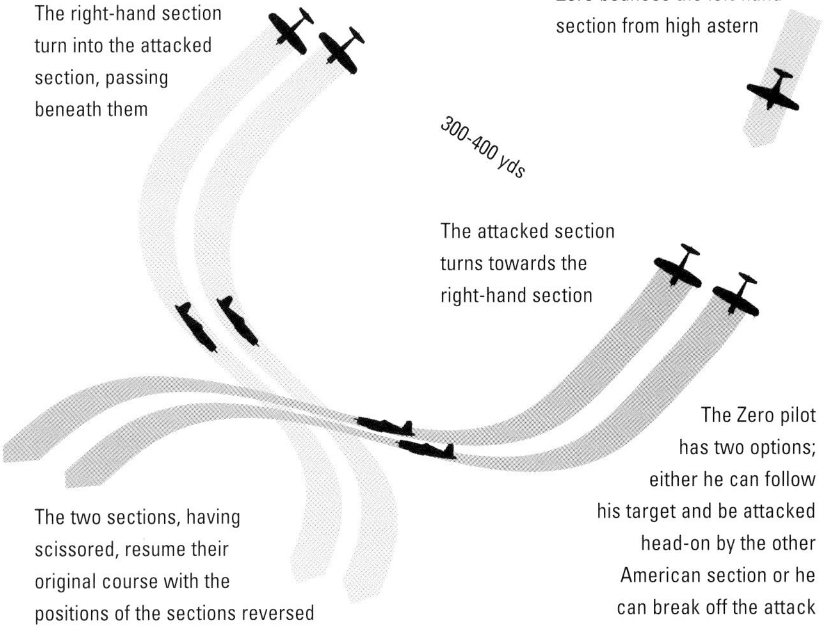

ABOVE: The beam defense maneuver known as the 'Thach Weave.' This was a maneuver invented by USN fighter pilot, LCdr John S. Thach in which a pair of aircraft would weave back and forth, providing the wingman the opportunity to shoot at an enemy fighter on his wingman's tail and vice versa. This tactic proved highly successful at the Battle of Midway.

Suddenly, the high-flying 'Grummans' which Applegate had been eyeing, pushed over to attack and he saw the red 'meatballs' on their wings. According to Applegate's debriefing report: *"The green-painted Georges made excellent coordinated attacks by sections. The first section came straight in, in a steep dive from 10 o'clock hig*h."

He signalled to his wingman to start weaving. The two Corsairs were just completing the first turn of their weave when Muto and Komeda's flight swept in. Applegate pulled up into them and fired a burst and they broke away. Muto and Komeda should have had the upper hand with the all-important altitude advantage, but the flight leader failed to hit Applegate.

The team of Hori and Imai pushed over when they were above the Corsairs and came down fast. In the blink of an eye, Applegate saw his wingman going down in a ball of fire. Ens Robert Speckmann from St. Paul, Minnesota, was just 20-years-old.

The fiery explosion caught the attention of Lt Malcolm Cagle and his wingman, Lt(jg) Kenneth Neyer, of VF-88 from the *Yorktown*. Cagle had seen aircraft milling around in the distance and did not know this was a dogfight until an aircraft exploded. When he and his wingman went to intervene, WO Minoru Honda and his wingman came down on them, forcing the pair into a 'Thach Weave' for mutual protection.

"I approached an F6F from behind above," stated Honda. "When I saw the head of the pilot, I felt a merciful heart but, in reality, I was engaged in a cold-blooded fight and I fired 20 rounds. The right wing of the enemy aircraft flew off and it went down in a spin. From the many experiences of air fighting since 1942, this was the only fight where the figure of an enemy pilot remains in my memory."

Lt(jg) Neyer became one of WO Honda's 40-plus victories, leaving Cagle to fend for himself. The lone Hellcat pilot was unable to release his belly tank and would have to fight with it attached. Almost immediately, Cagle came under a head-on attack by Ens Muto while Komeda went after Applegate. He fired on the incoming George and believed he hit it between the engine and cockpit, but his opponent flashed past as though unscathed.

Bob Applegate made a 180-degree turn and fired a 60-round burst at 60 degrees deflection into the second aircraft of the section that had accounted for Ens Speckmann. His tracers struck the engine and the fuselage around the cockpit. PO1/c Susumu Imai rolled his fighter over on its back and dived, smoking. Applegate attempted to follow but his Corsair started to spin. When he recovered, he found CPO Komeda right behind him.

RIGHT: WO Minoru Honda was a veteran pilot and a very demanding instructor who taught his subordinates how to survive against Hellcats and Corsairs. He stopped counting his aerial victories after his tally reached 40. (Y. Izawa)

Applegate recalled: "A George was on my tail! He was right up my tail, trying to lead his fire into my plane! I could see the pilot behind me. I was wrapping my plane up so tight, his bullets weren't hitting me."

The George was less than an aircraft length behind, with wings vertical, firing 20 mm shells from all four cannon which flashed past the FG-1D's tail; Komeda was simply too close to lead his fire into Applegate.

"I looked back and saw pieces flying off the George," recalled Applegate. "Then I saw a Hellcat which was firing into it. The entire tail section just vanished and he went in."

Lt Malcolm Cagle had just saved Applegate's life.

There were five or six Georges high above, holding their altitude advantage as the duel played out below. Lt(jg) Applegate signaled to Cagle to commence weaving. The one remaining George, most likely flown by Muto, made a high side run from four o'clock. Having lost two men from his division, he must have sought revenge. Instead of pulling back up, the Japanese pilot continued on across the Corsair and the Hellcat and turned back in a beam run from ten o'clock. The weave placed Applegate in a position to make a head-on run against his aggressor. Both aircraft went head-to-head and began firing at about 1,000 ft out.

Applegate saw his tracers hit the nose of the enemy aircraft and he felt his own engine take five or six hits. When the two aircraft closed to around 500 ft, both stopped firing; they had run out of ammunition.

The Japanese pilot continued to close on his opponent and it appeared to Applegate that he was intending to ram. Applegate: "I pushed my plane down and we cleared by about ten feet. I saw the pilot slumped forward."

ABOVE LEFT: CPO Nobuya Komeda.

ABOVE: Ens Kaneyoshi Muto photographed in February 1945. (Y. Shiga)

The George rolled over and down, and Applegate attempted to follow, but his aircraft was in no condition to give chase. No one saw the George hit the water.

With a damaged engine, the FG-1D began losing oil and running rough. Clouds of black smoke trailed behind it. Applegate attempted to radio Cagle that he would have to leave him, but his transmission was not received. He started down through the clouds and a part of the engine cowling came off and shattered the windshield. At about 1,000 ft, the engine gave out and Applegate went over the side. In the process, he struck his leg on the tailplane, but suffered no broken bones. He pulled the D-ring and ripped out two panels from his parachute, but landed safely in the water about four miles northwest of Ukuru Shima.

Lt Malcolm Cagle failed to observe his unidentified comrade exit his aircraft, but he did see the Corsair crash into the water. The Georges up above began peeling off to pounce on him, so he jumped into clouds and headed for home.

After fighting for about ten minutes, most of the *Shiden-Kai* pilots began to disengage and head back to base individually. Lt Ryoichi Yamada, who was short on fuel and ammunition, beat a hasty retreat from the engagement, flying back alone just above the sea.

LEFT: The aftermath of the Kure Harbor attack as seen in a reconnaissance photograph. An unidentified carrier has been left heavily damaged. (USN)

ABOVE: Bob Applegate (right) cuts a "Welcome Back" cake with Ens Arlen C. Davis aboard the Bennington. Both had been declared missing in action on 24 July 1945. (R. Applegate)

CPO Takahashi, who once joined up with Ens Muranaka, saw a Hellcat about 1,000 m below and made a solo pass on it; his flight leader could not dive. The Grumman fled into the clouds. Soon, two F6Fs were nosing up at him. Since Muranaka was out of sight, Takahashi wisely avoided a dogfight and headed for home.

CPO Hideo Nakao fought violently and "...felt good responses" but could not confirm enemy aircraft destroyed. He made an emergency landing at Oita Airfield due to lack of fuel on the way home.

"The groundcrewman had packed my gear wrongly!" recalled Applegate. "I tried to get the rubber raft out, but I had a lot of difficulty. Finally, I got my knife and tried to cut it out. In the process, I stabbed myself in the left arm, right through to the bone. I got into the raft. I figured I'd be seen because we were in the first strike of the day and there would be others. Then the next day, another strike. I put the dye marker in the water and used my signal mirror, but nothing happened."

Lt(jg) Applegate spent the night off Okino Shima. At around 22.00 hrs, some yellow flares were fired a short distance seaward from his position. Thinking that rescue was near, he fired off two red flares; it was a mistake. A 20 mm gun started firing in his direction from the island.

A short time after dawn the next morning, two F6F Hellcats from VF-34 off the *Monterey* were seen searching an area from about 1,000 ft altitude. They were not looking for Applegate, but for another pilot. Applegate caught their attention by firing red flares and this time it worked. The submarine *Dragonet* (SS-293) surfaced and hauled him aboard. The Corsair flyer would end the war on submarine duty until he was returned to Guam. He would later shake hands with Lt Cagle, the Hellcat pilot who saved his life. It was Cagle who confirmed Applegate's victory when he saw PO1/c Imai crash into the sea.

The 343 *Kokutai* had lost six pilots. The deaths of Lt Takashi Oshibuchi and Ens Kaneyoshi Muto were a very severe blow to the unit. In particular, Muto had been considered 'invincible'; his loss sent shock waves through the JNAF.

The 343 *Kokutai* claimed a total of 16 enemy aircraft shot down. American aircraft losses on this mission included three Grumman Hellcats, six Corsairs, 13 SB2C Helldivers, and six TBM Avengers. Most were shot down by anti-aircraft fire over Kure. Research indicates that the carriers *San Jacinto, Yorktown,* and *Bennington* certainly lost at least one F6F Hellcat and three Corsairs to the 343 *Kokutai.*

Chapter 21

"This man won't die whatever he may do!"
The Death of the last Squadron Leader: 1 August 1945

「この人は、どんなことをしても死なない！」

In his post-war memoir Capt Minoru Genda wrote: "*There was a man in the unit who could read the palms of individuals very well when we were at Matsuyama. He saw the lines in Lt Kanno's palm and declared 'This man won't die whatever he may do!' I was very pleased with those words and wished that this might be true. He was brave and intrepid. He had an eye for tactics and was unrivaled in fighting skills.*"

August 1st marked the beginning of the final two weeks of the Pacific War and the last week of July had proved devastating for the Japanese. Task Force 38 inflicted severe damage to warships and installations at Kure between 24 and 28 July. It returned repeatedly to mop up what was left. Two out of three carriers were reportedly badly damaged; the *Amagi* overturned and the *Katsuragi* sustained medium damage. The *Ryuho,* however, damaged earlier at Kure on 19 March, was well concealed and evaded attack. All the cruisers and destroyers in the harbor were reportedly sunk. Japanese fighters were conspicuously absent.

Adm 'Bull' Halsey was very pleased with the results and commented with obvious glee: "The Commander-in-Chief of the Combined Japanese Fleet could reach his cabin in his flagship, the light cruiser *Oyodo,* only in a diving suit!"

The 343 *Kokutai* mourned the loss of Lt Takashi Oshibuchi, but there was no respite from the war. If anything, morale was still high and all the pilots could think about was getting back at the Americans. They would get their chance on 1 August 1945.

B-24 Liberators were detected approaching Kyushu and the scramble order was given. Lt Naoshi Kanno was about to taxi forward when his engine began to malfunction. His take-off was delayed for several minutes. He abandoned the crippled fighter and jumped into 343-A-28, one of two aircraft assigned to Maintenance Petty Officer Wakashige Kondo. Finally, approximately 20 *Shiden-Kais* took off from Omura at 08.12 hrs, some armed with rockets and cluster bombs.

Lt Kanno's S301 and the S407 cruised at 6,000 m while Lt Keishichiro Hattori was down below with S701 at 5,000 m. Even at this late stage of the war, the 343 *Kokutai* was still a formidable unit with capable fighter leaders taking over from those who had died; Lt Hattori was transferred from S407 to S701 as a division leader following the death of Lt Oshibuchi.

Lt Kenzo Nakanishi from Kanno's division had returned to base due to engine trouble. The 23-year-old squadron leader was now missing a wingman. When three members of WO Mitsuo Hori's division aborted, Hori simply moved up to act as Kanno's wingman. The lead division now consisted of Kanno and Hori, with CPO Fukukichi Masago and PO2/c Tsuneharu Tamura. Hori was a seasoned veteran with over ten victories and 1,700 flight hours. It now became his duty to insure Kanno's life.

The Japanese knew that the Americans were using Yaku Shima as a staging area for their bombers attacking Kyushu. Aircraft from Okinawa would circle the island, assemble, then head for the mainland which was less than 74 km away.

Lt Naoshi Kanno (Maru)

Approaching Yaku Shima were 19 B-24s of the 494th Bomb Group out of Yontan, Okinawa. They had failed to link up with their fighter escort for a bombing run against targets on Amami Oshima. They were now proceeding to their rendezvous point at Yaku Shima, cruising between 8,000 and 11,300 ft over the thunderheads and the rain squalls below.

Lt Kanno was well acquainted with the rugged B-24s with their ten .50-caliber machine guns; he had fought them over Yap Island, east of the Philippines as a division leader attached to the 201 *Kokutai* and he had perfected a method of attacking them head-on, which was not for the faint-hearted. During a frontal approach, the pilot would execute a quick roll and dive vertically at the target's nose. Many of his men hated this maneuver and had no choice but to follow their commander. Kanno always flew at the front.

The *Shiden-Kais* arrived near Yaku Shima at 6,000 m altitude and spotted two Liberators circling the island below at 5,000 m. Lt Kanno alerted his men by radio; they were able to receive his transmissions, but they could not transmit unless in an emergency – a policy intended to eliminate needless radio chatter (Japanese pilots communicated mostly by hand signals). When Lt Kanno dropped his belly tank, others fell in unison.

"Receiving his warning on the wireless, I watched the two B-24s and found another B-24 appearing to join up with them," Hori recalled. "Our lead aircraft headed straight for the enemy group. The B-24s were now in two formations of three and two aircraft, starting southward. All the turret guns were directed toward us. Our formation changed course about 50 degrees south and made its way in a descent to take a position forward of the enemy. Reaching left forward 1,000 m distant from the B-24s, Lt Kanno glanced ahead and lowered the nose of his plane, going in from ahead. I followed him immediately. My target was the second aircraft at the right end of the enemy formation. The five B-24s opened fire all at once. Red flames grazed the windshield of my aircraft."

Gunners of the 866th Bomb Squadron were furiously firing bursts at the two Georges as they came streaking in. According to their mission report: "*Several passes out of sun and a few phosphorus bombs were dropped. One orange ball followed by puffs of smoke observed – believed to be a rocket... Two instances of B-24 gunners observing bursts of their tracer fire bounce off engines and bellies of enemy aircraft apparently causing no damage.*"

Lt Kanno was in the midst of his firing pass when the center section of his left wing suddenly exploded, tearing a one meter hole by his outside gun. But the Americans had not hit him, rather a defective 20 mm round had detonated in the breech of his cannon. This was not the first time such an incident occurred using the imported Oerlikon ammunition. Undoubtedly, he was claimed as destroyed by the gunners when he momentarily lost control of his fighter.

"After making one pass, I dived out of the range behind the enemy aircraft," Hori remembered. "Suddenly, I heard a radio call in a dive: '*This is Kanno, number one! My shell exploded in the barrel!*' Pulling up, I looked ahead but my leader's aircraft was nowhere to be seen. I leveled my aircraft and looked around. He was not there. I banked and looked down, then I saw his aircraft flying level and far below me. I dived and followed him."

Flying alongside Lt Kanno, WO Hori could see the gaping hole in his leader's wing and decided to escort him. However, Kanno would have none of it and signaled to continue the attack on the bombers. When Hori refused, Kanno gestured at him and pointed to the bombers but his wingman still would not leave. Then the squadron leader gave him an icy stare. Hori reluctantly peeled off toward the action when he saw B-24 formations escaping south in tight formation 7,000 to 8,000 m ahead at an altitude of over 1,000 m.

WO Hori chased after the B-24s; it took more than 15 minutes for him to take up position for another run. As he did so, he noticed that his men were attacking ineffectively, opening fire from too great a distance and breaking off too early. Then Hori began his attack. "I made a vertical diving attack," he recalled. "Tracers streamed right and left as well as ahead and behind the canopy of my aircraft. Though I was frustrated with the other men's indecisive attack, my aiming was not precise and my attack was not effective either, perhaps because I was caught by anxieties."

Hori went back to the area where he had last seen Kanno, but could not locate him.

"My heart thumped. I looked around carefully. I looked for him everywhere on the surface of the sea. I approached Yaku Island as I thought he might have crash-landed on the coast. I flew along the shores but

LEFT: Mitsuo Hori, Kanno's last wingman, is shown standing to far right. This photograph was taken at Rabaul at the end of 1942, when Hori served with the 582 Kokutai. (Masao Yamashita)

sighted no one. I looked up and around, calling '*Kanno, number one! Kanno, number one!*' There was no response. While I was searching, six *Shiden-Kais* gathered nearby. I thought that Kanno might have turned back toward Kagoshima before our assembly. I made up a formation with them and took a course to Omura. The remaining fuel did not allow us to make a further search."

While the rest of the pilots continued their attacks on the bombers over Take Island, they were jumped by four P-51s from the 348th Fighter Group. LCol William D. Dunham and Maj Edward S. Popek were leading two young lieutenants on an orientation flight and sweep to Kyushu when they saw the Japanese fighters attacking the bomber formation. They were unfamiliar with the Georges and misidentified them as Franks.

Dunham, a veteran Thunderbolt pilot with 15 victories, ordered his division to dive from 16,800 ft to the enemy below. The Americans reported 16 Japanese fighters from 10,000 to 17,000 ft in no particular formation. Apparently due to the absence of Lt Kanno, the Japanese were attacking the bombers in a disorganized fashion. They were observed making repeated individual overhead passes. In their initial pass, Dunham and 1/Lt Thomas Sheets each claimed a victory.

BELOW: LCol William D. Dunham (R.F. Toliver)

Maj Popek was about four miles north of Take Island when he selected his victim and came in from the rear. "I do not believe that the pilot knew that I was coming up from behind," Popek recalled. "He took no evasive action before or after he was hit."

A burst from Popek's six machine guns hit the George and knocked its right landing gear down. It spiraled to the left leaving a trail of smoke. His second victory came when he teamed with Tom Sheets to chase the enemy into clouds. He came in from behind and shot into the cockpit of a George. His gun camera film confirmed his two victories.

The action moved to the north where the Georges dived into clouds to escape and the Mustangs were unable to pursue them further.

The four P-51s returned to their base at Ie Shima without a scratch. For the Mustang jocks, these would be their last victories. Maj Ed Popek received credit for two victories that elevated his final score to seven. LCol William Dunham and 1/Lt Thomas Sheets each received one. Dunham ended his fighting career with 16 victories, while Sheets tallied three.

ABOVE: Maj Edward Popek (R.F. Toliver)

The mission report of the 348th Fighter Group stated: "*Favored evasive action and attempted to regroup for escape to the north. However, at least three separate attacks were made on our planes by individual Franks, although none were pressed closely. Evasive action was routine and almost exclusively to the left, nothing unusual being noted in tactics employed.*"

The Mustang pilots gave the following evaluation of their counterparts: "*Enemy pilots were not very aggressive and failed to regroup fast enough to make use of their numerical superiority against our attacks. Our pilots believe the method of attacking the bombers which Franks were using makes them very vulnerable to friendly fighter attacks. Routine evasive action was unspectacular. The Frank appears to be an excellent fighter and able to absorb considerable punishment although pilot protection is weak.*"

Gunners of the 866th Bomb Squadron claimed two enemy aircraft destroyed and three damaged. Their report stated: "*The upper turret gunner Sgt Merle Combs fired 30 bursts into the Zeke which climbed, went into a stall, fell away smoking in a spin and was observed to hit the water. Confirmed by pilot of A/P 747, Captain Junian G. Van Curren and S/Sgt Antone Bruns, Seventh Air Force aerial photographer.*"

The 343 *Kokutai* lost three pilots in combat: Lt Naoshi Kanno and PO1/c Sakutaro Moriyama of Squadron 301, and CPO Sukeo Yoshioka from Squadron 701.

The *Shiden-Kai* pilots claimed one P-47 destroyed and a B-24 as a probable. However, there were no P-47s involved in this action and all the Liberators returned to base. Added Edward Popek: "There were no P-47s in the area, to my knowledge, and our flight of four P-51s were the only US fighters there... I do not think that any B-24s were lost other than one that was badly damaged with a couple of engines shot out. It barely made it back to base by throwing everything that was loose out of the aircraft in order to maintain flight."

RIGHT: LCol Bill Dunham poses for a photograph on the wing of his P-51, which was named after his wife. (Bonnie Dunham)

B-24 '761 was hit hard. The pilot, 2/Lt Elmer L. Gladson and his upper turret gunner, Sgt Joseph L. Bushey, were killed when the Georges made two high passes out of the sun. Six other members of the crew were wounded, three seriously. 2/Lt Raymond L. Sturm took the controls and brought them home.

WO Mitsuo Hori landed back at base and was anxious to know if Kanno had returned. When his aircraft drew to a stop, he asked a member of the groundcrew if the lieutenant had returned. "*No, sir!*" came the answer. He rushed over to see LCdr Yoshio Shiga and reported that the squadron leader had a cannon barrel explosion. Incredulous, Shiga quickly telephoned Capt Genda at the mountain command hut and also instructed his staff to inquire with bases in the south.

"A few aircraft not belonging to our formation appeared in the sky over the base," recounted Hori. "Cdr Shiga, myself, and others gazed at them with our binoculars, praying that one of them might be Kanno. Our nerves were frayed every time the phone rang. We waited and waited, but his aircraft never showed up."

For many years, some believed that Lt Kanno was shot down by P-51s, but now this theory can be laid to rest. He was last seen away from the combat zone at low altitude. None of the Mustang pilots claimed a victory over a crippled Japanese fighter at low altitude. His loss was an operational accident during combat.

Before the Pacific War, the Japanese Navy had imported batches of 20 mm cannon ammunition from the Oerlikon Company in Switzerland. However, some rounds were found to be defective and exploded in the breech. Most of the defective ammunition was isolated but unfortunately a number of crates of these dangerous rounds were accidentally shipped to combat units in Kyushu by the *Gunjubu* (Munitions Department) and sporadic cases of breech explosions were reported. A team was dispatched to dispose of the inferior and dangerous batches, but the war ended before the team reached Omura.

Lt Naoshi Kanno, the last fighting squadron leader, a 25-victory ace with a bulldog spirit, was now gone. His death shocked the air group to its core. Capt Genda remembered: "*I lost three other squadron leaders one after another, and then the last one, Lt Kanno. I felt sorrow as if I had lost my own brother...*"

Chapter 22

The Final Days
Early August 1945

終末の日々

By early August 1945, American fighters and bombers roamed through Japanese airspace at will. The Fifth Air Force had arrived on Okinawa and Ie Shima at the end of June. It joined with the Seventh and Twentieth Air Forces to neutralize targets in Japan in preparation for the invasion of Kyushu. American pilots complained about the lack of opposition but they did not realize that the enemy was saving its aircraft for the one last-ditch *Kamikaze* attack against the impending Allied invasion.

The situation looked bleak for the 343 *Kokutai*; three of its original squadron leaders were now dead and it had lost 80 pilots. Yet Lt Kanno's untimely death only strengthened the resolve to fight on.

On 6 August, three B-29s embarked on a historic bombing mission to Japan. A single bomb fell from one Superfortress onto Hiroshima. In a flash, the city evaporated. Rumors soon reached Omura Airfield; high-ranking officers suspected that it was an atomic bomb.

On 8 August, the Twentieth Air Force dispatched an armada of B-29s on a daylight firebombing mission to Yawata, Japan's great steel manufacturing center. The 58th, 73rd, and 313th Bomb Wings cruised toward the Japanese mainland with 245 Superfortresses, escorted by more than 100 P-47s from the 318th, 413th, and 507th Fighter Groups from Ie Shima. The bombers flew at 20,000 ft while their escorts provided top cover at 1,000-3,000 ft higher. The Thunderbolt pilots were extremely confident in their aircraft on such long-range missions to Japan. With four drop tanks, the P-47 could stay airborne for up to nine hours.

At 07.30 hrs that morning, 23 *Shiden-Kais* – an apparently irregular formation because of the shortage of aircraft – under the leadership of Lt Takao Mitsumoto began rolling down the airfield. The first off was S407, followed by the 2nd *Chutai* of S301. A little later, the 3rd *Chutai* followed. CPO Shigeo Suzaki of S301 was the 24th and last man to sortie. S407's division leaders were Lt Sadao Oshio and Ens Mitsuo Ishizuka. Lt Keishichiro Hattori led the fighters from S701 and flew left behind S407. Lt Mitsumoto took his men eastward from Omura Airfield over the Ariake Sea where he planned to block the path of the enemy heading north.

En route to Yawata at around 09.30 hrs, 1/Lt Edward Freedman of the 333rd Fighter Squadron spotted an aircraft which he identified as a Frank. It was, in fact, a *Shiden-Kai* flown by CPO Yoshinori Matsumoto of S407. Freedman moved away from the bombers and his flight gave chase. Matsumoto was in a dire situation. As he dropped his belly tank, his engine stopped, having absorbed too much air in the fuel line, and he had to dive to about 1,000 m. As the Japanese fighter nosed down, the flight broke off its pursuit to return to the bombers.

RIGHT: Ens Mitsuo Ishizuka (Y. Matsumoto)

FAR RIGHT: From June 1945, CPO Yoshinori Matsumoto flew as Ens Mitsuo Ishizuka's wingman. On 8 August, Matsumoto was forced to make an emergency landing. As he dropped his belly tank, his engine absorbed too much air and stopped and he was forced to abort his attack on an American bomber formation. (Y. Matsumoto)

The Americans spotted two more enemy fighters up ahead and level. Freedman got behind the 'Tail-

End Charlie' and brought it down with two bursts. The lead Japanese split-S'ed with 2/Lt Joseph Irwin scoring damaging hits on the wing and fuselage. Just west of Yawata, they encountered another lone Japanese fighter. Freedman and FO Robert Coolbaugh damaged it in a tight spiral, but the enemy pilot escaped. These aircraft were also *Shiden-Kais* from the 343 *Kokutai*.

About five minutes after their first contact with the Georges, 11 more were spotted 2,500 ft below. The mission report of the 318th Fighter Group stated: "*The Japs were evidently climbing to attack the B-29s. The flight made a mutual support pass and succeeded in driving the Franks off to the right, away from the bombers. Ten minutes later, after the flight had resumed its escort, four Franks made two quick feints and broke off. The pilots believe that these Franks were attempting to draw the P-47s away from the B-29s. The first two Japs split-S'ed away and Lt Rasmussen got a short 90 degrees deflection burst at the third Frank, scoring a few hits behind the cockpit.*"

The B-29s, now numbering 221, arrived over Yawata with their Thunderbolt escorts and Lt Mitsumoto's Georges in pursuit. Around 10.00 hrs, the pilots of the 19th Fighter Squadron went after the Georges. The only pilot to score a victory was 2/Lt William J. Cuneo when he attacked a division of Georges from astern. They scattered like pigeons, but he stayed with one and smoked him. The flaming aircraft was last seen diving to earth.

"*After this encounter,*" stated the mission report, "*Lt Cuneo observed a George and attacked him from astern, again observing hits on the Jap's fuselage. The Jap used rolls, S'ing, half rolls, tight turns, and other acrobatic maneuvers in an effort to shake off pursuit. Lt Cuneo broke off his attack momentarily, and as he did so he began to have engine trouble; he therefore immediately headed for the rally point. While headed south over Shimabara Wan, Lt Cuneo spotted another George going in the same direction. He pulled behind the George... The Jap began racking his plane around in tight turns, and as he did so, Lt Cuneo managed to get a 90 degree deflection shot on him but did not observe any hits. Lt Cuneo began to lose power as the Jap made another tight turn so he put his plane into a steep dive and headed for the rally point at Yaku. (Lt Cuneo was forced to land with a dead engine.)*"

1/Lt Elmo Foster and 2/Lt Walter P. Germain each claimed a damaged in their pursuit attacks. They had spotted a group of 13 enemy aircraft flying parallel with a B-29 formation and each made a stern attack. Foster hit one, but his opponent escaped. He lost his wingman in the pursuit and being alone in enemy territory, he called it a day and headed back to Yaku Shima, the rally point. Germain also damaged an enemy aircraft despite suffering a malfunctioning gunsight.

Ens Mitsuo Ishizuka, the veteran pilot of Squadron 407, called "Dad" by his men, launched his rockets at a B-29 in a large formation over Mikawa Town in Kumamoto Prefecture. He was apparently hit by an escort fighter and his aircraft went down smoking, as seen by witnesses on the ground. He plunged into a mountain. Haruki Iijima and his comrades from the civilian Youth Corps searched but failed to find him. The following day, Ishizuka was found, still strapped into his seat. Pieces of his aircraft were scattered over 1,000 m. The engine was buried 30 m away from his body. The pilot was identified and cremated and his ashes deposited in a temple.

The 318th Fighter Group lost four Thunderbolts on this mission, three due to mechanical problems with their pilots posted missing. F/O Jesse B. Hill, whose aircraft was hit by a phosphorus bomb, bailed out and was rescued by a PBM Mariner.

The 507th Fighter Group lost F/O Paul Lynn when his aircraft lost power and spun into the sea in the Tokara Island group whilst flying to Kyushu.

The 318th Fighter Group pilots described the *Shiden-Kais* as "*...dark green in color with a white diagonal stripe just aft of the canopy. They appeared new and could not be identified.*" As for the Japanese pilots manning these fighters, the report stated: "*Jap pilots seemed experienced and used excellent individual tactics. Their formation tactics were poor, however, and they scattered when tapped. They were not aggressive and avoided our fighters when possible.*"

The 1st and 34th Fighter Squadrons were only three minutes over Yawata when they encountered the 343 *Kokutai*. In almost every case, the Thunderbolt pilots simply chased down their opponents and shot them down from astern. In all, the group claimed five enemy aircraft destroyed and two probables.

RIGHT: CPO Hideo Nakao photographed in a jovial mood in March 1945, though he endured a narrow escape from P-51 Mustangs five months later. Of interest is the white-painted box underneath the tail number. The names of the two aircraft maintenance men assigned to this aircraft are Kitagawa (left) and Nakao (right). Although the pilots did not have their own assigned aircraft, the chief mechanics did. (K. Osuo)

The American escort fighters also included P-51s. In a rare encounter with the Mustangs, CPO Hideo Nakao remembers the one-sided dogfights that followed: "Information was severely delayed on the day. When we took off from Omura, B-29s were already in sight and they had P-51 fighter escorts. The enemy bombers and fighters had altitude advantage and we could not get near them. They fell over us when we were at an altitude of 4,000 meters. After attacking us, the P-51s climbed limitlessly with tremendous climbing power. Our *Shiden-Kais* had no such climbing power. When we leveled off, they dived at us again. I slid and avoided the enemy bursts. I noticed an enemy fighter climbing from below. He was not fast and I shot at him in an instant. He had not noticed me, but I saw another enemy aircraft turning around on my tail. Momentarily, I evaded him. B-29s were separated in the directions to Oita and West Honshu. I chased a group heading toward West Honshu but I was not in time to climb to the B-29s' level, giving up the chase. By orders from base, I landed on Hofu Army Airfield. My classmate CPO Shigeyori Ina was with me. One of the wheels was punctured while I was taxiing to a revetment."

The 343 *Kokutai* suffered appalling losses. In all, nine pilots were killed in action and one wounded. Squadron 407 casualties were Ens Mitsuo Ishizuka, CPOs Sakae Masumoto and Masayuki Kutami, and PO2/cs Nobuyuki Nishimoto and Tatsuo Kuse. Kutami fought four Thunderbolts and was hit in the engine. He bailed out, but was shot dead whilst using his parachute.

Witnesses on the ground saw a friendly fighter being chased by a "Grumman" (actually a P-47 Thunderbolt) like a pigeon pursued by a hawk at low altitude. PO2/c Tatsuo Kuse, aged 18, trailed black smoke and crashed into a coal mine at Namazuda, Iizuka City, in Fukuoka Prefecture. The locals were so impressed with his valiant fight, they initially raised a wooden monument in his honor six years after the war (later replaced with a stone monument). Kuse had been the 3rd man to the late Lt Keijiro Hayashi since their days at Balikpapan in Borneo, and he also flew as wingman to Lt Takao Mitsumoto.

Squadron 701 lost three pilots: CPOs Yokio Taura and Kaemon Yokobori, and PO2/c Sueji Aburada. Yokobori fell to 12 Mustangs near Tsuiki Airfield when he attempted to intercept incoming B-25s. Aburada reportedly shot down a Liberator over the Ariake Sea and engaged three P-47s alone over Tara on his way back to Omura Airfield. He was hit and parachuted, but died shortly after he was rescued. Lt Keishichiro Hattori had his arm shot off and his aircraft was engulfed in flames. He bailed out and was rescued in the mountains with burns to his entire body. He was hospitalized.

BELOW: "Nip Clipper" had an admirable record of having flown the second-highest number of missions in the 9th Bomb Group. On its 35th mission, it was rammed by CPO Shigeo Suzaki. (Maurice Ashland)

CPO Shigeo Suzaki from S301 rammed the B-29 *Nip Clipper* (42-63512) from the 482nd Bomb Squadron, 9th Bomb Group over Dazaifu and crashed fatally while the stricken bomber plunged into the sea east of Iki Shima. The pilot, 2/Lt George F. Keller, was killed when his parachute failed to open. However, ten members of his crew were picked up by a Japanese

KELLER CREW: Rear: Levine, Ross, Keller, Correll, & Holden. Front: Zapf, Blake, Nikitas, Conley, Fowler, & Harman. Keller killed in bail out 8/8/45. All others POW and repatriated. A/C42-6352

ABOVE: The crew of "Nip Clipper". The only casualty as a result of CPO Shigeo Suzaki's ram attack was the pilot, 2/Lt George F. Keller (rear, center). The rest bailed out, were captured, and released at the end of the war. (Maurice Ashland)

fishing boat in the Sea of Japan off Shimane Prefecture on 14 August. They were sent on to Tokyo and returned home after the war.

The 343 *Kokutai* failed to thwart the fire raid on Yawata. Numerous fires were started and over 20 percent of the city was destroyed. One B-29 was rammed and three others were lost due to mechanical problems. Anti-aircraft fire damaged a further 20.

"*It was a failure!*" summarized Yoshinori Matsumoto of the intercept mission against the B-29s. His fear that enemy escort fighters with altitude advantage would thwart their plans was proven right.

On the following day, 9 August 1945, LCdr Yoshio Shiga and a few of his squadron and division leaders decided to engage in mountain hiking behind Omura. The unit was short on aircraft and fuel, and the outdoor excursion would help alleviate the stress his men felt. A single B-29 lumbered high in the distance.

Lt Ryoichi Yamada knew about the atomic bombing of Hiroshima three days earlier and felt uneasy about this lone intruder. He suspected that it was carrying another atomic bomb and he told his men to take cover if it dropped anything. Suddenly, one man shouted: "*Look, there's a parachute!*" A moment later there was a great flash and LCdr Shiga felt warmth on his cheek. In succession, a giant mushroom cloud rose beyond the mountains across Omura Bay in the direction of Nagasaki.

Nagasaki was leveled. More than 35,000 citizens were killed and over 60,000 injured. Wireless news reports sent shock waves across Japan. And still the war raged on.

LEFT: The mushroom cloud of the atomic bomb dropped on Nagasaki as seen by the men of the 343 Kokutai during their mountain hike around Omura on 9 August 1945.

Chapter 23

Surrender
15 August 1945

By mid-August 1945, most of 343 *Kokutai's* aircraft were grounded due to lack of fuel and spare parts but as long as the unit had one airworthy fighter left, it would continue to fight. On 12 August, WO Isamu Miyazaki was ordered to take his division and conduct a patrol to Yaku Island. The division consisted of Miyazaki, his wingman PO1/c Ken Okimoto, with section leader PO1/c Shoichi Handa, and his wingman PO2/c Hideo Izawa. The patrol was uneventful until they neared Karatsu, over Saga Prefecture.

Trigger-happy Japanese Army gunners opened up on the four unidentified aircraft. Amazingly, the anti-aircraft fire was accurate and three *Shiden-Kais* were hit. PO1/c Okimoto crash-landed in a rice paddy while Izawa ditched in the sea, sustaining serious injury. One of Miyazaki's landing gears came down and he was forced to make an emergency landing on Gannosu Army Airfield in Fukuoka Prefecture. It was not a perfect landing, and the pilot crashed his aircraft and was knocked unconscious. When he came to, the irate warrant officer made a protest to Army headquarters; it was an inglorious end for Miyazaki, who achieved over 13 victories, but who had almost been killed on his last mission by his own countrymen.

However, with the Americans now dominating the skies over Japan, it is easy to understand why such an incident took place. The *Shiden-Kai* looked very much like a Grumman Hellcat at high altitude. Squadron Leader Kanno was well aware of this and once flew a *Shiden-Kai*, a *Shiden*, and a Zero 52 over to Chiran Army Airfield in June, to instruct Army antiaircraft gunners in the different appearances of the aircraft.

The last dogfight involving the 343 *Kokutai* occurred on this same day and its last casualty was Lt Sadao Oshio of Squadron 407. Oshio had been attached to a *Raiden* interceptor squadron with the 302 *Kokutai* at Atsugi. As a newcomer to the 343rd, he probably entered combat around May.

BELOW: The 343 Kokutai's last casualty was Lt Sadao Oshio of Squadron 407.

Lt Oshio took off from Omura Airfield to test fly a *Shiden-Kai.* Meanwhile, 1/Lt Kermit A. Allen of the 460th Fighter Squadron, was flying cover in a P-51D at 11,000 ft. At 08.30 hrs he spotted a lone George and according to the squadron report: "*He opened the attack, closing rapidly to 200 yards dead astern and a burst was fired, scoring hits on the belly tank. The explosion tore the tank from the aircraft. The George split-essed and made a tight turn, using flaps. Staying with the Jap and still at 200 yards, Lt. Allen fired again on an 80-30-degree deflection shot scoring hits on the cockpit. The George once more split-essed, recovering momentarily on the deck while trailing heavy smoke. Then he suddenly snapped and split-S'ed into the water. He did not explode.*"

The poignant events of the last two days of the war, as seen through the eyes of the 343rd's personnel, are recorded by aviation writer Naoki Kodachi in his book *Reisen Saigo No Shogen.* On 14 August, a transport aircraft landed at Omura Airfield. It carried a very important dignitary on a special mission to visit Capt Genda. Lt Ryoichi Yamada of Squadron 701 met the official on the field and was irritated by the man's rude attitude. Yamada called LCdr Shiga on the telephone and told him to expect a visitor. Shiga, in turn, telephoned HQ and alerted the staff. It was their duty to screen all visitors who wanted to see the commander.

"*One rude man coming to see the captain!*" warned Shiga. "*Don't introduce him until you have carefully checked who he is!*"

The 'mystery man' was Toyosaku Kawanami, President of the Kawanami Shipbuilding Company and an advisor to the Naval General Staff. He had come to deliver the shocking news to Capt Genda that the Emperor would make a surrender announcement by radio the following day.

By the time Shiga arrived at HQ, Capt Genda was saying farewell to his visitor. After the car had departed, the CO informed his subordinate of the next day's surrender announcement. "I remember that I said to the captain, *'Really? That sounds good!'*" recalled Shiga. "But the captain did not respond."

At noon on 15 August, about 100 men of the 343 *Kokutai* stood at attention on the airfield apron in front of the loudspeaker. It had been a very difficult task to assemble every member of the air group in one location at any given time. At Omura, there were staff, pilots, and aircraft servicing crews. Medical and accounting departments were at Matsuyama. In addition, groundcrews were scattered at the revetments around the airfield.

Popular NHK Radio announcer, Shinken Wada, came on the air. "*This will be a broadcast of the gravest importance. All listeners please rise. His Majesty the Emperor will now read the Imperial rescript to the people of Japan. We respectfully transmit his voice.*"

The *Kimigayo,* Japan's national anthem, was played. After a slight pause, the Emperor began speaking. In this unprecedented event, many Japanese heard their leader's voice for the first time.

The assembled JNAF personnel at Omura strained to hear Hirohito, but his words were difficult to comprehend because of the poor transmission. Additionally, the Emperor used the Imperial language, which most people did not understand. However, most understood that Japan was going to formally surrender. Even though Genda was warned in advance about the surrender announcement, it still came as a shock to him. "I was quite surprised when the Emperor's address was delivered," remembered Genda. "I could not believe it… It was my thought we could continue war on the soil of Japan for as long as the Japanese nation existed, even if it took ten, 20, 50 or a 100 years!"

Japan had never lost a war in over 2,600 years and defeat was unthinkable. The 343 *Kokutai* had fought long and hard for its nation and too many comrades had died. It seems that some men were overcome with surprise while others were stoic and kept their emotions in check.

Aviation writer Naoki Kodachi has recorded the following reminiscences from some 343 *Kokutai* pilots: "I never thought we would surrender," recalled Lt Ryoichi Yamada from Squadron 701. "Should we stop fighting? No! We fight!"

WO Yoshio Nakamura, also of S701 remembered: "I could not understand the meaning of the broadcast. Though we heard from the command that Japan was defeated, I did not think that the war would end."

"Japan had lost the war, but I thought we, the pilots, had not lost anything. We did as much as we could. Most of the pilots simply showed an expression of relief on their faces." recalled WO Masao Sasakibara of S701.

"I first knew of the end of the war when the Emperor made the broadcast. We were ordered to go home," remembered WO Isamu Miyazaki of Squadron 301.

CPO Tomokazu Kasai of Squadron 301: "I was in the air on a test flight when the Emperor broadcast. I could not believe that Japan surrendered. I thought it was foolish!"

The reality was stunning: the war was over and they had somehow survived. Most were relieved. Now they started thinking about their families and the great unknown that was ahead.

In order to confirm the surrender announcement, Capt Genda flew to the headquarters of the 5th Air Fleet located at Oita Naval Base, escorted by Lt Masaji Matsumura, WO Minoru Honda, and CPO Miyuki Shimotsuru. While the captain was away, LCdr Shiga ordered all airworthy *Shiden-Kais* into the air. He led 18 fighters around the field to maintain morale and to bid farewell to the *Shiden-Kai.* Some mistakenly believed that it was a demonstration in support of keeping up the fight.

On the same day, a warrant officer from the 302 *Kokutai* at Atsugi arrived in a twin-engined aircraft to deliver a message. He carried word from Capt Yasuna Kozono, the fiery CO of the 302nd, that the war should be continued in violation of the Emperor's order. LCdr Shiga told the officer: "No! 343 *Kokutai* will not join you! Go away!"

Lt Ryoichi Koga, the servicing officer in the air group, had mixed feelings about the surrender. He was certainly happy that the war was over, but he and his men had given their heart and soul to keep the *Shiden-Kais* operational; it had been all for naught.

"Some days after the Emperor's address," reminisced Koga, "Capt Genda ordered us to assemble all the aircraft in one place after removing propellers and taking out gasoline. We assembled all the *Shiden-Kais* and other aircraft in the ruins of Omura Naval Arsenal. At that time the three squadrons had about 20 aircraft each, totaling about 60 aircraft including other types. It was so sorrowful for us to make the *Shiden-Kais* unairworthy, since we had done our best to keep them serviced for so long. One of the service crew said to me, '*I would rather burn with the aircraft than deliver them to the enemy!*' I persuaded him with every effort not to burn the aircraft. I had been ordered by Capt Genda to avoid trouble and deliver all weapons and ammunition to the Americans."

Capt Minoru Genda returned to base on 19 August. He was pleased to see that peace and order had been maintained. He gave orders for all unit records to be burned. He delivered an impressive address and thanked the men for their service and the dead for their sacrifice. Then he dismissed his men for the last time. With the exception of LCdr Yoshio Shiga, Lt Ryoichi Koga, and the accounting officer, personnel began to leave Omura to make their way back to their families.

The Americans arrive

On 14 September 1945, a reconnaissance team headed by Col Daniel W. Torrey, commanding Marine Air Group 22, landed at Omura Airfield in a Douglas transport aircraft. A jeep was unloaded from the aircraft and driven by Torrey. LCdr Shiga was surprised to see a colonel driving the jeep, particularly since only a few men were permitted to drive cars in a Japanese unit, and high-ranking officers were always chauffeured.

The Americans inspected the airfield and its facilities. In accordance with the surrender terms, mechanics had disarmed the fighters, removed the propellers, and flattened the tires. Everything seemed in order.

Sgt Ed Talty, a Marine combat correspondent, filed the following dispatch from Omura on 23 September 1945: "*This bomb-battered base on Kyushu bears a marked resemblance to an abandoned industrial area on the outskirts of an American city during the depths of The Depression.*"

"That's the first comparison which occurred to the majority of the initial echelons of the Second Marine Air Wing's Marine Air Group 22 when they landed on the field September 20. They were the first Marines to occupy Kyushu, southernmost of the Japanese home islands.

"The Omura base, about 25 miles from what the atomic bomb left of Nagasaki, presents a dreary appearance. Nothing has been done to repair the bomb damage to shattered barracks and hangars. Uncut

RIGHT: Marine Air Group 22, led by Col Daniel W. Torrey (5th from right), was the first American unit to occupy Omura Airfield. (E.S. Dew)

LEFT: A view of the aircraft hangar on Omura Airfield. At the end of the war, US Marines found 80 Shiden-Kais there, most of them non-operational.

BELOW: In accordance with surrender directives, the propellers and spinners were removed from the Shiden-Kais. Note the aircraft tails, which have been painted over to remove their 343 Kokutai designations.

RIGHT: Two Shiden-Kais of the 343 Kokutai with faded numbers in their Hinomaru. The aircraft in the foreground bears a flight leader's band. The tail numbers have been erased.

RIGHT: Col Daniel Torrey and LCdr H. Schwartz, group medical officer, walk toward the airfield's command post.

LEFT: Unloading supplies for the occupation of Omura.

RIGHT: Aerial view of Omura Airfield.

BELOW: Douglas transport aircraft with green surrender crosses sit on the perimeter of Omura airfield.

ABOVE Capt Hardy Wilcoxon stands by a newly erected sign at the former base of the 343 Kokutai which was transformed into a US Marine Airbase.

LEFT: An aerial view of Omura Airfield.

BELOW: Wrecked aircraft litter Omura airfield.

LEFT: Hardy Wilcoxon (standing middle) and fellow servicemen gather for a snapshot in front of a Shiden-Kai. Note that, contrary to Allied surrender directives, the aircraft's 20 mm guns and propeller have not been removed. The aircraft tail numbers have been painted out. (H.C. Wilcoxon)

RIGHT: Two US servicemen check out the cockpit of the Shiden-Kai. (H.C. Wilcoxon)

LEFT: A Shiden-Kai on Omura Airfield with the number 7 applied to the red Hinomaru. Hardy Wilcoxon is in the cockpit. (H.C. Wilcoxon)

LEFT: These VMF-113 Corsairs were the first American fighters to land at Omura.

weeds and grass contribute to the bedraggled appearance of the base. A large swimming pool in the rear of the officers' quarters is covered with moss. A large number of crumpled Jap aircraft litter the edge of the airfield and the hangars.

"But the major portion of the aircraft found at the base are in very good condition and many of them had apparently never been flown. One hangar contained several score fighter aircraft. Their propellers had been removed and piled neatly in the center of the building.

"There had been considerable damage to the facilities and material on the base before the arrival of Marine Air Group 22. Power connections to the buildings had been cut, telephone and telephone lines had been ripped out and the water system had been damaged. In one Jap transport aircraft, the cushions of the seats had been ripped open.

"But within 48 hours of the arrival of the Marines, the field was beginning to operate again and more than 100 Jap sailors and civilian workmen were restoring the administration and barracks building to a livable condition."

On 23 September, Corsairs of VMF-113 flew from Okinawa and landed on the airfield. The same day, Company C of the 27th Marines arrived at Omura to establish security. The airfield was selected as the base for Marine air operations in southern Japan.

The last flight of the Shiden-Kais

The Americans at Omura were very interested in the *Shiden-Kai* and wanted to ship examples back to the US. Six aircraft were to be test flown to determine airworthiness. From the six, three would be chosen for export.

At the end of September, Col Torrey requested Cdr Tatsuto Yamada, former commander of the 352 *Kokutai*, to fly these aircraft under American supervision. Yamada gave the assignment to LCdr Yoshio Shiga. Several pilots and servicing crewmen who had been demobilized were called back to Omura. The pilots were Shiga, CPOs Toshio Tanaka and Hideo Nakao (Harada) of Squadron 407, Masamori Ono of Squadron 701, and two other pilots.

The six *Shiden-Kais* were selected for flight. The tires were pumped up and propellers reattached. The engines were serviced and the fuel tanks filled with American aviation gasoline. The disarmed aircrafts' *Hinomaru* (red sun) markings were erased and replaced with white American 'stars and bars'.

Testing commenced on 13 October 1945. The first Japanese pilot airborne over Omura in the *Shiden-Kai* following the cessation of hostilities was CPO Nakao. He was told the following conditions for the test flight:

RIGHT: LCdr Yoshio Shiga briefs his men prior to a test flight in the Fall of 1945. From left to right: CPO Toshio Tanaka, LCdr Shiga center, unidentified (wearing cap with his back toward camera), Cdr Tatsuto Yamada (background, facing camera) and CPO Masamori Ono to right. The pilots are all wearing life jackets and armbands. (H.C. Wilcoxon)

LEFT: Another view of the briefing, with Shiga center.

BELOW: LCdr Yoshio Shiga is seen to right. The pilot in the white cap on the left is unidentified. The pilot in sunglasses (center) is CPO Toshio Tanaka.

BELOW: LCdr Yoshio Shiga poses in front of the Shiden-Kai he is about to test fly.

The American plan was to test six Shiden-Kais from which the most suitable three would be selected for transfer to the USA. Accordingly, six aircraft were test flown between 13 and 15 October, and three machines finally selected.

ABOVE: LCdr Yoshio Shiga about to enter the cockpit for the test flight.

ABOVE: Lt Willard Pedrick gives last-minute instructions to LCdr Shiga. Pedrick was a Marine intelligence officer. He later went on to become a distinguished professor of law.

BELOW: In the cockpit, LCdr Shiga goes over his pre-flight checklist, while mechanics prepare the aircraft for take-off. Note that the 20 mm cannon had been removed.

RIGHT: American and Japanese personnel cluster around a Shiden-Kai at Omura, October 1945.

LEFT: "This is a pretty interesting picture," wrote Wilcoxon. "George was a rather flimsy aircraft, as I understand it, and probably overpowered. Hence, when the mech pours on the revs during his engine checkout, he has to put a couple of guys on the tail end to keep the front end from tipping down." (H.C. Wilcoxon)

RIGHT: Testing engines.

- Start take-off when four F4U escorts, loaded with full ammunition, circle above in a traffic pattern
- Fly level at 200 knots
- Do not fly higher than 2,000 meters
- Only make left turns
- The flight range should be between Nagasaki and Sasebo

Flying any further north or south, or flying over ships in Omura Bay was prohibited. Nakao was warned that if he violated any of the rules, he would be shot down.

Though probably a little nervous at first, Nakao put his aircraft through its paces. When he made a steep turn, which was near a vertical turn, the Corsairs overran him, but the American pilots only smiled. Tension was eased. Nakao finished his task and made a perfect landing. When he stopped his aircraft, 20 to 30 Americans gathered, some with cameras. One of them pulled out a Japanese banknote and asked him for his autograph. He signed it *Hideo Nakao.* The Americans were amazed that he could write his name in English. Others joined in asking for autographs and Nakao was amused by his sudden 'celebrity' status.

Shiga discussed his task with Lt Willard Pedrick, a Marine intelligence officer. Shiga wanted to put on a demonstration for the Americans to show them what the *Shiden-Kai* was capable of. However, Pedrick warned him to fly no higher than 300 m and to fly straight and level.

BELOW: Shiden-Kais taxi out for their test flights.

ABOVE: LCdr Shiga confers with Lt Willard Pedrick, USMC, before the ferry flight on 16 October 1945. Shiga is wearing an American flight helmet and goggles.

BELOW: Belly tanks are attached for the long flight to Yokosuka.

ABOVE: Drop tanks are attached and the aircraft are ready to roll.

LEFT: A George pulls out.

ABOVE: Hollywood actor Tyrone Power (wearing leather jacket) in the 'chow line' at Omura Airfield. (USMC)

The *Shiden-Kai* took off and was quickly followed by a pair of Corsairs. Although the Japanese fighter was unarmed, the Americans were still worried about a possible suicide dive into one of their ships. They watched Shiga cautiously. There was a thin whisper of white spray trailing the *Shiden-Kai*, an indication of a fuel leak. Shiga quickly landed his aircraft to the applause of Marine onlookers.

Necessary repairs were soon made and Shiga took off again under the watchful guard of F4Us. Since this was his last flight, he decided to take liberties. He climbed to 500 m, dived, and rolled his aircraft. A Corsair approached and the American gestured for Shiga to take his aircraft down immediately and he complied.

When Shiga landed, he was politely told of his breach of regulations. "But you didn't say anything about a roll!" responded Shiga. The Americans saw the humor in his response and told him that everything was fine. The testing continued for three days.

On 16 October 1945, the time came for the ferry flight to Yokosuka. LCdr Shiga met with Cdr Yamada and several Marine officers to discuss plans. One of the officers, a handsome man, appeared very familiar to Shiga. The plan was for the flight of three *Shiden-Kais* to be escorted by four Corsairs, with the *Shiden-Kais* in the middle, flanked by a pair of F4Us on the outside. The Americans ordered the Japanese pilots to fly straight and level, otherwise they would be shot down. The 20 mm cannon were reinstalled on the fighters, but were obviously not loaded. Mechanics attached belly tanks for the long journey.

When all the aircraft were airborne, the three Japanese pilots decided to unleash the full power of the *Shiden-Kai*. With no ammunition, the aircraft was a little lighter. Using high octane American aviation fuel, the Homare engine exhibited its maximum power. As if on cue, the three ferry pilots gave full throttle and virtually left their escorts behind.

"The Corsairs raced after us at full speed!" recounted CPO Toshio Tanaka. "It was very thrilling and delightful! I gestured to them to speed up and try to catch us."

The flight reached Yokosuka Airfield without any problems and the three *Shiden-Kais* were turned over to the Technical Air Intelligence Center. Shortly afterwards, the Marine officer in charge of the ferrying mission arrived by transport aircraft. It was not until the next day that Shiga learned that the handsome major was Hollywood film star, Tyrone Power. After ferrying the fighters, Power asked Shiga if he would show him around Kyoto and Shiga agreed to do so. They fast became friends and continued communicating with one another for many years afterwards.

The three *Shiden-Kais* from the 343 *Kokutai* were ferried to the United States where they were test flown and evaluated. One was refurbished and is now currently on display at NAS Pensacola, sporting the original tail number 343-A-19. The National Air and Space Museum restored another for display. The whereabouts of the third 343 *Kokutai* aircraft are not known for certain, although Wright-Patterson AFB possesses one.

BELOW: Maj Tyrone Power in the pilot's seat with Col Elkin S. Dew. They followed the three 343 Kokutai Shiden-Kai fighters to Yokosuka. (E.S. Dew)

At the end of November 1945, all the Japanese aircraft were brought together on Omura Airfield to be destroyed. The aircraft were bulldozed into a long heap and surrounded with wooden boxes and crates which were then doused with gasoline and ignited. Within a few hours, the last of the 343 *Kokutai's* proud *Shiden-Kai* fighters were reduced to a pile of blackened metal. At that moment, the 343 *Kokutai* passed into history. Scrap metal merchants would eventually recycle 'Genda's Blade' into 'plowshares' for the post-war rebuilding of Japan.

LEFT & BELOW: Black smoke drifts into the air over Omura and Matsuyama as the last Shiden-Kai fighters of the 343 Kokutai meet their end.

LEFT: Smoking fighters at Matsuyama Airfield, September 1945.

RIGHT: Yoshio Shiga (left) with Jiro Yoshida and Gen Minoru Genda (right). This photograph was taken about a month before Gen Genda passed away in 1989. (J. Yoshida)

RIGHT: The Shiden-Kai at the USAF Museum in Dayton, Ohio.

LEFT: A well-documented veteran, this restored Shiden-Kai carries the original tail code 343-A-19. It is on display at the National Museum of Naval Aviation in Pensacola, Florida. (NMNA)

RIGHT: The Shiden-Kai belonging to the National Air and Space Museum was restored at Champlin Fighter Museum in Mesa, Arizona. It will be displayed at the NASM Dulles Facility near Washington DC. (Robert Mikesh)

Chapter 24

A Ghost from the Past

海底に眠っていた紫電改

On 15 November 1978, a diver from the Johen Marine Club looking for an anchor lost by a fishing boat made a startling discovery in Hisayoshi Bay, which is located at the north end of Sukumo Bay in Ehime Prefecture. An intact *Shiden-Kai* was found under 130 ft of water, encrusted with barnacles, with its left wing buried in the sand. News of this remarkable find was broadcast all over Japan.

At the time this aircraft was salvaged, no one could determine its origins. Before the authors present their analysis, it is important to acquaint the reader with the following facts.

1. Ditching of the aircraft and date

It was assumed initially that the *Shiden-Kai* belonged to the 343 *Kokutai* because the unit was known to have operated in the area at the time. However, proof was needed. The Matsuyama NHK-TV station located Sakue Hamami, a housewife, who saw this aircraft on the day it ditched. In the interview, she clearly remembered the date as 24 July 1945 because she was preparing for her father's funeral that day: "I was in the house when I heard this loud noise," she recalled. "I went outside and saw this aircraft coming over the mountains. Then it disappeared from my view, going low toward the sea."

Thus, the date of the ditching is confirmed as 24 July 1945.

After the initial news broadcast, Ryuzo Inoue of Hiroshima City came forward to say that he saw this fighter ditch in the bay. Inoue was, at the time of the incident, a young flight student who was stationed in the area to build defensive positions. He and two others were out on the cove in their small Navy cutter when the aircraft appeared. "It made a right turn and landed with its tail contacting the water at first," he explained. "It sank with its nose down."

According to Inoue, the engine was silent as it came in. The three men in the boat raced to rescue the pilot. However, just as they reached it, the aircraft sank before their eyes. Looking directly down into the water, they could clearly see the pilot sitting motionless in the cockpit. Soon, the fighter vanished into deeper, murkier, green water. All agreed that the pilot had made a skillful landing in the bay, only 200 m from shore. They stayed over the area for a while, but the pilot never surfaced.

The date 24 July 1945 and the submerged *Shiden-Kai* were now linked. On this date, there were fierce dogfights over the Bungo Straits involving *Shiden-Kais* of the 343 *Kokutai,* and six such aircraft did not return from operations. Only the 343 *Kokutai* lost *Shiden-Kais* on this date in this area.

2. Cockpit remains

When the aircraft was salvaged in the presence of Gen Minoru Genda, LCdr Yoshio Shiga, and other air group veterans on 14 July 1979, there were no pilot's remains found in the cockpit, though a local man claimed that he had found two items. At the request of Asahi Broadcasting Company, underwater cameraman Tadashi Murashima dived to the wreckage on 15 December 1978, one month after the aircraft was discovered. According to Murashima's article 'Pacific War Recovery, Kawanishi Shiden-Kai' in the British publication *After the Battle*, he "*...examined the cockpit thoroughly for evidence of human remains but found only a pair of flying boots and helmet.*" However, these items were missing when the aircraft was brought up. Murashima made no mention of whether he left the two items in the cockpit or took them out with him.

There was no trace of the pilot's parachute harness. This was standard equipment for all naval fighter pilots.

3. Aircraft identification number

"*We also tried to learn the identity of the aircraft from the number inscribed on the vertical tail fin,*" wrote Murashima. "*But this part had already reached a state where the lettering was indecipherable to the naked eye. By using infrared photography, we were able to make out something, but it proved impossible to read.*"

Typically, the mission reports include the aircraft number and the name of the pilot who flew it. The tail numbers of the six aircraft for this mission were not known except those of Lt Oshibuchi and CPO Hatsushima.

4. Battle damage

1. Tadashi Murashima states in his article that there were holes, which were apparently caused by bullets: "*Holes which appeared to have been caused by gunfire could be seen all over the fuselage...*"
2. Susumu Koyama, a former Army fighter pilot living in Matsuyama, took part in the recovery operation as an advisor to the salvaging company. As Sgt Susumu Kajinami, he fought in New Guinea with the Army's 68th *Sentai* in 1943-44. By his own reckoning, he had shot down 24 American fighters. Very familiar with bullet damage, he is of the opinion that the numerous holes in the aircraft were, in fact, caused by corrosion.
3. LCdr Shiga states that there were areas in the engine and cowling where metal had fallen off, but it was not clear if the resultant holes were caused by bullets or corrosion. Only one 'streak' was observed on the left topside of the fuselage, which seemed to be a trace of a bullet having come from behind.
4. Co-author Sakaida visited the salvaged aircraft in 1981 and noticed numerous holes, but nothing resembling bullet damage.
5. Co-author Takaki also inspected the aircraft in July 2001 and stated: "At least to my eyes, there were no holes that could be interpreted to have been caused by explosives... I saw only one hole of about three centimeters in diameter near the edge of the windshield on the left top of the nose section. The small hole looked to be simple with no edges bent."

5. Ammunition

1. There were 500 rounds left in the aircraft, according to LCdr Shiga. He also stated that some armor-piercing bullets of inferior quality were carried.
2. According to the article in the *Sankei Shimbum* newspaper dated 16 July 1979, there were 515 rounds found in the magazines.
3. One of the four ammunition magazines (left, outside) was empty while the ammunition in the other three magazines remained unused, according to Koyama.

Note: The *Shiden-Kai* had an ammunition capacity of 800 rounds (200 rounds per gun). If 500 or 515 rounds were found, then 300 or 285 rounds had been fired, with each of the three guns firing about 33 or 28 rounds before stopping. Unlike the Zero, which could selectively fire its cannon and machine guns separately, all four cannon in the *Shiden-Kai* fired in unison.

6. Adjustment of the foot bar and seat positions

1. According to aviation historian, Yoshiro Ikari, the foot bar position was adjusted for a pilot with a height of about 160 cm.
2. Veteran 343 *Kokutai* pilot Chitoshi Isozaki stated that the seat setting was for a pilot taller than 160 cm.

7. Canopy

1. According to Koyama, when the aircraft was *first* discovered, the canopy was open. Some said that it was *partially* opened. Such information should be based on the observations of the *first* discoverers, but there is also an observation that it was closed. Tadashi Murashima was not one of the *first* discoverers.
2. According to Murashima: "When the aircraft was first discovered, the canopy was closed." However, it is not clear if he said this after he heard of it from the original discoverers. He slid back the canopy to search inside.

Note: When the aircraft was salvaged, the canopy was closed.

8. Seat belt on the pilot's seat

The seat belt was unbuckled when the cockpit was inspected.

9. Other observations

The four propeller blades were bent backwards. The throttle was in the full open position. Propeller pitch was low-speed revolving. The flaps were fully down. The landing gears were retracted.

10. Casualties on 24 July 1945

On 24 July 1945, six pilots of 343 *Kokutai* did not return from operations. They were Lt Takashi Oshibuchi and his wingman, CPO Jiro Hatsushima, of Squadron 701; PO1/c Kenshin Mizoguchi of Squadron 407; Ens Kaneyoshi Muto, CPO Nobuya Komeda, and PO1/c Susumu Imai, all from Squadron 301.

Analysis

In order to ascertain the identity of the ditched aircraft, it was necessary, through a process of elimination, to endeavor to identify each pilot casualty and the location where they were shot down.

Lt Takashi Oshibuchi and CPO Jiro Hatsushima, S701

It is a known fact that Lt Oshibuchi left the dogfight on 24 July 1945 due to engine problems and was escorted by CPO Jiro Hatsushima. In combing through the vast amount of US Navy mission reports, only the report of VF-49 mentions such an encounter. Not only did Lt(jg) Jack A. Gibson and his wingman, Lt(jg) George M. Williams spot a pair of Georges hightailing it away from the action, but Gibson especially noted a large "...*white band about a foot wide around the fuselage just forward of the horizontal stabilizer.*"

Gen Ryoichi Yamada, Ret., of the Japan Air Self Defense Force and former Squadron 701 division leader states: "A white two-band marking was painted on the fuselage of Lt Oshibuchi's plane, while that of the division leaders like me had only one white band. Depending upon the servicing situation, however, division leaders did not always use aircraft with the marking, but Lt Oshibuchi, who was the commander of all the squadrons, almost always flew the aircraft with the white two-band marking."

Oshibuchi, who often flew 343-C-45, was not flying this aircraft on his last mission; he was flying 343-C-13. This aircraft had one band.

The action report of VF-49 uses the term "*white band*" (singular), but Gibson uses also the plural form "*white bands*" in his letter to the author dated 24 July 2000. The aircraft with the white marking could not have been CPO Hatsushima's plane. Thus, Gibson's opponent was Lt Takashi Oshibuchi, while George Williams shot down Jiro Hatsushima. The two Georges crashed in the vicinity of Mizunoko Shima Lighthouse.

PO1/c Kenshin Mizoguchi, S407

The three squadrons of the 343 *Kokutai* initially went into the combat area together as a large group. Oshibuchi's Squadron 701 went in first, with Squadron 407 behind them. Squadron 301 separated from the two squadrons and fought independently, but in the same area around Mizunoko Shima.

Ens Walter A.Yancy of VF-49 reported sighting seven unidentified aircraft high on the Kyushu side and to the rear of his group. This would have been the six *Shiden-Kais* of S407. At this time, the S701 formation had broken up. Lt Oshibuchi and his wingman, CPO Hatsushima, were already in retreat or had been shot down. Ens Kazuo Muranaka was alone or formed up with CPO Yoshio Takahashi, who lagged behind. The Yamada/Ono and Yagi/Kurita sections were dogfighting; they did not form a division of four aircraft.

The two divisions from S407 were led by veteran pilots Lt Takashi Mitsumoto (1st division) and Ens Mitsuo Ishizuka (2nd division). Mitsumoto initially led the squadron, but he had to flee when his belly tank would not release. Ens Minoru Honda, another veteran, had to take over a three-aircraft division.

Ens Yancy hit a George and saw it crash into the sea. He was also given credit for a probable on a second aircraft, but no one saw this aircraft crash. PO1/c Kenshin Mizoguchi was the only casualty from S407. According to LCdr Yoshio Shiga's official report of Mizoguchi's death, the young pilot was unable to drop his belly tank, but stayed in the dogfight while lagging behind. The co-authors believe that Mizoguchi was downed by Ens Walter A. Yancy. After this encounter, Yancy joined up with Lt(jg) George Williams somewhere at the mouth of the Bungo Straits. It suggests that Mizoguchi was shot down on the Kyushu side between Mizunoko Shima Lighthouse and the mouth of the Bungo Straits.

Ens Kaneyoshi Muto, CPO Nobuya Komeda and PO1/c Susumu Imai, S301

The three remaining casualties were all from Squadron 301. The co-authors place these pilots in action against Lt Malcolm Cagle and Lt(jg) Robert Applegate near Mizunoko Shima Lighthouse. Lt(jg) Applegate initially saw a division of four aircraft crossing their group up ahead. Of the two divisions of Squadron 301 aircraft, only Muto's division was intact at this time.

The 500 or 515 rounds of 20 mm ammunition in the salvaged George now come into consideration. Muto was the senior veteran pilot who led the division composed of Muto/Komeda and Hori/Imai. As the division leader, he would have initiated the attack. Lt(jg) Applegate reported being bounced by the first section of Georges, but they failed to hit any of the American aircraft. However, Muto was an expert aerial marksman. If three of his cannon malfunctioned in the diving attack, this may account for the fact that they hit nothing. Sometimes in an attack, the wingman would withhold fire to cover his leader.

Muto and Komeda broke away when Applegate pulled up and fired a burst. It also suggests that Muto first noticed in his diving attack that only one of his cannon was functioning and he instantly broke away, with his wingman following.

WO Mitsuo Hori and PO1/c Susumu Imai of the second flight came down after Muto/Komeda had made their pass. Apparently, Ens Robert J. Speckmann was shot down by Hori, flying the lead aircraft of the second section, who was a veteran ace, while his wingman covered him. Applegate reported hitting the second aircraft of the second flight; this would have been PO1/c Susumu Imai. Imai's aircraft was seen to crash by Lt Malcolm Cagle of VF-88.

There are two pilot casualties remaining: Muto and Komeda. If it is assumed that Ens Muto's cannon had failed him, then it was Komeda who was shot down by Lt Malcolm Cagle. Applegate could see the flashes of all four cannon right behind him as a George was sitting just one aircraft-length behind him. An eight-year veteran with close to 30 victories, Muto probably would not have clung to his opponent in such an absorbed way, firing with wild abandon. Lt Cagle jumped in and blew the George's tail away.

The remaining opponent made a head-on run against Applegate. Both pilots stopped firing when they ran out of ammunition. In the salvaged *Shiden-Kai*, one magazine was empty. This means that in the head-on run against Applegate, the Japanese pilot had only one functioning cannon!

The enemy pilot was slumped over when he cleared Applegate by ten feet, a sure indication that he had been wounded. When Applegate turned to go after him, the George disappeared. No one saw this aircraft crash into the sea.

It became imperative that this aircraft land immediately. It is hard to believe that the pilot was so short of fuel that he had to land right away. Most pilots, including Lt Yamada, who engaged in combat, managed to return to base without refueling on the way back. The area around Hisayoshi Bay is mountainous and the most outstanding feature of the area is the ragged coastline. The pilot was most likely wounded. His aircraft may have been damaged, although there is no clear evidence. He flew his aircraft eastward looking for an airfield or clearing to land. There was a small airfield at Uwajima 30 km north of Johen, where a new Navy training air group was formed in March 1945. It is not known if the pilot of the ditched aircraft knew this.

Given the circumstances, Hisayoshi Bay was probably the only option left for a ditching. It was also natural that any Navy pilot should choose this procedure. The pilot made a masterful landing on the water but he was motionless after the aircraft stopped. This suggests that the pilot was skilled but physically unable to quickly exit the cockpit.

A former 343 *Kokutai* veteran, Isamu Miyazaki, comments: "It is certain that such a beautiful landing on that topography, from the mountainside to the narrow cove, could be made by a pilot with considerable skill."

The standard procedures for ditching were as follows:

1. Open cockpit canopy and lock it in place.
2. Remove flight helmet and accompanying radio earphones and cords. There were cases where American pilots drowned because they became entangled in their own radio cords.
3. Unhook parachute harness at chest and legs prior to touching down, thus allowing for quick exit without being encumbered by heavy gear. Japanese Navy parachutes had a quick release mechanism at chest level. The pilot simply pinched two levers together and three of the four restraining straps fall away.
4. Ditch the aircraft with tail in contact with the water first. Dragging the tail on the water helps reduce the aircraft's speed.

It is unknown if the canopy was opened or closed at the time the fighter came in contact with the water, as there were two different observations. Most likely the cockpit was closed. Had it been open, the pilot would have drowned immediately, still buckled into his seat. His remains would have been found in the cockpit, along with the remains of his parachute harness.

When the aircraft sank, the pilot must have survived for a period. He could not have slid back the canopy until the cockpit filled entirely with water, to equalize the pressure. The pilot must have survived for a few minutes as the cockpit slowly filled with water.

There is a case where a pilot survived in a similar situation. In December 1944, CPO Takumi Sugitaki, of the 302 *Kokutai* (later with the 343rd), was flying back to base after shooting down a B-29. He ran out of fuel and prepared to ditch. However, the nose of his Jack was so heavy, he dived into Tokyo Bay. He was able to breathe for a while after hitting the seabed as water slowly rose through the floorboard. Unable to slide back the canopy, the desperate pilot kicked out the Plexiglas and squeezed through the canopy to reach the surface, half-drowned.

The pilot of the *Shiden-Kai* had no chance of survival. He was simply too deep and probably wounded. He removed his flight boots and helmet, and using his whole strength, pulled back the canopy. He managed to squeeze through, but undoubtedly became caught somehow, and drowned. His kapok-filled life jacket would have floated him to the surface, but none of the three young men in their boat above saw anything surface.

A few days after the ditching, the mangled upper torso of a Japanese Navy pilot in a flight suit was washed ashore at Yoko Shima, a small island located to the west of Hisayoshi Bay. His face was missing and there was no identification on the flight suit or life jacket.

An examination of the corpse by a military physician and the local *Kempei Tai* revealed that the pilot was about 20 years old, and that he had been in an aircraft crash. The remains of this pilot were cremated and his ashes deposited in Seirinji Temple near Johen Town. Of the six missing pilots, their ages were: Lt Takashi

Oshibuchi, 26; CPO Jiro Hatsushima, 22; PO1/c Kenshin Mizoguchi, 21; CPO Nobuya Komeda and PO1/c Susumu Imai 20-years-old. Ens Kaneyoshi Muto was the eldest at 29. The pilot of the submerged *Shiden-Kai* cannot be confirmed as the unidentified corpse found on the shores of Yoko Shima.

The foot bar position suggests that the pilot was a short man of about 160 cm, and both Muto and Komeda were short. The other four missing pilots were a little taller. The identity of the pilot cannot be determined solely by this information.

After the last engagement, Lt(jg) Applegate could not fly with Lt Cagle due to engine trouble caused by battle damage and he parachuted. He landed in the water "...four miles northwest of Ukuru Shima." This small island is located near the mouth of Sukumo Bay in Southwestern Shikoku and about 15 km south-southwest of Hisayoshi Bay. Hisayoshi Bay is a part of Sukumo Bay and is located at its north end. After the two opposing fighters each sustained bullet damage in their head-on run, the pilot of the *Shiden-Kai* ditched at the north end of Sukumo Bay, while the Corsair pilot parachuted from his mount near the mouth of the same bay.

After the *Shiden-Kai* was salvaged, the barnacles were sandblasted away and the entire aircraft was repainted to prevent further deterioration; no attempt was made to repair the aircraft. As there was no conclusive evidence, no attempt to identify the pilot was made by the survivors from the 343 *Kokutai*. Japanese historians also lacked the critical information necessary to undertake such a seemingly impossible task. Thus, the question had remained unsolved.

The *Shiden-Kai* is now displayed in a modern building, small but firmly built, atop Mt. Base, overlooking the cove where the aircraft was found. It rests as a monument for the six pilots who died in the air battle on 24 July 1945. It is also a symbol of all the *Shiden-Kais* that fought to defend the homeland at the last stage of the war.

Reliable information has been obtained here after a lapse of more than 20 years since the *Shiden-Kai* was pulled from its watery grave on 14 July 1979. The two authors are convinced that the pilot of the ditched *Shiden-Kai* was Ens Kaneyoshi Muto, but readers are urged to draw their own conclusions from the real circumstances of the day's combat.

RIGHT: The view overlooking the cove where the Shiden-Kai was found. The area is used for fish farming.

LEFT: July 1979: the vague, but unmistakable shape of an aircraft can be seen just below the water as a large ship crane is used to bring the Shiden-Kai to the surface. (Yoshinori Matsumoto)

LEFT: Carefully placed lifting straps help to bring the delicate wreck out of the water. Note the snagged fishing net over the tail. (Yoshinori Matsumoto)

ABOVE: The salvage operation made national headlines in Japan. Veterans from the 343 Kokutai were on hand to watch the proceedings. (Yoshinori Matsumoto)

LEFT: The aircraft is placed gently on a barge to be brought ashore. (Yoshinori Matsumoto)

ABOVE: The interior of the cockpit is encrusted with barnacles and marine growth.

LEFT: The seat was examined for clues as to the pilot's stature. However, the seat belt was unbuckled and there was no parachute pack, a clear indication that the pilot had made his exit from the cockpit.

BELOW: Next-of-kin of the six 343 Kokutai pilots who died on 24 July 1945 were invited to watch the salvage operation.

ABOVE: The Shiden-Kai as it appears today in the small museum overlooking the cove where it was found. (Koji Takaki)

LEFT: There are many holes in the aircraft due to corrosion. Co-author Koji Takaki inspected the wreck for bullet-hole damage, but there were no definite signs. (Koji Takaki)

LEFT: A view of the tail section. The fabric-covered surfaces had rotted. The aircraft tail number would not have helped identify the pilot since flight records were burned at the end of the war. (Koji Takaki)

BELOW: The salvaged Shiden-Kai was never fully restored. The barnacles were sandblasted and the surfaces were repainted, but no repairs were made. Photographs of the six pilots reported missing on 24 July 1945 can be seen in a frame just below the engine. (Koji Takaki)

Appendix

三四三海軍航空隊

Aircraft Colors and Markings of the 343 *Kokutai*

One of the most controversial topics concerning the *Shiden-Kais* of 343 *Kokutai* is their markings. Color profiles in various Japanese aviation publications portray subtle differences. Some are based on monochromatic photographic evidence, thus rendering the colors speculative. However, several extant relic samples and three surviving *Shiden-Kai* do reveal the camouflage colors used by Kawanishi.

OVERALL COLOR

The *Shiden-Kais* were painted in a glossy dark green/olive finish on the upper and side surfaces. According to James Lansdale, an expert in Japanese aircraft markings, they were in the color range of the Federal Standards FS-34052, 34077, and 34079. The *Shiden-Kai* owned and refurbished by the National Air and Space Museum is painted in a semi-gloss black-green (Munsell 10 GY 2/2).

Other than prototypes, the undersurfaces were usually left natural metal with grey or aluminum-colored lower portions on the control surfaces. This was to save cost and time during production.

The *Hinomaru* (red sun) markings were painted in FS-11136 on the wings and fuselage. The 'meatball' on the fuselage measured approximately 85 cm with a 75 mm white border. However, in some photographs of 343 *Kokutai* aircraft, the white border is absent on the fuselage. The *Hinomaru* on each upper wing measured approximately 110 cm and was not bordered in white.

Wing leading edges were yellow-striped for identification of friendly aircraft. It was also useful as a warning marking to maintenance crews when the propellers were spinning. The example at the NASM is painted in deep yellow (Munsell 10 YR 7/14).

Propeller blades were painted in a semi-gloss dark brown (Munsell 10 R 2/2). The spinners were dark green. The propeller tips were painted in the same color as the wing leading edges, and were some 5 cm in width about 5 cm inset from the tip.

It is interesting to note that in a few American mission reports the *Shiden-Kais* of the 343 *Kokutai* have been described as brownish in appearance or mistaken for bare-metal P-47 Thunderbolts. Weathering may account for the first description and sun glint from a distance may account for the second.

TAIL NUMBERS

The unit designation was painted on the tail. The 343 *Kokutai* used the Roman alphabet to differentiate each of the three fighter squadrons: A=Squadron 301; B=Squadron 407; C=Squadron 701. The tail code, 343-A-15, denoted 343 *Kokutai*, 301 Squadron, Aircraft No.15. The squadron letter code was painted on the top of the tail and was not placed between the unit and aircraft number. The tail codes were painted in deep yellow as per the leading edges and propeller tips.

It is interesting to note that the refurbished *Shiden-Kai* at the NASM was numbered 35 to avoid markings of a known pilot. The aircraft number was selected after extensive research. However, the co-authors located the 343 *Kokutai* mission report for 12 April 1945, which shows that 343-A-35 was flown by CPO Katsue Kato, who was forced to abort from the sortie to Amami/Kikaiga Shima due to engine problems. Kato made a name for himself in the unit's baptism of fire on 19 March 1945 over Kure when he claimed nine victories. He was killed on 16 April on the second mission to Amami.

AIRCRAFT FUSELAGE BANDS/STRIPES

Aircraft bands or stripes were common on JNAF fighters. They served as quick recognition guides to identify flight leaders. Interestingly, the purpose of painting band markings on the aircraft for the squadron leaders of the 343 *Kokutai* was to have enemy pilots identify their aircraft. As recounted in *Kamikaze Operations* by aviation historian Yoji Watanabe, Lt Naoshi Kanno asked his groundcrew to paint two diagonal yellow bands on the fuselage so that he could lure and destroy enemy aircraft. The other two squadron leaders approved of Kanno's idea and followed suit. It is said that these markings were applied after the unit's baptism of fire on 19 March and before its move to Kanoya Airfield in April 1945.

According to Gen Ryoichi Yamada, who served as *buntaicho* in S701, the same diagonal band markings were painted at the same position, forward of the *Hinomaru*, on the fuselages of the three squadron leaders' aircraft: two bands for the squadron leaders and one band for the *buntaicho*. The color was white in S701 and S407.

Yamada remembers that he flew his assigned aircraft with the one-band marking only on his first or second mission. In most cases afterwards, he flew aircraft with no band because of unavailability due to maintenance problems. However, he further states that his squadron leader, Lt Takashi Oshibuchi, flew his aircraft with the two-band marking more frequently.

Lt Goro Ichimura, *buntaicho* in S407, states that he does not clearly remember the color of the band.

He flew his assigned aircraft (343-B-25) for a long time. Some surviving members of his squadron remember the color as white.

There apparently was a color difference in S301. Gen Masaji Matsumura (*buntaicho*) states that Lt Kanno's two-band marking was definitely yellow. Yet he cannot remember the band color on his own aircraft since he rarely flew his assigned machine.

The foregoing represents the general aspects of the flight leaders' band markings in the 343 *Kokutai* as remembered by the three surviving *buntaicho* officers over half a century after the end of the war.

According to Gen Yamada, and CPO Tomokazu Kasai of S301, no band markings were painted on any aircraft used by pilots other than those of the squadron leader and the *buntaicho*. Their view is that no markings could be painted because there were no assigned aircraft for the rank and file pilots. CPO Seiichiro Sato (post-war name Yamamoto) states: "I never flew the same aircraft on every mission." He participated in many battles as the 2nd aircraft to Lt Masaji Matsumura from 12 April to 28 May 1945, until he was severely wounded.

The position of the bands on the fuselage also invites controversy. There are some photographs of S407 aircraft showing a diagonal band *between* the *Hinomaru* and the tail. This is possibly because the marking of aircraft was left to the discretion of each squadron.

In less-recent works on Japanese aviation and on model aircraft boxes (*circa* 1970-1980s), there are inaccurate profiles of two famous 343 *Kokutai* aircraft. One is that of Lt Naoshi Kanno. His 343-A-15 is shown with a yellow-white-yellow band, with the squadron letter 'A' placed between the unit and aircraft number.

In the case of Lt Takashi Oshibuchi's 343-C-45, the error in his aircraft profile is the double red diagonal bands. His aircraft did not have red bands. Although there is no photograph of his aircraft available, remarkably, there was an American eyewitness who confirmed the existence of a band(s). He was Lt(jg) Jack A. Gibson who shot down Lt Oshibuchi. In his debriefing report, which was written into the VF-49 mission report for 24 July 1945, Gibson noted: "*White band about a foot wide around fuselage just forward of the horizontal stabilizer.*"

Bands on aircraft made identification easy. However, maintenance problems played havoc with this system. The tried and tested method of identifying flight leaders was to approach the aircraft to identify its pilot or read its tail number. Pilots were briefed before the sortie on the formation composition. CPO Kasai believes that bands were not absolutely necessary. He states that it was risky for the flight leader to carry such bands because enemy pilots could easily identify him. Except in very rare circumstances, only the *buntaicho* flew banded aircraft. An exceptional case is observed in S407. Ens Mitsuo Ishizuka flew 343-B-25, a *buntaicho* aircraft, as the mission leader of 12 fighters on 17 May 1945.

The co-authors believe that it is difficult to make definitive statements on the bands simply because no former members of the 343 *Kokutai* can agree.

NUMBERS IN THE *HINOMARU* ('MEATBALL')

Some of the 343 *Kokutai Shiden-Kais* were photographed with large white numbers painted inside the *Hinomaru*. Aviation historian Osamu Tagaya explains that large white numbers were commonly used on aircraft for training purposes. In a letter to the NASM restoration staff in May 1994, Tagaya wrote: "*The practice of painting the individual aircraft number in the fuselage* Hinomaru *was a standard method of visual aid in the JNAF in the late-war period for operational units during the time that they were in training and working up to full operational status. Once they were fully operational or deployed to the combat zone, these numbers, usually roughly hand-applied using water-soluble paint, were removed.*"

In a letter to co-author Henry Sakaida, LCdr Yoshio Shiga explained that the large numbers were used for maintenance purposes. However, Maintenance Petty Officer Wakashige Kondo of S301 states: "They were not painted for servicing purposes. Definitely not!"

Kondo believes they were used in training, while CPO Kasai states that they were not. It is possible the numbers were used to train the inexperienced 'C' Class pilots. Veterans such as CPO Kasai probably had no need for such numbers as formation training aids.

During the research for this book, Gen Yamada advised he does not remember if his squadron used such numbers. Responses from all the former pilots that co-author Koji Takaki contacted in regards to the numbers were either that they did not remember them, or they never used such numbers in training!

Were these numbers temporary or permanent? LCdr Shiga explained to the authors that they were painted with a water-soluble solution of ground chalk, and were never meant to be permanent. Maintenance Petty Officer Kondo, in his telephone interview with Koji Takaki on 29 December 2001, stated: "They were painted with a solution of lime and water with a little amount of adhesive agent so that they could be easily erased." He went on to state that they were ordered to erase all the numbers in the *Hinomaru* after the combat of 19 March 1945. It is not known if such orders were issued to the other squadrons. Strangely, Lt Kanno's aircraft was photographed on 10 April 1945 with a large 15 painted in the *Hinomaru*. It is said that the number was eventually erased after the unit arrived at Kanoya, the combat base for the Okinawa Campaign.

VICTORY MARKINGS

Victory markings were painted on aircraft, the purpose of which was to raise morale. Only a few aircraft were known to have carried victory markings on the fuselage. CPO Tomokazu Kasai of S301 states that victory markings were definitely painted on Lt Kanno's *Shiden-Kai*, and that they took the shape of silhouettes of aircraft with arrows through them. Gen Yamada confirms that victory markings were found on Lt Kanno's aircraft, but no victory markings were painted on S701 aircraft, including Lt Oshibuchi's aircraft.

Aviation historian Kazuhiko Osuo once saw photographs of 343 *Kokutai* aircraft with victory markings in a private photo album. They included 343-A-15 flown by S301 leader Lt Naoshi Kanno, which had five; 343-A-11 showed 16 kills; and 343-B-17 had three.

Since the JNAF did not recognize nor award individual credit, and enlisted pilots did not have assigned aircraft, it meant that the aircraft recorded the victory, not the pilot. The victory markings took the shape of silhouettes of the enemy aircraft downed. They were located on the fuselage above the step, below the canopy, or between the *Hinomaru* and the tail.

Order of Battle and Flight Organization of the 343 *Kokutai*

The 343 *Kokutai* utilized the two-aircraft and four-aircraft division which copied US Navy fighter formations. This basic unit afforded each member protection since he had a wingman (*rekki*). The JNAF's earlier three-aircraft formation was found to be inferior when pitched against American fighters in the early part of the war. While two pilots could watch out for each other, they could not protect a third, odd man. Additionally, under the later formation, each of the two sections could protect the other.

Capt Minoru Genda chose the leaders of his three fighter squadrons very carefully; leadership and combat skills were paramount. His squadron leaders were all Naval Academy graduates who were respected veterans. All were aces, with Lt Naoshi Kanno being the top scoring fighter pilot of all the Naval Academy graduates.

The squadrons also had a high number of special duty (*Tokumu*) officers who helped guide the lower ranked enlisted pilots and bolstered the unit's level of morale. In the JNAF, officers who came up from the enlisted ranks (*Tokumu*) generally had more experience than an average lieutenant junior grade, as well as a newly minted ensign straight from the air training group following graduation from Naval Academy. However, *Tokumu* and reserve officers were promoted very slowly and usually did not lead formations containing commissioned officers. As observed in the combat formation compositions, this principle was applied in the 343 *Kokutai* also.

All squadrons of the 343 *Kokutai* comprised approximately 40 pilots and were made up of two *buntai* or groups. The term *buntai* can be translated as 'division,' but it does not have the same meaning as in the American military terminology. A *buntaicho* was the leader of a group (*buntai*=group, *cho*=leader) under a squadron leader called *hikotaicho* (*hikotai*=squadron). For example, in Squadron 701, which was led by Lt Takashi Oshibuchi, Lt Ryoichi Yamada served as the 1st *Buntaicho* while his Naval Academy classmate, Lt Kunio Matsuzaki, served as 2nd *Buntaicho* from the formation of the squadron (until Lt Matsuzaki was killed in action on 19 March 1945). The vacancy was filled by Lt Kazuchika Kinoshita in May, who was from the Reserve Student Course. He was killed in action on 5 July. Kinoshita's vacancy was taken by Lt Keishichiro Hattori, a Naval Engineering Academy graduate from Squadron 407 in August. In S407, which was led by Lt Yoshishige Hayashi (KIA 21 April), Lt Goro Ichimura served as the 1st *Buntaicho*, and his classmate, Lt Kozo Shima (KIA 19 March), was 2nd *Buntaicho*. Shima's vacancy was filled by Lt Tsuneyasu Hayami, a Naval Academy graduate who came from the Yokosuka *Kokutai* in June. He was Lt Ichimura's classmate.

The organizational position of the *buntaicho* in the squadron meant that he had the right to command after the *hikotaicho*. If the position of the *hikotaicho* was vacant, the senior *buntaicho* would take over leadership both in the air and on the ground. For example, Lt Goro Ichimura (*buntaicho* of S407) was appointed acting squadron leader until May when Lt Keijiro Hayashi was named to replace squadron leader Lt Yoshishige Hayashi who was killed in action.

A veteran special duty lieutenant could also become a *buntaicho*. For example, Lt Chitoshi Isozaki, who had over 4,000 flight hours and more than ten victories, arrived in May and became *buntaicho* of S301 where Lt Masaji Matsumura had served as the only *buntaicho* since the squadron was formed at Matsuyama. Another veteran, Lt Mitsuo Toyoda, was the *buntaicho* of S401, a training unit within the 343 *Kokutai*.

The basic flight formation was called a *kutai* (*ku*=section). This term corresponds to the American flight term 'division.' It appears that the 343 *Kokutai* had no specific formal term for 'section' which consisted of two aircraft. It was often called a 'pair' – from the English. It was the combination of the 1st and 3rd aircraft, or of the 2nd and 4th aircraft.

Kutaicho (division leader) was a leading position in a *kutai* formation. There were specific pilots who could fly in this position due to their rank (generally above chief petty officer), combat experience, and flying hours. However, these pilots did not always fly as *kutaicho*. For example, CPO (later WO) Mitsuo Hori, one of the veteran pilots in S301, flew as the 2nd aircraft to Lt Masaji Matsumura on 16 April. LCdr Yoshio Shiga, the XO, says that some of those veteran pilots who would usually lead a division, flew as the 2nd aircraft on 24 July 1945.

As a general rule, fixed aircraft were assigned only to *hikotaicho* (squadron leader) and *buntaicho*. However, as seen in the formation compositions, those fixed aircraft were not always used by them because of airworthiness problems.

The terms *chutai* and *shotai* are also confusing and were used by the 343 *Kokutai*. As a basic understanding, it can be said that the terms were used in the flying formation compositions. A *chutai* (*chu*=mid or middle) was formed when a mission was flown by more than three *kutai* (divisions) from more than two squadrons and one *chutai* was made up of more than two *kutai*. They often flew missions in a formation of aircraft from the three squadrons, or from two of the three squadrons. An aircraft group from one squadron was called a *chutai* and was usually led by a squadron leader or a *buntaicho* when the group consisted of more than two divisions.

The order of the *chutai* depended upon the seniority of each squadron leader. When three squadron leaders were on duty, Lt Takashi Oshibuchi (Naval Academy 68th Class) of S701 assumed overall command and he headed the 1st *Chutai*. Lt Yoshishige Hayashi (Naval Academy 69th Class) of S407 commanded the 2nd *Chutai*. Lt Naoshi Kanno (Naval Academy 70th Class) led the 3rd. This *chutai* formation composition was adopted on the 16 April mission. When Lts Oshibuchi and Hayashi were off duty, Lt Kanno led two aircraft groups from two squadrons as the mission leader. On the 12 April mission, Lt Kanno led the 1st *Chutai* from S301 and Lt Ryoichi Yamada (Naval Academy 71st Class), the *buntaicho* of S701, led the 2nd consisting of two divisions. A Naval Academy classmate never participated as a *chutaicho* on a mission led by another classmate of the same rank.

On 2 June, Lt Keijiro Hayashi led the 1st *Chutai* as mission leader with Lts Ryoichi Yamada and Masaji Matsumura, both *buntaicho*, leading the 2nd and 3rd *Chutai*. Lt Kanno, the mission leader's classmate, was off duty.

A veteran officer pilot whose rank was below lieutenant could become a *chutaicho*. For example, Ens Mitsuo Ishizuka (S407), who was promoted from the enlisted ranks, led a *chutai* made up of three *kutai* (two from S407 and one from S701) on 17 May 1945.

It is often stated that a *shotai* (*sho*=small) was a term for a two-*kutai* (division) formation, but this is not accurate. The number of *kutai* in a *shotai* varied from one to three, according to the composition of a flying formation. For example, the mission on 17 May 1945, as mentioned, was flown by a *chutai* consisting of two *kutai* from S407 and one from S701. The two divisions from S407 formed a *shotai* directly led by the mission leader, the *chutaicho*, who was also the *shotaicho* (namely, the leader of the two-division formation). The other division from S701 was also a *shotai*, led by a chief petty officer who was the *kutaicho* (division leader) and *shotaicho* (leader of the *shotai* consisting of one division). A *shotai* could be led by a *kutaicho* of any rank, provided he was the senior pilot in the *shotai*.

The above terminologies were not peculiar to the 343 *Kokutai*, but were common to the JNAF air groups during the late war period when the four-aircraft formation was adopted. The 343 *Kokutai* pursued formation fighting using the four-aircraft more than any other unit.

Japanese Naval Air Force Ranks

Commissioned Ranks

Gensui	Fleet Admiral
Taisho	Admiral
Chujo	Vice Admiral
Shosho	Rear Admiral
Taisa	Captain
Chusa	Commander
Shosa	Lieutenant Commander
Taii	Lieutenant
Chui	Lieutenant (junior grade)
Shoi	Ensign

Enlisted Airmen (ranks as of 1 November 1942)

Hiko heisocho	Flight Warrant Officer
Joto hiko heiso	Flight Chief Petty Officer
Itto hiko heiso	Flight Petty Officer 1st class
Nito hiko heiso	Flight Petty Officer 2nd class
Hiko heicho	Flight Leading Airman
Joto hiko hei	Flight Chief Airman
Itto hiko hei	Flight Airman 1st class
Nito hiko hei	Flight Airman 2nd class

Notes:

1. To differentiate Navy from Army, the word *Kaigun* (Navy) was prefixed to each rank such as *Kaigun Taisa* (Navy Captain) or *Kaigun hiko heisocho* (Navy Flight Warrant Officer). For enlisted pilots, the prefix *Hiko* (Flight) was added.
2. The Imperial Navy held the following courses to train fighter pilots:
   - i. Flight Student to train officer pilots for graduates from the Navy Academy or Navy Engineering Academy.
   - ii. Flight Reserve Student to train officer pilots for graduates from universities and colleges.
   - iii. Pilot Trainee (*Soju Renshu-sei* or *Soren*) for ranks from *Kaiheidan* (Seamen's training corps).
   - iv. A Flight Trainee (*Ko-hi*) for volunteers above the 4th grade of middle schools (over about 17 years old) to train petty officer pilots for future special duty (*Tokumu*) officers.
   - v. B Flight Trainee (*Otsu-hi*) for volunteers above the 2nd grade of advanced primary schools or middle schools (over about 15 years old) to train future primary-class commanding pilot.
   - vi. C Flight Trainee (*Hei-hi*) for ranks from Kaiheidan. Pilot Trainee Course was abolished at the start of this course in 1941.
   - vii. Special B (*Toku Otsu*), an intensive short course, for seniors out of B Flight Trainees, started from December 1942.
   - viii. Reserve Trainee for graduates from civilian pilot training schools at Ehime and Nagasaki.
3. There was a great difference under the Military Orders Rule in the power or authority between the officers from the Navy Academy and special duty officers promoted from the ranks (*Tokumu*). Officers who were not Naval Academy graduates held *Tokumu* ranks.
4. The lowest rank for front line pilots who were posted to the 343 *Kokutai* was Leading Airman. They were from the Special B course 1st and 2nd term, graduating from the advanced fighter training courses in July or August 1944.

Glossary

US military terminology

AC	Aircraft commander
ACI	Air Combat Intelligence
AMM	US Navy Aviation Machinest Mate
AOM	US Navy Aviation Ordnance Man
ARM	US Navy Aviation Radio Man
AV	US Navy Seaplane Tender
CAG	Commander Air Group
CAP	Combat Air Patrol
CBI	China-Burma-India Theater of Operations
CO	Commanding Officer
CPO	Chief Petty Officer
Deck	Ground (as in "... dived for the deck")
Ditch	To make a landing in the sea
Division	Element of four aircraft
Dumbo	Martin PBM Mariner
Flak	Antiaircraft fire – from Ger. '*Flieger Abwehr Kanone*'
Flight	Element of two aircraft (A flight was also known as a 'Section')
Flattop	Slang for aircraft carrier
George	Allied codename for the Kawanishi N1K1-J Shiden or N1K2-J *Shiden-Kai* fighter

HVAR High Velocity Aerial Rocket
ICS Internal Communication System
JAAF Japanese Army Air Force
Jack Allied codename for the Mitsubishi J2M *Raiden* fighter
JNAF Japanese Navy Air Force
Jock Slang for 'pilot'
Jug Slang for P-47 Thunderbolt
Myrt Allied codename for the Nakajima C6N carrier reconnaissance aircraft
NAS Naval Air Station
PBM Martin Mariner patrol flying boat
Privateer Consolidated-Vultee PB4Y-2 land-based patrol bomber
Rufe Nakajima A6M2-N floatfighter
SB2C Curtiss Helldiver, US Navy dive-bomber, two-man crew
Scramble When all pilots on duty take off immediately for interception
Split S or Split-S, or Split-Ess An aircraft maneuver in which an aircraft makes a half-roll on its back, and then dives in a curve, usually leveling off on a reversed heading at much lower altitude.
SNAFU "Situation normal – all fucked up"
Tally-ho! A code word derived from old an English hunting expression, meaning "enemy sighted!"
TBM Grumman Avenger torpedo bomber
TF Task Force
TG Task Group (Task Groups made up a Task Force.)
Tony Allied codename for the Kawasaki Ki-61 fighter
USAAF US Army Air Force
VBF US Navy Fighter-Bomber Squadron
VF US Navy Fighter Squadron
VMF US Marine Corps Fighter Squadron
VPB US Navy Patrol Bomber Squadron
VT(N) US Navy Night Torpedo Bomber Squadron
W Warrant Officer
XO Executive Officer
Zeke Allied codename for the Mitsubishi Zero fighter

Japanese military terminology

Buntai Group (not the same as an American air group; comprised of about 20 pilots. In the 343 Kokutai, two *Buntais* made a *Hikotai* or Squadron)
Buntaicho Group leader
Chutai Mid-sized flying group comprised of three or more *Kutais*
Chutaicho Chutai group leader
Etajima Naval Academy (at Etajima)
Hikotai Squadron (in the 343 *Kokutai* approximately 40 pilots formed a squadron)
Hikotaicho Squadron leader
Kamikaze "Divine Wind" or Special Attack Suicide aircraft, pilot, or operation
Kempei Tai Military Police
Kikusui "Floating Chrysanthemum" – name given to a series of Kamikaze suicide attacks
Kokutai Naval Air Group
Kugisho Air Technical Arsenal
Kutai A Divison of four aircraft
Kutaicho A leader of four aircraft
Rekki Wingman
Sango Bakudan Aerial burst bomb
Sentai Army Air Combat Group
Shotai Usually consisted of one to three *Kutais* or from four to 12 fighters

Formation compositions from 343 Kokutai Action Reports (Sento Shoho)

12 April 1945 intercept mission against carrier aircraft over Amami/Kikai from Kanoya

Lt Naoshi Kanno, S301, Mission Leader

1st *Chutai*, Lt Naoshi Kanno leading
  1st *Shotai*, Lt Naoshi Kanno leading
    1st *Kutai*
      A15 Lt Naoshi Kanno
      A35 CPO Katsue Kato (aborted)
      A26 PO1/c Toshinobu Shimizu (canceled take-off)
      A43 PO Mikio Mitsuishi, missing
    2nd *Kutai*
      A11 CPO Shoichi Sugita
      A23 CPO Tomokazu Kasai
      B06 PO2/c Toyomi Miyazawa (aborted)
      C25 LA Tsuneharu Tamura (canceled take-off)
  2nd *Shotai*, WO Shoji Shibata leading
    1st *Kutai*
      A37 WO Shoji Shibata
      A41 CPO Nobuya Komeda
      A43 PO2/c Hirotoshi Miyata, missing
      C31 PO2/c Toru Otsubo
    2nd *Kutai*
      A13 CPO Itaru Shikano
      C38 CPO Yoshio Aoyama, missing
      B35 PO2/c Sakutaro Moriyama
      C35 PO1/c Osamu Omori, missing
  3rd Shotai, Lt(jg) Tatsutoshi Hashimoto leading
    1st *Kutai*
      A7 Lt(jg) Tatsutoshi Hashimoto, missing
      B40 CPO Eiichiro Sakurai
      C37 CPO Koichi Niizato, missing
      C41 PO1/c Mahito Yoshihara

2nd *Chutai*, Lt Masaji Matsumura leading
4th *Shotai*, Lt Masaji Matsumura leading
1st *Kutai*
A02 Lt Masaji Matsumura
A38 CPO Seiichiro Sato
C07 PO2/c Susumu Imai
A40 PO1/c Makoto Nishimura, missing
2nd *Kutai*
B48 CPO Mitsuo Hori
B32 CPO Hiroshi Tanaka
B47 PO1/c Teruo Kiriyama
B46 LA Harutoyo Naka (aborted)
5th *Shotai*, WO Isamu Miyazaki leading
1st *Kutai*
B27 WO Isamu Miyazaki (aborted)
A06 PO1/c Shoichi Handa
B38 PO2/c Ken Okimoto
A33 PO1/c Rokuro Asama (aborted)
2nd *Kutai*
A14 CPO Wataru Tomisugi (aborted)
A12 CPO Ichiro Muraki
B31 PO2/c Takeshi Ishikawa, missing
B53 PO2/c Toshio Shimemoto, missing

3rd *Chutai*, Lt Ryoichi Yamada leading
6th *Shotai*, Lt Ryoichi Yamada leading
1st *Kutai*
C36 Lt Ryoichi Yamada
C32 CPO Takumi Sugitaki
A45 CPO Sukeo Yoshioka
A39 LA Ryoji Niwa
2nd *Kutai*
C40 CPO Yasuo Matsumoto
B52 PO1/c Tsutomu Kimura
C44 PO2/c Tsutomu Yamada
C16 PO2/c Hiroyoshi Yoshida, missing

(*Note: The tail code A=Squadron 301's aircraft, B=S407 and C=S701, but this rule was not applied on this mission because of the confused servicing situation following the 343* Kokutai *transfer to Kanoya from Matsuyama to support Kamikaze Operations.*)

16 April 1945, intercept mission against carrier aircraft over Amami/Kikai from Kanoya

Lt Takashi Oshibuchi, S701, Mission Leader

1st *Chutai*, Lt Takashi Oshibuchi leading
1st *Shotai*, Lt Takashi Oshibuchi leading
1st *Kutai*
C45 Lt Takashi Oshibuchi
C34 CPO Jiro Hatsushima
C25 PO2/c Toru Kurita
C41 LA Kikuo Fujiki
2nd *Kutai*
C49 Lt(jg) Tetsuya Omura
C51 WO Narunobu Ibusuki
C33 PO1/c Jiro Funakoshi
C48 PO1/c Kiyoshi Toyohara

2nd *Chutai*, Lt Yoshishige Hayashi leading
1st *Shotai*, Lt Yoshishige Hayashi leading
1st *Kutai*
B38 Lt Yoshishige Hayashi
B43 CPO Teigo Ishida, missing
B57 LA Hitoshi Kotake, missing
B45 LA Eikichi Nishitsuruzono, missing
2nd *Kutai*
B03 WO Hiroshi Ohara (canceled take-off)
B35 PO1/c Shojiro Ishii (canceled take-off)
B32 PO2/c Kenshin Mizoguchi (canceled take-off)
B40 PO1/c Tomio Yamamoto

2nd *Shotai*, Lt(jg) Itaru Kawabata leading
1st *Kutai*
B55 Lt(jg) Itaru Kawabata
B15 PO1/c Sueo Ishizue
B27 LA Shokichi Kurumoto (aborted)
B52 PO1/c Kazuo Yotsueda, missing
2nd *Kutai*
B46 CPO Katsuyoshi Tanaka, missing
B43 CPO Shigeyori Ina
B47 PO2/c Zenji Asai, missing
B50 PO1/c Hisaichi Kubo

3rd *Chutai*, Lt Naoshi Kanno leading
1st *Shotai*, Lt Naoshi Kanno leading
1st *Kutai*
A15 Lt Naoshi Kanno
A02 CPO Katsue Kato, missing
A01 Lt(jg) Hiromu Tanaka
A30 PO1/c Toshinobu Shimizu
2nd *Kutai*
A23 WO Shoji Shibata
A14 CPO Itaru Shikano (aborted)
A13 CPO Nobuya Komeda
A21 PO2/c Sakutaro Moriyama

2nd *Shotai*, Lt Masaji Matsumura leading
1st *Kutai*
A11 Lt Masaji Matsumura
A33 CPO Mitsuo Hori
A40 CPO Seiichiro Sato
A19 PO2/c Susumu Imai
2nd *Kutai*
A37 WO Isamu Miyazaki
A47 CPO Wataru Tomisugi, missing
A22 PO2/c Ken Okimoto
A27 PO1/c Rokuro Asama, missing

21 April 1945, intercept mission against B-29s from Kokubu

Lt Takashi Oshibuchi, S701, Mission Leader

1st *Chutai*, Lt Takashi Oshibuchi leading
1st *Shotai*, Lt Takashi Oshibuchi leading
1st *Kutai*
C51 Lt Takashi Oshibuchi
C41 PO1/c Toshio Komabashiri (aborted)
C25 CPO Takamoto Kashima
C26 LA Kikuo Fujiki

2nd *Kutai*
C33 CPO Tsutomu Kimura
C43 PO2/c Toru Kurita
C47 PO2/c Masao Hasegawa

2nd *Shotai*, Lt(jg) Tetsuya Omura, leading
1st *Kutai*
C49 Lt(jg) Tetsuya Omura
C48 PO1/c Kiyoshi Toyohara
C40 PO1/c Jiro Funakoshi
C50 PO1/c Akira Takahashi

(*Note: The 2nd* Shotai *was made up of only one division.*)

3rd *Shotai*, Lt Yoshishige Hayashi, S407, leading
1st *Kutai*
B30 Lt Yoshishige Hayashi, crashed
B10 PO1/c Shojiro Ishii
B36 CPO Shigeyori Ina
2nd *Kutai* (made up of two aircraft)
B50 Lt(jg) Itaru Kawabata
B27 LA Shokichi Kurumoto

2nd *Chutai*, Lt Goro Ichimura leading
1st *Shotai*, Lt Goro Ichimura leading
1st *Kutai*
B51 Lt Goro Ichimura
A04 CPO Tsuneo Ozeki
A48 LA Magoshichi Koma
B18 LA Kinzo Kasuya
2nd *Kutai*
C53 Lt Toshio Tanaka
B17 CPO Hisamitsu Watanabe
B07 LA Akira Hoshino

(*Note: The 2nd* Chutai *was a two-division formation led by Lt Ichimura airborne from Matsuyama. This chutai encountered B-29s and the 2nd* Kutai *engaged them on the way to Kokubu. Lt Yoshishige Hayashi's shotai was under command of the 1st Chutaicho, Lt Oshibuchi, without forming another* chutai *probably because his two divisions were composed of only five aircraft.*)

3rd *Chutai*, Lt Naoshi Kanno, S301, leading
1st *Shotai*, Lt Naoshi Kanno leading
1st *Kutai*
A11 Lt Naoshi Kanno
A06 Lt(jg) Hiromu Tanaka
A22 PO1/c Toshinobu Shimizu, crashed
A47 PO1/c Yoneyasu Tsuneishi
2nd *Kutai*
A19 Lt Masaji Matsumura, cancelled take-off
A13 CPO Seiichiro Sato
A05 PO2/c Susumu Imai
A50 PO1/c Mahito Yoshihara
3rd *Kutai* (made up of 2 aircraft)
A40 CPO Mitsuo Hori
A37 PO2/c Ken Okimoto

(*Note: This* shotai *was made up of 3* kutai *of 10 aircraft from the same squadron. The* shotai *formed a* chutai.)

22 April 1945, intercept mission against B-29s from Kokubu

Lt Takashi Oshibuchi, S701, Mission Leader

1st *Chutai*, Lt Takashi Oshibuchi leading
1st *Shotai*, Lt Takashi Oshibuchi leading
1st *Kutai*
C45 Lt Takashi Oshibuchi
C41 PO1/c Toshio Komabashiri
C26 CPO Takamoto Kashima
2nd *Kutai*
C53 WO Narunobu Ibusuki
C50 CPO Takeo Yamada
B12 PO2/c Toru Kurita
2nd *Shotai*, Lt Naoshi Kanno, S301, leading
1st *Kutai*
A48 Lt Naoshi Kanno
A19 Lt(jg) Hiromu Tanaka
A04 PO2/c Ken Okimoto
A14 PO1/c Yoneyasu Tsuneishi (aborted)
2nd *Kutai*
A37 Lt Masaji Matsumura (aborted)
A05 CPO Nobuya Komeda
B10 PO2/c Susumu Imai
B07 PO1/c Mahito Yoshihara

11 May 1945, intercept mission against flying boats from Omura

Lt Masaji Matsumura, S301, Mission Leader

1st *Kutai*,
A52 Lt Masaji Matsumura
A29 CPO Seiichiro Sato
A08 CPO Yoshikazu Miyamoto
A17 CPO Mahito Yoshihara

15 May 1945, intercept mission against flying boats from Omura

Ens Akio Matsuba, S701, Mission Leader

1st *Kutai*
C55 Lt (jg) Akio Matsuba
C53 CPO Takumi Sugitaki
C58 CPO Takeo Yamada
C66 PO1/c Nobumitsu Minoura, crashed
2nd *Kutai*
C62 CPO Ryuji Yagi
C49 CPO Takamoto Kashima (aborted)
C65 PO1/c Toru Kurita
C21 PO2/c Sueji Imamura (aborted)

16 May 1945, intercept mission against flying boats from Omura

Lt Naoshi Kanno, S301, Mission Leader

1st *Chutai*, Lt Naoshi Kanno leading
1st *Shotai*, Lt Naoshi Kanno leading
1st *Kutai*
A15 Lt Naoshi Kanno
A27 Lt(jg) Hiromu Tanaka
A47 CPO Tomoichi Yokoyama
A08 PO1/c Susumu Imai

2nd *Kutai*
A19 WO Mitsuo Hori
A03 CPO Hiroshi Tanaka
A21 CPO Tesshin Kamei
A01 PO2/c Mutsue Naka

2nd *Shotai*, Lt Goro Ichimura, S407, leading
3rd *Kutai*
B25 Lt Goro Ichimura
B21 CPO Tsuneo Ozeki
B10 LA Shokichi Kurumoto
B30 LA Yoshio Takagi

(*Note: This* shotai *was made of up one* kutai.)

17 May 1945, intercept mission against flying boats from Omura

Ens Mitsuo Ishizuka, S407, Mission Leader

1st *Chutai*, Ens Mitsuo Ishizuka leading
1st *Shotai*, Ens Mitsuo Ishizuka leading
1st *Kutai*
B25 Ens Mitsuo Ishizuka
B36 PO1/c Shigeru Aoyagi
B34 CPO Tomio Yamamoto
B01 PO1/c Yoshinori Matsumoto
2nd *Kutai*
B10 CPO Toshio Tanaka
B07 CPO Hideo Nakao (aborted)
B32 CPO Hisamitsu Watanabe
B06 PO2/c Akira Hoshino, crashed

2nd *Shotai*, CPO Sanpei Shiono, S701, leading
3rd *Kutai*,
C64 CPO Sanpei Shiono
C65 CPO Tomeshiro Hiro, missing
C21 CPO Junichi Miyake
A19 PO2/c Masayuki Tashiro

(*Note: The* chutai *was made up of two shotai because the 3rd* Kutai *was from another squadron, S701. The 2nd* Shotai *was made up of only one* kutai. *Even though a* kutai *was led by a chief petty officer, it formed a* shotai *as the* kutai *was from another squadron.*)

Remarks:

1. A *Shiden-Kai* with the tail code 343-A-19 is on display at the National Museum of Naval Aviation in Pensacola, Florida, USA. As listed above, the aircraft was piloted by PO2/c Susumu Imai on 16 April, Lt Masaji Matsumura who canceled take-off on 21 April, Lt(jg) Hiromu Tanaka on 22 April, WO Mitsuo Hori on 16 May, all from S301, and PO2/c Masayuki Tashiro, S701, on 17 May 1945. It is not known who flew the aircraft in the latter period of June-August. Of the five pilots, only PO2/c Susumu Imai was killed in action, later, on 24 July 1945. It was a 'lucky' aircraft.
2. The aircraft with the tail code 343-A-11 was flown by CPO Shoichi Sugita, S301, on 12 April. After Sugita was shot down on 15 April, this aircraft was flown by Lt Masaji Matsumura on 16 April and also Lt Naoshi Kanno on 21 April 1945.
3. Note the aircraft No. 343-B 25 in the formation compositions for 16 and 17 May. Lt Goro Ichimura, *Buntaicho* of S407, states that he flew B-25 for a long time. It is thought that the aircraft was assigned to Lt Ichimura and probably had a one-band marking on the fuselage. The assigned aircraft with the code B-25 was flown by Ens Mitsuo Ishizuka, the mission leader on 17 May. It may have been rather exceptional, but this is an example where a *buntaicho* aircraft with the marking was flown by another pilot.
4. As seen above, a *shotai* was led by a chief petty officer, warrant officer, lieutenant junior grade, or lieutenant. A *shotai* could be led by a pilot of any rank on condition that he was a division leader and senior to the other pilots in the formation.

343 *Kokutai* Flight Personnel Casualties (Killed in Action)

Date (1945)	Unit		Location
19 March	S301	Lt(jg) Isaburo Inoue	Kure
	S301	CPO Yasuharu Nikko	Hojo
	S301	PO1/c Noriyoshi Kubo	Kure
	S407	Lt Kozo Shima	Iyo Nada
	S407	CPO Shiro Endo	Iyo Nada
	S407	PO1/c Haruhiko Takeshima	Iyo Nada
	S407	PO2/c Tokimasa Shinoda	Iyo Nada
	S407	LA Yoichi Saiki	Iyo Nada
	S407	LA Yoshihiro Nakatani	Iyo Nada
	S407	CPO Mitsuo Nakajima	Strafed on base
	S701	Lt Kunio Matsuzaki	Iyo Nada
	S701	Lt(jg) Yukihiro Watabe	Nakayama
	S701	CPO Kikuichi Ishikawa	Iyo Nada
	S701	LA Itsuo Waki	Iyo Nada
	T4	Ens Mitsuru Takada	Kochi-Ken
	T4	CPO Minoru Endo	Kochi-Ken
	T4	CPO Hiroshi Kageura	Kochi-Ken
12 April	S301	Lt(jg) Tatsutoshi Hashimoto	Amami/Kikai Is.
	S301	CPO Yoshio Aoyama	Amami/Kikai Is.
	S301	CPO Koichi Niizato	Amami/Kikai Is.
	S301	PO1/c Makoto Nishimura	Amami/Kikai Is.
	S301	PO1/c Osamu Omori	Amami/Kikai Is.
	S301	PO1/c Mikio Mitsuishi	Amami/Kikai Is.
	S301	PO2/c Takeshi Ishikawa	Amami/Kikai Is.
	S301	PO2/c Hirotoshi Miyata	Amami/Kikai Is.
	S301	PO2/c Toshio Shimemoto	Amami/Kikai Is.
	S701	PO2/c Hiroyoshi Yoshida	Amami/Kikai Is.
15 April	S301	CPO Shoichi Sugita	Kanoya
	S301	PO2/c Toyomi Miyazawa	Kanoya
16 April	S301	CPO Wataru Tomisugi	Amami/Kikai Is.
	S301	CPO Katsue Kato	Amami/Kikai Is.
	S301	PO1/c Rokuro Asama	Amami/Kikai Is.
	S407	CPO Teigo Ishida	Amami/Kikai Is.

Date (1945)	Unit		Location
16 April	S407	CPO Katsuyoshi Tanaka	Amami/Kikai Is.
	S407	PO1/c Kazuo Yotsueda	Amami/Kikai Is.
	S407	PO2/c Zenji Asai	Amami/Kikai Is.
	S407	LA Hitoshi Kotake	Amami/Kikai Is.
	S407	LA Eikichi Nishitsuruzono	Amami/Kikai Is.
21 April	S301	PO1/c Toshinobu Shimizu	Izumi
	S407	Lt Yoshishige Hayashi	Izumi
4 May	S301	CPO Teruo Kiriyama	Kikai Island
	S301	CPO Yoneyasu Tsuneishi	Kikai Island
	S407	Lt(jg) Itaru Kawabata	Kikai Island
	S407	LA Magoshichi Koma	Kikai Island
	S701	Lt(jg) Hirohisa Yasui	Kikai Island
	S701	PO2/c Kikuo Fujiki	Kikai Island
5 May	S407	PO2/c Kinzo Kasuya	Taketa, Oita
	S701	Ens Teruo Sasaki	Omura
12 May	S407	WO Hiroshi Ohara (wounded, died in Aug.)	Omura
15 May	S701	PO1/c Nobumitsu Minoura	Uku Shima
16 May	S407	CPO Tsuneo Ozeki	Iki Straits
17 May	S407	CPO Tomeshiro Hiro	Onna Jima
	S407	PO2/c Akira Hoshino	Onna Jima
28 May	S407	CPO Motoo Isokawa	Kanoya
	S701	CPO Takeo Yamada	Kanoya
	S701	PO2/c Masaru Morikawa	Kanoya
2 June	S301	CPO Eiji Mikami	Kagoshima
	S701	CPO Jiro Funakoshi	Kagoshima
3 June	S407	PO2/c Kiyoshi Miyamoto	Kanoya
10 June	S701	PO2/c Yukio Muramoto	Miyakonojo
22 June	S407	Lt Keijiro Hayashi	Amami Is.
	S407	CPO Sueo Ishizue	Amami Is.
	S701	CPO Takashi Yanagisawa	Amami Is.
	S701	PO1/c Susumu Tsuchiya	Amami Is.
2 July	S407	=CPO Shojiro Ishii	Satsuma Pt.
	S407	PO1/c Ichimaru Naruse	Satsuma Pt.
	S701	CPO Takashi Sakuma	Satsuma Pt.
5 July	S701	Lt Kazuchika Kinoshita	Nishisonogi
	S701	CPO Akira Takahashi	Nishisonogi
	S701	CPO Kiyoshi Toyohara	Nishisonogi
9 July	S701	Lt Teishi Ueshima	Omura
24 July	S301	Ens Kaneyoshi Muto	Bungo Straits
	S301	CPO Nobuya Komeda	Bungo Straits
	S301	PO2/c Susumu Imai	Bungo Straits
	S407	PO1/c Kenshin Mizoguchi	Bungo Straits
	S701	Lt Takashi Oshibuchi	Bungo Straits
	S701	CPO Jiro Hatsushima	Bungo Straits
1 Aug	S301	Lt Naoshi Kanno	Yaku Is.
	S301	PO1/c Sakutaro Moriyama	Take Is.
	S701	CPO Sukeo Yoshioka	Take Is.
2 Aug	S407	CPO Seiji Noma	Take Is.
8 Aug	S301	CPO Shigeo Suzaki	Dazaifu
	S407	Ens Mitsuo Ishizuka	Kumamoto
	S407	CPO Sakae Masumoto	N.Kyushu
	S407	CPO Masayuki Kutami	N.Kyushu
	S407	PO2/c Nobuyuki Nishimoto	Fukuoka
	S407	PO2/c Tatsuo Kuse	Iizuka
	S701	CPO Yokio Taura	N. Kyushu
	S701	CPO Kaemon Yokobori	Tsuiki
	S701	PO2/c Sueji Aburada	Tara, Saga
12 Aug	S407	Lt Sadao Oshio	Sasebo

Wounded In Action

Date (1945)	Unit		Location
19 March	S301	Lt Naoshi Kanno*	Imabari
	S301	LA Hideo Izawa*	Iyo Nada
	S701	Ens Akio Matsuba	Iyo Nada
	S701	PO1/c Tsutomu Murayama	Iyo Nada
7 May	S701	LA Hisao Sugita	Oita
15 May	S701	Ens Akio Matsuba	Uku Shima
28 May	S301	CPO Seiichiro Sato*	Kagoshima
2 July	S701	CPO Takumi Sugitaki*	Satsuma Pt.
5 July	S701	PO1/c Teruichi Sato*	Omura Wan
8 Aug	S701	Lt Keishichiro Hattori*	Fukuoka

*Bailed out

US Strategic Bombing Survey Interrogation No.479 of Capt. Minoru Genda

(Authors note: Verbatim from original. The following is a partial transcript of the interrogation report as it pertained to the homeland defense and the 343 *Kokutai.*)

Date: 28-9 November 1945
Place: Tokyo, Meiji Building
Interrogators: Majs Edwin McElwain and Robert Braucher
Interpreter: LCdr Walter Nichols, USNR
Allied officers present: 1/Lt Donald Meiklejohn, MI

(*Note: Capt Genda was cooperative and intelligent; he was at once willing to venture his best estimate in cases where his information was not complete but at the same time, he showed great discrimination in refraining from unfounded expressions of opinion*)

Q: What was the largest number of JNAF planes ever to intercept B-29s over Tokyo?
A: My guess would be about 100 planes. This was the biggest anywhere in the Empire and occurred when B-29s based in China hit the Empire in November 1944.
Q: What system did the JNAF use to control its fighters in the air?
A: Radio. There were differences between the units. Generally, the defense command post at Yokosuka would get advance information and order the planes to take off. This information was generally reliable, except that altitude could not be determined from our radar reports and had to be estimated.
Q: When was the 343 *Kokutai* organized?
A: It was organized in December 1944. I assumed command in January. This was not strictly a defense unit, but did take part in these operations. Our main duty was to contact American planes and attack them.
Q: What happened to your unit?
A: We trained at Matsuyama until early April. In April, we moved to Kanoya. When we were at Matsuyama, we undertook to intercept American Naval planes making a carrier strike on Kure. While we were at Kanoya, we did not have enough range to hit Okinawa so we staged through Kikai before hitting Okinawa.
Q: Did your group take part in the *Kikusui* Operations?
A: Yes. We did not participate in the first one, but we were in the others. We always covered the suicide planes. They went straight to Okinawa, while we veered toward Kikai to intercept American fighter planes.
Q: Were the suicide attack planes under your command or under their own commander?
A: Under their own separate command.
Q: What were your losses in the *Ryukyus* operations?
A: From April until August, we lost about 100 planes and 100 pilots.
Q: What was the average flying time of your pilots?
A: About 500 hours on the average. All of our planes were Georges and we had the very best pilots. Many of these men were lost. As a matter of fact, by the end of the war, all of our commanders had been killed in action. I was on the verge of assuming command in the air myself and expect that I would have been killed had the war lasted another month.
Q: Did B-29s have any effect on the Okinawa campaign because of their raids in April?
A: They had a big effect. Not very many planes were lost in the bombings, but the dropping of time bombs on the airfield resulted in our not being able to use the fields for much longer periods of time than heretofore.
Q: Did these April raids have any effect on maintenance?
A: Except for one particular occasion when we were caught completely by surprise and lost fifty or sixty men, our personnel losses were very few. We did have considerable damage inflicted on our installations—hangars, sheds, etc.—but they were not in use at the time so that the effect was comparatively negligible.
Q: What was your serviceability rate at the beginning of the Okinawa campaign?
A: About 70% at first, but after the heavy and continuous B-24 raids beginning in July, our serviceability rate went down continuously. At the end, I estimate that it was somewhere around 50%. B-29 raids had very little effect on our serviceability rate.
Q: Sometime during this period, it was decided to not concentrate on B-29s, and to concentrate on KETSU operations. Who decided this? When?
A: This decision was made by the Combined Fleet. In July, I received orders to cease B-29 interception. The reason given was the lack of fuel. Upon the receipt of this order, I sent a written protest to headquarters. I submitted my contrary opinion quite frequently. We gradually stepped up our interception again, but it never reached the degree that it had attained…(this section missing—co-authors)
Q: (The question is missing –co-authors)
A: No. We limited our training somewhat, but no plane was ever kept on the ground during operations because of a shortage of fuel. We had plenty of fuel on hand, but had orders not to use it. It was being saved for Ketsu Operations.
Q: Was the landing gear on the George (*Homari* #21) any good?
A: No. It was very poor. Much worse than on the average Japanese plane. In dive-bombing or strafing, if the plane reached a speed of 420 M.P.H., the gear was apt to fall apart in the air. About one-fourth of the pilots we lost in combat were lost for this reason. It got to be a better plane toward the end.
Q: Was the George a better fighter than the Jack?
A: Yes. It was designed for a different purpose than the Jack, but I think that it was more effective against the same target.
Q: Did you ever consider using alcohol in a Jack #21?
A: No. I never thought of it. We always used 90-100 octane gasoline in them.
Q: How was pilot training at the end of the war?
A: There was very little pilot training and there were no first-class replenishments. Men had from 150-200 hours, but they were not used in operational flying. At the end of the war, those pilots flying in operations had at least 300 hours in the air. Some of our pilots had up to 3,000 hours in the air (These men had generally come up thru the ranks). I should say that about 5% of our pilots had 2,000 hours or more.
Q: How many had between 300 and 400 hours?
A: I would say about 40-50%.
Q: Did the 343rd Air Group stay in Kanoya?
A: It was there until April and then went to Omura via Kokubu. One of our four airplane units went to Matsuyama for interception.
Q: How did the idea of Kamikaze arise in the Naval air force?
A: Younger officers and enlisted flyers in operational areas, after the losses in the Marianas, tried to get Imperial General HQ to start the use of suicide planes, saying they were ready to make the crashes. Imperial GHQ said, however, that they should wait until a heavy enough bomb was developed. Also, OKA (i.e., the BAKA bomb) was being worked on at this time too. Vice Admiral Onishi, however, finally gave the first order in October 1944 to start the suicide attacks and he merely reported to Imperial General HQ that they had been started. HQ did not know about it until then. They did not object, however.
Q: Was the decision to begin made by the Navy alone or as a joint Army-Navy matter?
A: By the Navy alone.
Q: What cooperation was there between the Army and the Navy during the Philippines as to suicide attacks?
A: The only cooperation came in the exchange of plans as to how many suicide planes were being sent out on each attack. There were no exchanges on tactics, etc.
Q: Did the Navy set up any special suicide organization or were suicide planes within the normal air units (*kokutai*)?
A: Usually volunteers from various air units were put together into a single new air unit (*kokutai*). There were no other suicide organizations.

Q: Did Navy suicide planes ever fly on missions with any crewmembers along, other than the pilot that is?

A: Yes, in the larger planes, a crewmember sometimes went with the pilot. In some cases, two went along. Frequently when the pilot announced that he was volunteering for a suicide mission, his crewmembers – the radio operator and the observer, for example – said that since they had been together this far, they would like to go along on the final suicide mission so they all went together. No effort was made to stop them from going with the pilot.

Q: During the Philippines Campaign we had a report from a U.S. carrier that a Japanese suicide plane had crashed on the carrier deck and that when it was examined, it was found that the pilot, dressed in some sort of ceremonial robe, had his feet tied to the pedals of his plane. Can you tell us anything about such cases?

A: Such a thing might have happened in isolated cases but there never was any such established policy in the Navy air force.

Japanese Bibliography

Fukubayashi, Toru *Downed US Aircraft and Flyers in the Kyushu Area, Downed US Aircraft and Flyers in the Chugoku and Shikoku Areas*, privately published, with 1947 GHQ/SCAP Investigations Reports No.65

Genda, Minoru *Kaigun Kokutai Shimatsu Ki* (History of the Rise and Fall of the Japanese Naval Air Force). Tokyo: Bungeishunju Publishers, 1962

Hakiri, Matsuo *Ozora No Kessen* (Showdown in the Sky). Tokyo: Asahi Sonorama Publishers, 1994

Hata, Ikuhiko and Izawa, Yasuho *Nihon Kaigun Sentoki Tai* (Japanese Navy Fighter Units). Tokyo: Kantosha Publishers, 1975

Hata, Ikuhiko and Izawa, Yasuho *Nihon Rikugun Sentoki Tai* (Japanese Army Fighter Units). Tokyo: Kantosha Publishers, 1977

Hata, Ikuhiko *Taiheiyo Senso Koku Shiwa* (Historical Stories of the Pacific Air War). Tokyo: Tojusha Publishers, 1980

Hori, Mitsuo *Shiden-Kai Kusen Ki* (History of the Shiden-Kai in Aerial Combat), appearing in *Taiheiyo Senso Dokyumentari* (The Pacific War Documentary), Vol. 5. Tokyo: Konnichino Wadaisha Publishers

Ikari, Yoshiro *Sentoki Shiden-Kai* (The Shiden-Kai Fighter Aircraft). Tokyo: Shirakane Shobo Publishers, 1976

Ikari, Yoshiro *Shiden-Kai No Rokki* (Six Shiden-Kai Pilots). Tokyo: Kojinsha Publishers, 1987

Ikari, Yoshiro *Saigo No Gekitsuio* (The Last Fighter Ace). Tokyo: Kojinsha Publishers, 1991

Ikari, Yoshiro *Saigo No Sentoki Shiden-Kai* (The Last Shiden-Kai Fighter). Tokyo: Kojinsha Publishers, 1994

Kato, Kan'ichiro *Zerosen No Himitsu* (The Secrets of the Zero Fighter). Tokyo: Kodansha Publishers, 1995

Kimata, Jiro *Nihon Kubo Senshi* (Combat History of the Japanese Carriers). Tokyo: Tosho Shuppansha Publishers, 1977

Kimata, Jiro *Kure Gunko Dai Kushu* (The Great Air Raid on Kure Harbor), appearinig in *Taiheiyo Senso Kaisen Jiten* (Encyclopedia of Pacific Sea Battles), Maru Extra Edition No.14, Tokyo: Kojinsha Publishers, 1995

Kodachi, Naoki *Reisen Saigo No Shogen* (The Last Witnesses on the Zero Fighter). Tokyo: Kojinsha Publishers, 1999

Kodachi, Naoki *Reisen Saigo No Shogen II* (The Last Witnesses on the Zero Fighter II). Tokyo: Kojinsha Publishers, 2000

Kodaka, Noritsura *Ah Seishun Zerosen Tai* (Ah! The Zero Fighter Squadron of My Youth). Tokyo: Kojinsha Publishers, 1969

Kojima, Noboru *Senkan Yamato* (The Battleship Yamato). Tokyo: Bungeishunju Publishers, 1977

LeMay, Curtis E. and Yenne, Bill *Superfortress: The B-29 and American Power* (translated by Yoji Watanabe). Tokyo: Asahi Sonorama Publishers, 1991

Morimoto, Tadao *Tokko* (Special Attack). Tokyo: Bungeishunju Publishers, 1992

Murashima, Tadashi *Pacific War Recovery, Kawanishi Shiden Kai* appearing in *After The Battle No.28*, London: Battle of Britain Prints International, 1980

Okumiya, Masatake *Kaigun Kokutai Zenshi* (The Complete History of the Japanese Naval Air Force). Tokyo: Asahi Sonorama Publishers, 1988

Shiga, Yoshio *343 Ku Taishi* (An Anthology of the 343 Kokutai). Tokyo: 343 Kokutai Association, 1981

Shiga, Yoshio *Shiden-Kai No Jitsuyosei* (Efficiency of the Shiden-Kai), appearing in *Umiwashi No Koseki* (Wakes of the Sea Eagles), edited by the Kaiku Kai Association. Tokyo: Hara Shobo Publishers, 1982

Sugiyama, Riichi *Kaigun Kokutai Shirei No Ki* (A Memoir of a Navy Air Group Commander), appearing in *Zoku Kantai Kokutai* (Carrier Air Groups Continued). Tokyo: Konnichino Wadaisha Publishers, 1978

Takashi, Sakurai *Rikugun Hiko Dai 244 Sentai Shi* (History of the Army 244th Fighter Group). Tokyo: Sobunsha Publishers, 1995

Tanaka, Etsutaro *Yokosuka Kokutai Koboki* (Memoir of the Rise and Fall of the Yokosuka Air Group), appearing in *Shogen Showa No Senso 13*. Tokyo: Kojinsha Publishers, 1991

Terada, Chikao *Kaitei Kara Kikan Shita Shiden-Kai No Nazo* (The Riddle of the Shiden-Kai that Returned from the Seabed), appearing in *Shiden, Shiden-Kai, Type 94 Recon Seaplane*. Tokyo: Kojinsha Publishers, 1992

Tono, Toshio *Omei* (Disgrace). Tokyo: Bungeishunju Publishers, 1985

Ugaki, Matome *Sensoroku* (Ugaki Matome's Diary). Tokyo: Hara Shobo Publishers, 1993

Umemura, Takeshi *Wanpaku Zerosen Taicho Netto Shimatsu* (The Combat Story of a Errant Zero Fighter Squadron Leader), appearing in *Kojinsha NF Bunko N-90 720*. Tokyo: Kojinsha Publishers, 1995

Warner, Denis and Peggy, and Seno, Sadao *The Sacred Warriors* (Japanese translated edition). Tokyo: Tokuma Shoten Publishers, 1989

Watanabe, Yoji *Omoi Hikoki Gumo* (Heavy Contrails). Tokyo: Bungeishunju Publishers, 1999

Watanabe, Yoji: *Kyokuchi Sentoki Raiden* (Intercept Fighter J2M Raiden). Tokyo: Asahi Sonorama Publishers, 1992

Yanagida, Kunio *Zerosen Moyu* (The Burning Zero Fighter), 6. Tokyo: Bungeishunju Publishers, 1993

Yasunobu, Takeo *Ah Kamikaze Tokko Tai* (Ah! Kamikazes). Tokyo: Kojinsha Publishers, 1977

OTHER REFERENCE SOURCES

Hohai No Seishun (The Youth in the Surging Billows), the record of the Navy Academy 70th Class. Yokohama: Edited and published by the Navy Academy 70th Class Association, 1995

Kaigun Sentoki Tai Shi (History of the Imperial Navy Fighter Force), *edited by Zero Fighter Pilots Association, 1987.*

Kyokuchi Sentoki Shiden-Kai *(Intercept Fighter Shiden-Kai). Tokyo: Gakusha Kenkyusha Publishers, 2000*

Maru Extra No.32 *(1973), containing memoirs by Tadashi Nakashima, Goro Ichimura, and Mitsuo Hori. Tokyo: Ushio Shobo Publishers*

Maru No.395 *(1979) and* Maru Extra Edition No. 117 *(1988), containing articles, combat diaries of the 343 Kokutai, and memoirs by Minoru Genda, Ryoichi Yamada, Ryoichi Koga, Isamu Miyazaki and Tomokazu Kasai. Tokyo: Ushio Shobo Publishers*

Maru Extra *(1990), containing a memoir by Hideo Harada. Tokyo: Ushio Shobo Publishers*

Maru Extra *(1990), containing an article by Hiroshi Goto* Senkan Ise Kure Gunko Boei Sen *(Combat of the Battleship Ise in the Defense of Kure Harbor). Tokyo: Ushio Shobo Publishers*

Maru Extra No.3 *(1997), containing a memoir by Goro Ichimura. Tokyo: Ushio Shobo Publishers*

Nippon No Kaigunki (Japanese Navy Planes) in *Koku-Fan*. Tokyo: Bunrindo Publishers, 1970

Sankei Shimbun *(Sankei Newspaper, evening edition 16 July 1979, containing the story of the salvage operation of the Shiden-Kai)*

WWII Imperial Japanese Navy Aircraft, Air World Extra Edition. *Tokyo: Air World Publishers, 1994*

Zerosen No.35 *containing a memoir by Toru Kurita. Tokyo: Zero Fighter Pilots Association*

American Bibliography

Bradley, F.J. No Strategic Targets Left. *Paducah, KY: Turner Publishing, 1999*

Erickson, Roy D. "Eric" *Tail End Charlies! Navy Combat Fighter Pilots at War's End*. Published by author, 1995

Francillon, Rene J. *Japanese Aircraft of the Pacific War*. London: Putnam and Co., 1970

Francillon, Rene J. *US Navy Carrier Air Groups Pacific 1941-1945*. London: Osprey Publishing, 1978

Hata, Ikuhiko and Izawa, Yasuho *Japanese Naval Aces and Fighter Units in World War II*. Translated by Don Cyril Gorham. Annapolis, MD: US Naval Institute Press, 1989

Hess, William N. *Pacific Sweep, The 5th and 13th Fighter Commands in World War II*. Garden City, NY: Doubleday & Co., 1974

Lambert, John W. *The Pineapple Air Force*. St. Paul, MN: Phalanx Publishing Co., 1990

Morison, Samuel Eliot *A History of U.S. Naval Operations in World War II*, Boston: Little, Brown, and Co., 1969

Olds, Robert *Helldiver Squadron, The Story of Carrier Bomb Squadron 17 with Task Force 58*. Washington, DC: Zenger Publishing, 1945

Olynyk, Frank J. *USMC Credits for the Destruction of Enemy Aircraft in Air-to-Air Combat World War 2*. Published by the author, 1982

Olynyk, Frank J. *USN Credits for the Destruction of Enemy Aircraft in Air-to-Air Combat World War 2*. Published by the author, 1982

Olynyk, Frank J. *USAAF (Pacific Theater) Credits for the Destruction of Enemy Aircraft in Air-to-Air Combat World War 2*. Published by the author, 1985

Rust, Kenn C. *Seventh Air Force Story*, Temple City, CA: Historical Aviation Album, 1979

Rust, Kenn C. *Twentieth Air Force Story*, Temple City, CA: Historical Aviation Album, 1979

Shaw, Henry I. Jr. *The United States Marines in the Occupation of Japan*, Washington DC: Historical Research Branch, G-3 Division, 1961

Sherrod, Robert *History of Marine Corps Aviation in World War II*, Washington DC: Combat Forces Press, 1951

Sweet, Donald *Seaplanes at War*. Burnsville, NC: Celo Valley Books, 1999

Sweet, Donald *The Sailor Aviators, A Tale of the Men of Navy Patrol Bombing Squadron 21 in World War II*, Burnsville, NC: Celo Valley Books, 1996

Tillman, Barrett *Hellcat: The F6F in World War II*. Annapolis, MD: US Naval Institute Press, 1979

Tillman, Barrett *Corsair: The F4U in World War II and Korea*. Annapolis, MD: US Naval Institute Press, 1979

Index